Springer Series in Solid-State Sciences

Editors: M. Cardona P. Fulde H.-J. Queisser

Volume 40 **Semiconductor Physics** – An Introduction By K. Seeger

Volume 41 **The LMTO Method** Muffin-Tin Orbitals and Electronic Structure
By H.L. Skriver

Volume 42 **Crystal Optics with Spatial Dispersion, and Excitons**
By V.M. Agranovich and V.L. Ginzburg

Volume 43 **Resonant Nonlinear Interactions of Light with Matter**
By V.S. Butylkin, A.E. Kaplan, Yu.G. Khronopulo, and E.I. Yakubovich

Volume 44 **Elastic Media with Microstructure II** Three-Dimensional Models
By I.A. Kunin

Volume 45 **Electronic Properties of Doped Semiconductors**
By B.I. Shklovskii and A.L. Efros

Volume 46 **Topological Disorder in Condensed Matter**
Editors: F. Yonezawa and T. Ninomiya

Volume 47 **Statics and Dynamics of Nonlinear Systems**
Editors: G. Benedek, H. Bilz, and R. Zeyher

Volume 48 **Magnetic Phase Transitions**
Editors: M. Ausloos and R.J. Elliott

Volume 49 **Organic Molecular Aggregates,** Electronic Excitation and Interaction Processes
Editors: P. Reineker, H. Haken, and H.C. Wolf

Volume 50 **Multiple Diffraction of X-Rays in Crystals** By Shih-Lin Chang

Volume 51 **Phonon Scattering in Condensed Matter**
Editor: W. Eisenmenger, K. Laßmann, and S. Döttinger

Volume 52 **Superconductivity in Magnetic and Exotic Materials**
Editors: T. Matsubara and A. Kotani

Volume 53 **Two-Dimensional Systems, Heterostructures, and Superlattices**
Editors: G. Bauer, F. Kuchar, and H. Heinrich

Volume 54 **Magnetic Excitations and Fluctuations**
Editors: S. Lovesey, U. Balucani, F. Borsa, and V. Tognetti

Volume 55 **The Theory of Magnetism II,** Thermodynamics and Statistical Methods
By D.C. Mattis

Volumes 1 – 39 are listed on the back inside cover

Magnetic Excitations and Fluctuations

Proceedings of an International Workshop,
San Miniato, Italy, May 28 – June 1, 1984

Editors:
S. W. Lovesey, U. Balucani, F. Borsa,
and V. Tognetti

With 114 Figures

Springer-Verlag
Berlin Heidelberg New York Tokyo 1984

Professor Dr. Stephen W. Lovesey
Rutherford Appleton Laboratory,
Chilton, Oxfordshire OX11 OQX, Great Britain

Dr. Umberto Balucani
Istituto di Elettronica Quantistica, CNR, I-50127 Firenze, Italy

Professor Ferdinando Borsa
Dipartimento di Fisica "A Volta", I-27100 Pavia, Italy

Professor Valerio Tognetti
Dipartimento di Fisica dell' Università di Firenze, Largo E. Fermi,
I-50125 Firenze, Italy

ISBN-13:978-3-642-82371-8 e-ISBN-13:978-3-642-82369-5
DOI: 10.1007/978-3-642-82369-5

2153/3130-543210

Preface

An international workshop on Elementary Excitations and Fluctuations in Magnetic Systems was held in San Miniato, Italy, for five days beginning 28 May, 1984. The workshop comprised eight working sessions that contained a total of 43 invited talks, and 58 scientists were in attendance from 14 countries. Our aim was to review some topics of current interest in the statistical physics of magnetic materials and models, with an emphasis on theoretical studies and confrontations between these and experimental and computer simulation data.

This book contains summary papers written by the invited speakers, and the material will be of immediate interest to graduate students and researchers engaged in studies of magnetic properties. There is, perhaps, no effective way to record and convey the benefit of the numerous discussions between the participants that are a significant integral feature of a workshop.

The magnificent venue of the workshop, I Cappuccini, was made available to us by the Cassa di Risparmio San Miniato. Financial support for the workshop was received from Consiglio Nazionale delle Ricerche, Università degli Studi di Firenze and the Gruppo Nazionale Struttura della Materia. Our administrative load and the burden of preparing the proceedings for publication was made light by the talents of Carla Pardini (CNR, Florence), and Caroline Monypenny and Jane Warren (Rutherford Appleton Laboratory).

Finally, we wish to thank all the participants for their attendance and individual contributions to the success of the workshop.

Chilton, Great Britain — *S.W. Lovesey*
Firenze, Italy — *U. Balucani*
Firenze, Italy — *V. Tognetti*
Pavia, Italy — *F. Borsa*

Contents

Part I Experimental Studies of One-Dimensional Magnets

Specific Heat of TMMC due to Non-Linear Excitations: Comparison with Classical Models. By *F. Borsa*, M.G. Pini, A. Rettori, and V. Tognetti (With 3 Figures) 2

Narrow and Broad Solitons in the Antiferromagnetic Chains of $CsCoCl_3$ and TMMC. By *J.P. Boucher*, L.P. Regnault, A. Pires, J. Rossat-Mignod, Y. Henry, J. Bouillot, W.G. Stirling, and J.P. Renard (With 8 Figures) 6

Spin Waves in a One-Dimensional Antiferromagnetic Insulator By *M.F. Collins* and B.D. Gaulin (With 3 Figures) 12

Solitons in a XY (S=1/2) Quantum System. By *W.J.M. de Jonge*, A.M.C. Tinus, and K. Kopinga (With 3 Figures) 16

Mössbauer Relaxation Studies of π-Domain-Wall Dynamics in Pure and Impure Ising-Type Magnetic Chains. By *H.J.M. de Groot*, R.C. Thiel, and L.J. de Jongh (With 4 Figures) 21

k=0 Excitations in the Quantum Chain Systems $(C_6H_{11}NH_3)CuCl_3$ and $(C_6H_{11}NH_3)CuBr_3$. By *K. Kopinga*, W.J.M. de Jonge, C.H.W. Swüste, A.C. Phaff, R. Hoogerbeets, and A.J. van Duyneveldt (With 5 Figures) 27

Magnon-Phonon Coupling in Spin-½ Quasi 1D Antiferromagnets By D.J. Lockwood (With 3 Figures) 33

Spin Dynamics in $CsNiF_3$: Spinwaves, 2-Spinwaves, Solitons ... By M. Steiner (With 1 Figure) 39

Part II Spin Wave Dynamics in One Dimension

Classical Ferromagnetic Chain in a Field; Spin-Waves Versus Non-Perturbative Interactions. By *U. Balucani*, M.G. Pini, A. Rettori, V. Tognetti, and R. Vaia (With 1 Figure) 46

The Life Time of the Spin Waves of a Classical Ferromagnetic Chain in an Applied Magnetic Field. By U. Balucani, M.G. Pini, and *V. Tognetti* (With 1 Figure) 51

Quantum Spin Chains: Phase Behaviour and Dynamic Excitations By *J.C. Bonner* and J.B. Parkinson (With 7 Figures) 55

Lecturers name is written in *italic*

Nonlinear Excitations in the Classical XY-Model. By *R.W. Gerling* and D.P. Landau (With 5 Figures) 61

Field Dependent Spin Dynamics of Antiferromagnetic Heisenberg Chains By *G.P. Meech* and S.W. Lovesey (With 2 Figures) 67

Is $CsNiF_3$ a Classical System? By *M.G. Pini* and A. Rettori (With 4 Figures) 72

Spin Wave Theory for the Classical Easy Plane Ferromagnet in Zero Field: Exact Results at Low Temperature. By G. Reiter (With 1 Figure) 77

Part III Soliton Theories

Sine-Gordon Statistical Mechanics. By R.K. Bullough, D.J. Pilling, and *J. Timonen* 80

Solitons in Lattice Systems. By H.W. Capel 86

Thermodynamics and Correlation Functions of Non-Linear Excitations in the Easy-Plane Ferromagnetic Chain By H. Fogedby (With 2 Figures) 92

Is TMMC a Sine-Gordon System? By *M. Fowler*, N.F. Wright, and M.D. Johnson 99

Dynamics of Phase Transitions in Ferromagnetic Domain-Wall Structures. By *E. Magyari* and H. Thomas (With 1 Figure) 102

Part IV Mixed and Disordered Magnets

Stress Fluctuations and Local Structural Distortions in Ferromagnetic Metallic Glasses: Their Role on the Relaxation of the Magnetic Permeability. By *P. Allia* and F. Vinai (With 1 Figure) 110

Frustration on Fractals. By *M. Cieplak* and J.R. Banavar 115

A Weakly Anisotropic Ising Model in a Random Field By *R.A. Cowley*, M. Hagen, H. Yoshizawa, G. Shirane, and R.J. Birgeneau (With 4 Figures) 121

Spin Dynamics Associated with Local Magnons in FeF_2: Mn By D.W. Hone and *S.M. Rezende* (With 6 Figures) 126

2D Pure and Dilute Quantum Transverse Ising Model. By G. Kamieniarz . 137

The Random Field Ising Model. Equilibrium and Non-Equilibrium Properties. By J. Villain (With 2 Figures) 142

Part V Excitations in Ordered Magnets

Theory of the Raman and Brillouin Scattering on the Excited Electronic States and Magnons in Europium Chalcogenides. By W. Cieslewicz and *A. Stasch* 150

Effect of a Surface on Magnon-Magnon Interactions in Ferromagnets By D.E. Kontos and *M.G. Cottam* (With 2 Figures) 153

Correlation Theory of Magnetic Excitations
By P.-A. Lindgård (With 3 Figures) 159

A Rescaling Approach to Excitations in Magnetic Systems
By *B.W. Southern* and P.D. Loly (With 5 Figures) 165

Part VI **Novel Magnetic Models and Systems**

Monte Carlo Study of the 2-Dimensional Spin-½ XY Model
By *H. De Raedt*, B. De Raedt, J. Fivez, and A. Lagendijk (With 4 Figures) 172

Quenching of Spin Fluctuations in Highly Enhanced Magnetic Materials
By *K.A. Gschneidner, Jr.* and S.K. Dhar (With 2 Figures) 177

Quasiparticle Model for High Frequency Inelastic Neutron Scattering from Intermediate Valence Compounds. By D.L. Huber (With 4 Figures) 183

Singlet Ground-State Dimer Systems $Cs_3Cr_2Br_9$ and $Cs_3Cr_2Cl_9$
By *J.K. Kjems*, D. Petitgrand, R. Feile, B. Leuenberger, and H.U. Güdel (With 4 Figures) 189

Helimagnetism in the Heisenberg Model: New Features and Old Mechanisms
By *E. Rastelli*, L. Reatto, and A. Tassi (With 2 Figures) 195

Magnetic Excitations in the Quasi-2d Planar Magnets $BaM_2((XO)_4)_2$ (M=Co,Ni; X=P,As). By *L.P. Regnault*, J. Rossat-Mignod, J.Y. Henry, R. Pynn, and D. Petitgrand (With 8 Figures) 201

Spin Dynamics of Rare Earth Ions in Crystal Fields Probed by Muon Spin Rotation. By P.M. Richards (With 2 Figures) 207

Ground-State Properties of s = 1 Easy-Plane Ferromagnetic Heisenberg Chains. By T. Schneider (With 3 Figures) 212

Bound State of Phonons and CEF-Excitations in $CeAl_2$
By P. Thalmeier (With 5 Figures) 217

Local Magnetic Field Distributions in Ising Models. By *M.F. Thorpe*, M. Thomsen, D. Sherrington, and T.C. Choy (With 3 Figures) 222

List of Contributors 227

Part I

Experimental Studies of One-Dimensional Magnets

Specific Heat of TMMC due to Non-Linear Excitations: Comparison with Classical Models

F. Borsa
Dipartimento di Fisica dell'Università ed Unità GNSM, I-27100 Pavia, Italy
M.G. Pini
Istitudo di Elettronica Quantistica del CNR, I-50127 Firenze, Italy
A. Rettori and V. Tognetti
Dipartimento di Fisica dell'Università ed Unità GNSM, I-50125 Firenze, Italy

In recent years, conclusive experimental evidence has been obtained indicating that in certain ranges of temperature and external magnetic field, non-linear excitations not describable by single or multiple spin-wave excitations give a significant contribution both to equilibrium thermodynamic properties and dynamic properties of the one-dimensional antiferromagnet TMMC ($(CH_3)_4NMnCl_3$). Direct experimental evidence was obtained by neutron scattering [1] and specific heat[2] measurements and indirect evidence by NMR[3] and thermal conductivity[4]. However, no theoretical model has proved as yet completely satisfactory to explain the experimental results[5]. Following the original suggestion by Mikeska[6], TMMC should be mapped by a sine-Gordon classical chain for external magnetic fields perpendicular to the chain axis of magnitude less than about 5 Tesla.

The Hamiltonian for TMMC in an external magnetic field perpendicular to the c axis can be written as

$$H = 2J\sum_{i=1}^{N} (\vec{S}_i\cdot\vec{S}_{i+1} - \delta S_i^z S_{i+1}^z) - g\,\mu_B H \sum_{i=1}^{N} S_i^y \tag{1}$$

which in the classical limit can be reduced by a decimation procedure[7] to the effective Hamiltonian in order to calculate the partition function

$$H_{eff} = \sum_{i=1}^{N/2} \left[-\tilde{J}\ \vec{S}_{2i}\cdot\vec{S}_{2i+2} + 4\tilde{D}\,(S_{2i}^z)^2 + (\tilde{H}^2/8\tilde{J})\ (S_{2i}^y)^2\right] \tag{2}$$

where $|\vec{S}_{2i}|=1$, $\tilde{J}=JS^2$, $\tilde{D}=J\delta S^2$ and $\tilde{H}=g\mu_B HS$ and $\delta=0.016$ for TMMC.

The general Hamiltonian for a discretized sine-Gordon chain is

$$H = \sum_i Aa\left[\dot{Y}_i^2/2 + \omega_0^2(1-\cos Y_i) + (c_0^2/2a^2)(Y_{i+1}-Y_i)^2\right] \tag{3}$$

which reduces in the continuum limit to

$$H = A \int dx \left[\tfrac{1}{2}\,\dot{Y}^2 + \omega_0^2(1-\cos Y) + \tfrac{1}{2}c_0^2\ Y_x^2\right]. \tag{4}$$

In the continuum limit the Hamiltonian (1) can be mapped into a sine-Gordon problem (4) with

$$\begin{aligned} A &= (16Ja)^{-1} \\ c_0 &= 2aJS(1-\tfrac{1}{2}\delta)^{\frac{1}{2}} \\ \omega_0 &= g\mu_B H(1-\tfrac{1}{2}\delta)^{\frac{1}{2}}\ . \end{aligned} \tag{5}$$

In order for the mapping to be valid it is essential that TMMC can be treated as a classical spin chain, with negligible fluctuations along the chain axis and that the continuum limit hold. The last point can be best appreciated by comparing the numerical results for a classical SG chain[8] to the results obtained[5] by a classical transfer-matrix calculation for a 1D classical planar model described by the Hamiltonian

$$H = \sum_i \left[2JS^2\cos(\phi_i-\phi_{i+1}) - g\mu_B HS\cos\phi_i\right] . \tag{6}$$

As seen in Fig. 1 one obtains essentially the same values for the specific heat ΔC_S where ΔC_S is the contribution to the specific heat obtained by subtracting the term ascribed to linear spin waves and refers therefore only to the contribution of non-linear excitations. Also in Fig. 1 one can compare the numerical calculations for a classical SG chain[8] (or planar model[5]) with the experimental results obtained with the procedure outlined in ref. 5. The agreement is quite good considering that no adjustable parameter enters in the plot. (We used J=6.7K and S=5/2 pertinent to TMMC.)

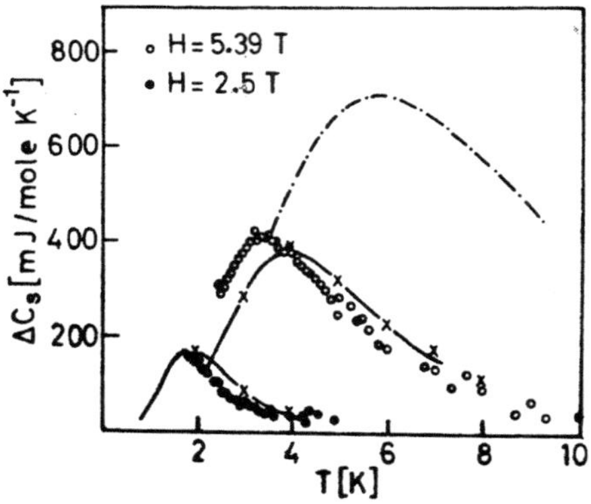

Fig.1 Experimental contribution of non-linear excitations to the specific heat of single crystal TMMC vs T for the magnetic field applied perpendicular to the c axis. ΔC_S was obtained by subtracting from the measured specific heat the value for H=0 after correcting for the linear spin-wave contribution (see ref.5). (——) numerical results for a classical SG chain from ref.8;(x) our numerical results for the planar model;(—·—·) theoretical curve for the soliton free-gas phenomenology

It should be mentioned that it is also of considerable theoretical interest to establish the nature of the non-linear excitations. One proposal[9] is that the main contribution in the SG model comes from thermally excited solitons. Transfer matrix calculation interpreted in terms of an ideal gas was developed yielding for TMMC

$$\Delta C_S = 6.63\, m\, (\beta E_S)^{\frac{1}{2}} \exp(-\beta E_S) \left[(\beta E_S)^2 - (\beta E_S) - 1/4\right] \tag{7}$$

where E_S is the soliton's formation energy and m is the soliton's effective mass. For TMMC: $E_S = g\mu_B HS = 3.35H$ (in K) and $ma = (E_S/2JS^2) = 0.04H$ (with H in Tesla). The result (7) agrees only qualitatively with the exact numerical calculations in Fig. 1. However, it appears that considering finite-temperature corrections[10] in the transfer matrix calculations, one can reach excellent agreement with the numerical results[8] (or planar model[5]) confirming the soliton-type character of the non-linear excitations in the SG chain.

Regarding the comparison with the experimental results, however, the main problem remains the recognition of the theoretical model which best approximates the real 1D system.

For magnetic fields close to the critical field H_c=8.9T for which $\tilde{H}/8\tilde{J}$= 4D the Hamiltonian (2) describes an almost easy-axis system [11] . Therefore for the two values of H (7.6T and 10T) for which experiments have been performed one can compare the experimental results directly with the transfer matrix calculations without further assumptions (see Figs. 2 and 3).

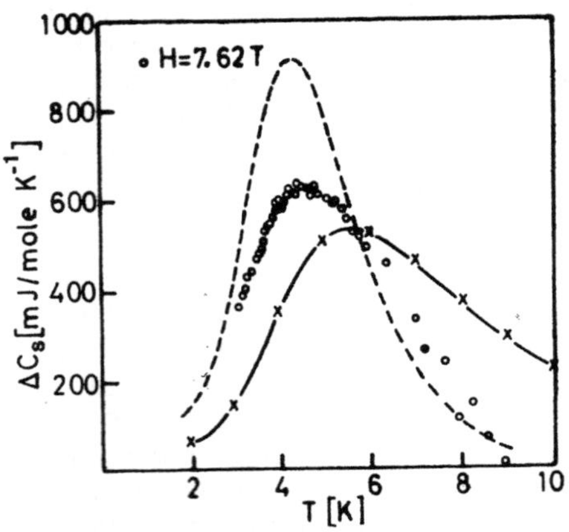

Fig.2 Experimental contribution of non-linear excitations to the specific heat of TMMC compared with theoretical results;(---) our numerical results for the quasi-uniaxial model. For the remaining symbols see caption to Fig.1

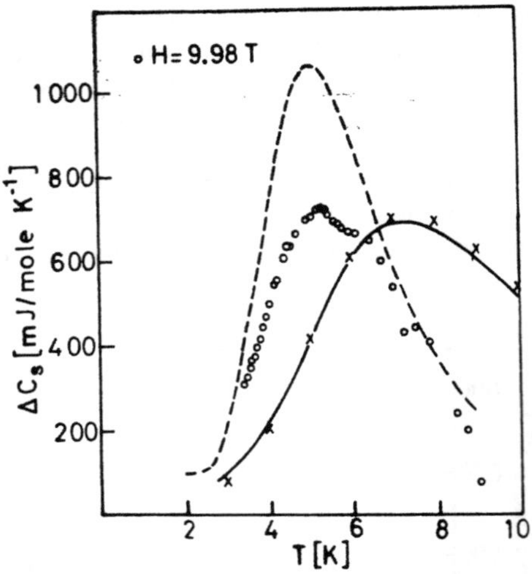

Fig.3 Experimental contribution of non-linear excitations to the specific heat of TMMC compared with theoretical results;(---) our numerical results for the quasi-uniaxial model. For the remaining symbols see caption to Fig.1

The agreement, without adjusted parameters, is excellent regarding the position of the maximum, while it appears that the classical calculation overestimates the magnitude of the non-linear contribution by about 30÷40%. For H= 5.4T and H=2.5T the transfer matrix calculations on Hamiltonian (2) are made difficult by lack of convergence. Thus the calculations were performed on the planar model and the results are compared with the experiments in Fig. 1. In this case the agreement about the position of the peak is less satisfactory while the magnitude is in much better agreement than for the calculations performed at higher fields with the model (2). However, it should be noticed that the agreement about the magnitude of the peak is probably fortuitous since in the planar model one neglects the out-of-plane contributions which are instead included in the complete Hamiltonian (2).

It should be stressed that the over all agreement obtained by comparing the experimental results in TMMC with classical numerical calculations is much better than the one obtained in the 1D ferromagnetic chain $CsNiF_3$ where recent transfer matrix classical calculations performed on the complete Hamiltonian are completely unable to reproduce the experimental results[12]. One can conclude that, contrary to $CsNiF_3$, quantum corrections are small in TMMC and they should affect only the magnitude of the specific heat peak.

References

1. J.P. Boucher, L.P. Regnault, J. Rossat-Mignod, J.P. Renard, J. Bouillot and W.G. Stirling, Solid State Comm. 33, 171 (1980)
2. F. Borsa, Phys. Rev. B25, 3430 (1982)
3. J.P. Boucher and J.P. Renard, Phys. Rev. Lett. 45, 486 (1980)
4. J.A.H.M. Buijs and W.J.M. de Jonge, J. Phys. C15, 6631 (1982)
5. F. Borsa, M.G. Pini, A. Rettori and V. Tognetti, Phys. Rev. B28, 5173 (1983)
6. H.J. Mikeska, J. Phys. C11, L29 (1978); ibidem C13, 2913 (1980)
7. J. Villain and J.M. Loveluck, J. Phys.(Paris) Lett. 38, 77 (1977)
8. T. Schneider and E. Stoll, Phys. Rev. B22, 5317 (1980)
9. R.M. Leung, D.W. Hone, D.L. Mills, P.S. Riseborough and S.E. Trullinger, Phys. Rev. B21, 4017 (1980)
10. K. Sasaki and T. Tsuzuki, Solid State Comm. 41, 521 (1982); J. Magn. Magn. Mater. 31-34, 1283 (1983)
11. I. Harada, K. Sasaki and H. Shiba, J. Phys. Soc. Japan 51, 3069 (1982)
12. M.G. Pini and A. Rettori, Phys. Rev. B29, (1 May 1984)

Narrow and Broad Solitons in the Antiferromagnetic Chains of $CsCoCl_3$ and TMMC

J.P. Boucher*[1], L.P. Regnault**, A. Pires*[2], J. Rossat-Mignod**, and Y. Henry**
Département de Recherche Fondamentale, RM* et DN** et Laboratoire d'Electronique et de Technologie de l'Information, CRM*** CENG
F-38041 Grenoble Cedex, France

J. Bouillot and W.G. Stirling
Institut Laue-Langevin, F-38042 Grenoble Cedex, France

J.P. Renard
Institut d'Electronique Fondamentale, Université Paris IX
F-91405 Orsay Cedex, France

I INTRODUCTION

The two quasi one-dimensional (1D) compounds $CsCoCl_3$ and $(CH_3)_4NMnCl_3$ (TMMC) are almost ideal systems in which to study soliton excitations. Both they have antiferromagnetic (AF) couplings in the chains and at low temperature they exhibit an Ising symmetry favourable for the occurrence of solitons. This symmetry is an intrinsic property of $CsCoCl_3$ while in TMMC it is achieved only by the application of an external magnetic field H perpendicular to the chains. In the 1D short range order regime two energetically equivalent configurations are expected for the spins. Solitons can be seen as Bloch walls separating ordered domains and allowing the spins to pass from one configuration to the other. In the case of a "strong" Ising symmetry ($CsCoCl_3$) the walls are reduced to one lattice unit ("narrow" solitons) (Fig.1a), while in the case of a "weak" Ising symmetry (TMMC) the walls extend over several lattice units (10 to 30) ("broad" solitons) (Fig.1b). To maintain a paramagnetic state, these walls move rapidly along the chains, inducing characteristic fluctuations. As an example, the fluctuation spectrum for narrow solitons as predicted by J. VILLAIN [1] is shown in Fig.2. Depending on the wave vector q to be considered with respect to the inverse of the correlation length, κ, two different types of fluctuations can be observed. For $q^* = |\pi-q| \gg \kappa$, we are concerned with the propagation of individual solitons (Soliton mode) while

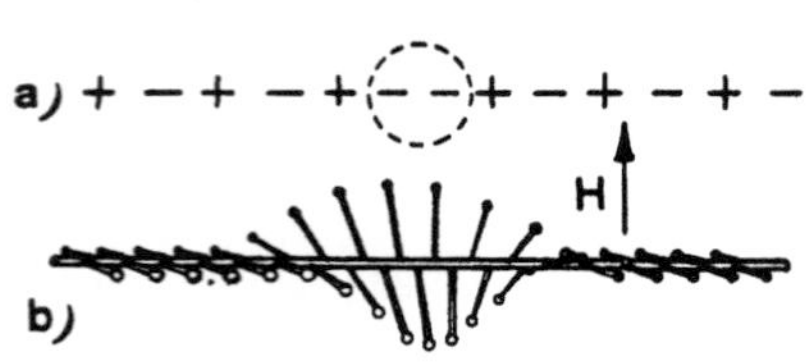

Fig.1 Narrow (a) and broad (b) solitons in AF chains

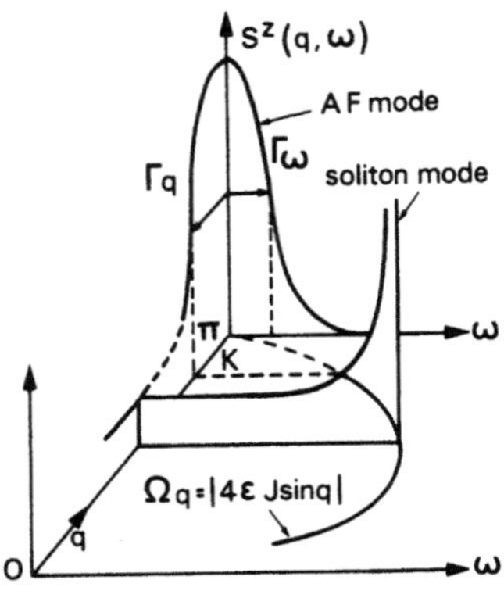

Fig.2 Fluctuation spectrum for narrow solitons

for $q^* \lesssim \kappa$, we observe essentially the fluctuations of the ordered domains which show up as a mode (AF mode) centred at $q = \pi$. In the soliton model, the width Γ_q which measures the short range order is simply the soliton density n_s :

$$\Gamma_q = \kappa = 2\, n_s \,. \tag{1}$$

Concerning the dynamics, each time a soliton goes through an ordered domain a flipping is induced which makes the two AF sublattices exchange their spin direction. The frequency broadening of the AF mode results from these flippings and Γ_ω measures the flipping rate :

$$\Gamma_\omega = 2\, n_s\, V_o = \Gamma_q\, V_o \tag{2}$$

where V_o is the average velocity of the solitons. In the present work we shall consider only the AF mode. It can be observed by neutron scattering (NS) measurements which allow both q and ω analysis. However, this requires a very good instrumental resolution together with a high flux of neutrons. These conditions are provided by the triple-axis spectrometer IN12 of the Institut Laue-Langevin. The AF mode can also be studied by nuclear magnetic resonance (NMR) measurements. The spin-lattice relaxation rate $1/T_1$ of nuclei located in the neighbourhood of the chains is governed by the flippings of the magnetic sublattices [2] :

$$1/T_1 \sim \Gamma_\omega \,/\, (\Gamma_\omega^2 + \omega_N^2) \;\; \simeq \;\; 1/\Gamma_\omega \tag{3}$$

if the nuclear Larmor frequency ω_N is smaller than Γ_ω. This technique is particularly interesting when NS measurements are not possible. An unusual example of this complementarity will be shown in the present work.

II EXPERIMENTS on $CsCoCl_3$

In this compound, the 1D short range order regime where the picture of moving solitons is meaningful extends from $T_M \simeq 75$ K (maximum of the susceptibility) down to the ordering temperature $T_{N1} \simeq 21$ K. The Hamiltonian for such AF chains (assumed to be along the Z axis) can be written as [3] :

$$\mathcal{H} = \sum_n 2\, J\, S_n^z\, S_{n+1}^z + \varepsilon\, J\, (S_n^+\, S_{n+1}^- + S_n^-\, S_{n+1}^+)$$

with $S = 1/2$, $J = 75$ K and $\varepsilon \simeq 0.12$. According to Maki [4], the shape of the AF mode is described approximately by the following expression :

$$S^z(q,\omega) = \frac{S^2}{2\pi}\, \frac{\Gamma_\omega^2 \,/\, \Gamma_q}{[\Gamma_\omega^2(1 + q^{*2}/\Gamma_q^2) + \omega^2]^{3/2}} \tag{4}$$

with Γ_q and Γ_ω defined by (1) and (2). In the present case of "narrow" solitons, the soliton density n_s and the velocity V_o are given by $n_s = \exp(-J/T)$ and :

$$V_o = \frac{2}{\pi}\, \frac{\sinh(2\varepsilon J/T)}{2\varepsilon J/T}\, \frac{4\varepsilon J}{I_o(4\varepsilon J/T)} \;\simeq\; \frac{2}{\pi}\, 4\varepsilon J \tag{5}$$

for $2\varepsilon J \ll T$. $I_o(X)$ is the modified Bessel function.

The NS measurements were performed between 13-60 K. In the fitting procedure using (4), the widths Γ_q and Γ_ω have been considered as adjus-

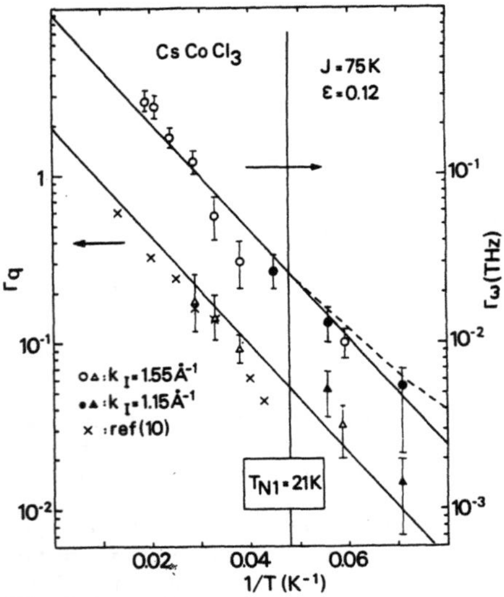

Fig.3 Experimental values of Γ_q and Γ_ω

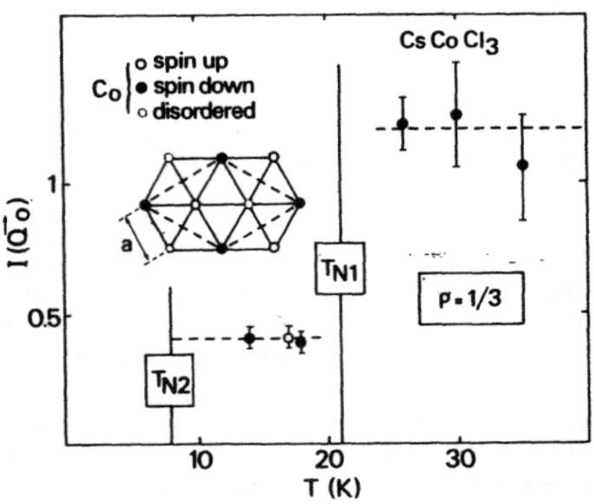

Fig.4 Integrated intensity of the neutron scattering

table parameters. The resulting values are plotted in Fig.3. The full lines are theoretical curves corresponding to (1) and (2) together with (5). We emphasize that they are not a fit to the data. The dashed line shows the influence of the temperature for $T \lesssim 2\varepsilon J \simeq 18$ K by using the exact expression for V_0. Since no adjustment of any physical parameters was made the agreement between the soliton model and the experiments is extremely good. At this stage it is worth noting that the temperature range explored in these measurements extends well below the 1D regime. This means that paramagnetic 1D correlations persist below T_{N1}. Looking first at the crystallographic structure of $CsCoCl_3$ we observe that in the planes perpendicular to the chains, the spins form a triangular lattice [5]. Secondly we note that with an AF coupling between nearest chains one is led to the simplest situation for frustration effects : one third of the chains should remain "disordered" as shown in the inset of Fig.4. To check this point, we have evaluated the number of spins involved in the NS. This is made by performing a double integration in both q and ω on the observed intensity. The resulting values $I(\vec{Q}_0)$ obtained at different temperatures are shown in Fig.4. We observe that they are temperature independent, as expected, but different below and above T_{N1} : the average intensity below T_{N1} is actually $\rho = 1/3$ of that above T_{N1}. This study of the soliton dynamics confirms that below T_{N1}, the chains of $CsCoCl_3$ undergo frustration effects which maintain 1/3 of them in a paramagnetic state. Fluctuations arising from individual solitons (soliton mode) have also been observed below T_{N1}. Their study is reported in a separate paper.

III RESULTS IN TMMC

The chains being along the Z axis, the 1D Hamiltonian for this compound can be written as :

$$\mathcal{H} = \sum_n 2\,J\,\vec{S}_n\vec{S}_{n+1} - 2\,D\,S_n^zS_{n+1}^z - g\mu_B\,H\,S_n^y \qquad (6)$$

with $JS^2 \simeq 42.5$ K, $DS^2 \simeq 1$ K and $S = 5/2$. At low temperature and in moderate fields ($20 \lesssim H \lesssim 60$ kOe), Equation (6) can be reduced to a sine-Gordon Hamiltonian. The corresponding soliton density is given by :

$$n_s = \sqrt{8/\pi}\ m \sqrt{E_s/T}\ \exp(-E_s/T) = \Gamma_q/2 \quad (7)$$

where E_s is the soliton energy and m the soliton mass ($m = E_s/4JS^2$). Assuming a Gaussian distribution for the velocities, one obtains for V_0 and Γ_ω :

$$V_0 \simeq \sqrt{2/\pi}\ 4JS/\sqrt{E_s/T} \quad , \quad \Gamma_\omega = (8/\pi S)\ E_s \exp(-E_s/T) \ . \quad (8)$$

In low field ($H < 60$ kOe), the agreement between theory and experiments is very good with E_s proportional to H, as expected [6], $E_s = \alpha H$, with :

$$\alpha = 0.26 \pm 0.02 \text{ K/kOe} \ . \quad (9)$$

For $H \gtrsim 60$ kOe, deviations from this model can possibly be attributed to the presence of defects in the chains as the length of the ordered domains becomes large [6]. It could also result from the intrinsic properties of the Hamiltonian (6)[7]. Since in antiferromagnets, a field H plays the role of an anisotropy at low T, we can conversely describe the effect of D in terms of an effective field $H_D \sim \sqrt{J\ D}$ applied along the chain axis (Z axis). In the classical spin description, we evaluate $H_D \simeq 90$ kOe. However, it has been observed that the magnon energy gaps (at $q = \pi$) for the Z and Y components of the spins are equal in a field of $H \simeq 70$ kOe [8]. This value could be a relevant parameter for TMMC. The soliton regime is determined by the competition between H and H_D. For $H \ll H_D$, as Z is the "hardest" axis, the spins inside the walls turn preferentially in the XY plane perpendicular to the chain axis (transverse solitons) (Fig.1b). On the other hand, for $H \gg H_D$, they will turn inside the ZX plane going through the chains (longitudinal solitons). Equations (7) and (8) can be used again but with $E_s \sim H_D$. For $H \simeq H_D$, a symmetrical configuration is reached (equivalent to an easy axis ferromagnet) [7]. The soliton density becomes :

$$n_s = 4\ m\ (E_s/T)\ \exp(-E_s/T) = \Gamma_q/2 \ . \quad (10)$$

The difference between (7) and (10), specifically in the E_s/T dependence, expresses the fact that in the walls, the spins are not confined anymore in a plane, but the π rotation of the spin can be made within a sphere. With the same V_0, we expect now :

$$\Gamma_\omega = 8\sqrt{2}\ /\ (S\sqrt{\pi})\ E_s\sqrt{E_s/T}\ \exp(-E_s/T) \ . \quad (11)$$

In order to probe simultaneously Γ_q and Γ_ω, NS measurements have been performed from 60 up to 100 kOe. The results obtained at T = 83 kOe are shown in Fig.5, plotted as a function of $\Gamma_q T$ and $\Gamma_\omega \sqrt{T}$, corresponding to (10) and (11). Both quantities display the same exponential behavoiur as a function of 1/T. In particular, no deviation is observed for Γ_q, showing that there is no appreciable effect due to defects in the chains. This confirms the high quality of the single crystals used in our experiments.

For higher fields ($H > 100$ kOe), we resort to NMR measurements. A special experiment has been set up at the Service National des Champs Intenses (S.N.C.I.) in Grenoble. The magnetic field is obtained with a Bitter coil, especially designed to reach a sufficient homogeneity (better than 10^{-5}

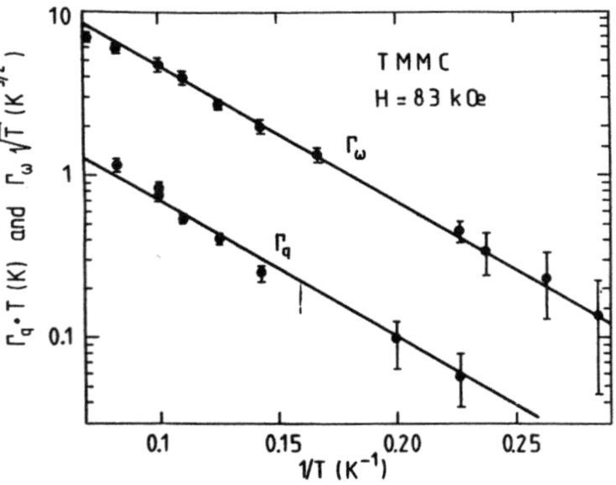

Fig.5 Experimental values of Γ_q and Γ_ω at H = 83 kOe

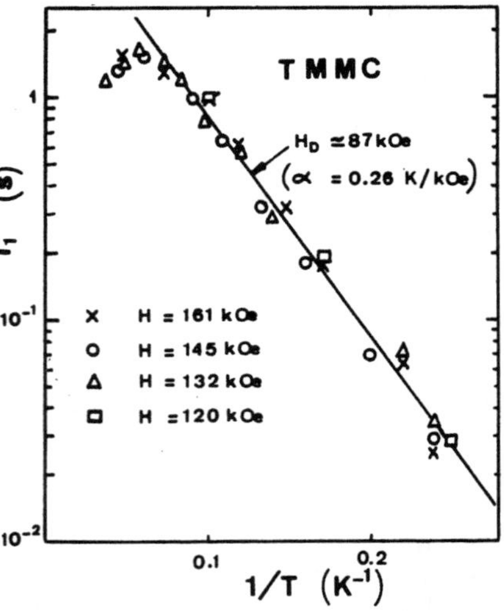

Fig.6 Spin relaxation time of ^{15}N for H > 100 kOe

over 1 cm^3) over the field range 100-170 kOe. To emphasise these exceptional (and probably unique) experimental conditions, we remark that a power of about 8 MW is required to obtain the nominal field. The data shown in Fig.6 give the values of the relaxation rate ($1/T_1$) measured on the spins of ^{15}N nuclei in an enriched single crystal of TMMC. They are independent of H and follow an exponential behaviour with 1/T. This is in agreement with the model of longitudinal solitons as E_S is governed only by the dipolar field H_D. Figure 7 summarizes the values obtained for the soliton energy E_s from 20 up to 170 kOe. The full line represents the results of previous work in low field [9]. A crossover occurs between 60 and 100 kOe. Below 60 kOe, the dynamics are governed by transverse solitons with $E_S \sim H$; for $H > 100$ kOe, they are governed by longitudinal solitons with $E_S \sim H_D$, independent of H. If the classical value $H_D \simeq 90$ kOe is used, we evaluate $\alpha = E_S/H_D \simeq 0.23$ K/kOe almost identical to the value obtained in low field [(9)]. However with $H_D \simeq 70$ kOe, deduced from the magnon gaps [8], we obtain $\alpha \simeq 0.33$ K/kOe, which is close to the classical prediction $\alpha = g\mu_B S \simeq 0.36$ K/kOe. In a recent publication, Mikeska has proposed a description in terms of two soliton branches [10] : in low field the Sine-Gordon picture is only an approximation with persistent out of plane fluctuations increasing with H (would this explain the small value $\alpha \simeq 0.26$ K/kOe ?). Instability of these solitons is expected at $H \simeq H_D$. Experimentally no drastic effect is observed. On the other hand, the Sine-Gordon description would be completely applicable in high field ($H > 100$ kOe) for the longitudinal solitons. Finally, we show in fig.8 the phase diagram of the different soliton regimes as determined experimentally for TMMC. At low T, it is limited by the 3D ordering. The high temperature limit corresponds to the condition that the number of spins inside a soliton ΔN_S remains smaller that the number N_S of spins between solitons. Experimentally it agrees with $\Delta N_S/N_S \lesssim 0.6$ [9].

The investigation of these two compounds, $CsCoCl_3$ and TMMC, illustrates the advantage of antiferromagnets as the AF mode yields an accurate determination of the soliton regime. Narrow and broad solitons are observed to behave very similarly.

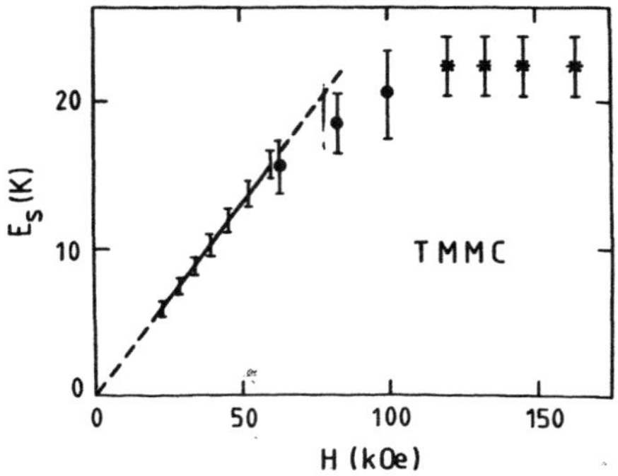

Fig.7 Soliton energy in the field range 20-170 kOe

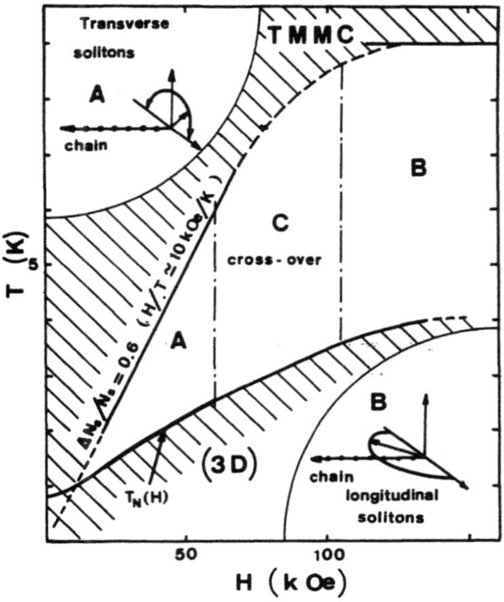

Fig.8 Phase diagram of the soliton regimes in TMMC

One of us, J.P.B., is most grateful to J.C. Picoche and P. Rub for the preparation of the high field magnet at the SNCI in Grenoble.

1 Also member of Equipe de Recherche CNRS 216
2 Permanent address : Departamento de Fisica, Universidade de Minas Gerais, Belo Horizonte, M.G. Brazil

1 J. Villain: Physica B 79, 1 (1975)
2 J.P. Boucher and J.P. Renard: Phys. Rev. Lett. 45, 486 (1980)
3 H. Yoshizawa, K. Hirakawa, S.K. Satija and G. Shirane: Phys. Rev. B 23, 2298 (1981)
4 K. Maki: Phys. Rev. B 24, 335 (1981)
5 M. Mekata and K. Adachi: J. Phys. Soc. Jap. 50, 3904 (1981)
6 J.P. Boucher, L.P. Regnault, J. Rossat-Mignod, J.P. Renard, J. Bouillot and W.G. Stirling: J. Appl. Phys. 52, 1956 (1981)
7 I. Harada, K. Sasaki, H. Shiba: Sol. St. Comm. 40, 29 (1981)
8 I.U. Heilmann and J.K. Kjems, Y. Endoh, G.F. Reiter, G. Shirane and R.J. Birgeneau: Phys. Rev. B 24, 3939 (1981)
9 J.P. Boucher, L.P. Regnault, J. Rossat-Mignod, J.P. Renard, J. Bouillot, W.G. Stirling and F. Mezei: Physica 120B, 241 (1983)
10 N. Flüggen and H.J. Mikeska: Sol. St. Comm. 48, 293 (1983)

Spin Waves in a One-Dimensional Antiferromagnetic Insulator

M.F. Collins and B.D. Gaulin

Physics Department, McMaster University
Hamilton, Ontario, L8S 4M1, Canada

The study of one-dimensional materials has proved fruitful because computational difficulties are reduced while the underlying physics is maintained. Effectively one-dimensional magnetic materials do exist in nature with magnetic interactions that can be well characterized [1]. The pathological nature taken on by fluctuations under these conditions prevents the appearance of a state with long range order although there is extensive short range order at low temperatures.

We describe in this work the measurement of spin fluctuations in the quasi-one-dimensional paramagnet $CsMnBr_3$ by neutron inelastic scattering. This material crystallizes into the hexagonal space group $P6_3/mmc$ [2] and has been characterized by susceptibility, specific heat [3] and EPR [4] measurements as well as by quasi-elastic neutron scattering [5], neutron inelastic scattering [6] and neutron diffraction [3] measurements. The last two of these were performed in the three-dimensionally ordered state which occurs below 8.3 K. The picture that emerges is one of chains of Mn^{2+} ions along the hexagonal c axis with magnetic moments that are coupled strongly and antiferromagnetically with Heisenberg exchange constant between neighbouring manganese atoms of -0.88 ± 0.01 meV. The coupling between spin in different chains is almost three orders of magnitude weaker [6].

Our own earlier measurement [7] in the paramagnetic state at 15K show that, except near the zone centre, the magnetic excitations are well described by sharp spin-wave modes. The dispersion relation is shown in fig. 1. This is as expected for a Heisenberg chain at limitingly low temperatures except that the finite intercept at the magnetic zone centre indicates the presence of small anisotropic terms arising from dipolar interactions and probably also single ion anisotropy [8]. These terms have a magnitude that is only a few percent of the Heisenberg intrachain coupling and should give a relatively insignificant contribution to the spin wave properties at reduced wavevectors, q_z, greater than 0.1. We thus interpret our spin wave measurements at these wavevectors in terms of a one-dimensional Heisenberg Hamiltonian. The experimental conditions for these measurements involve standard neutron triple-axis techniques, and are the same as described in references 7 and 8.

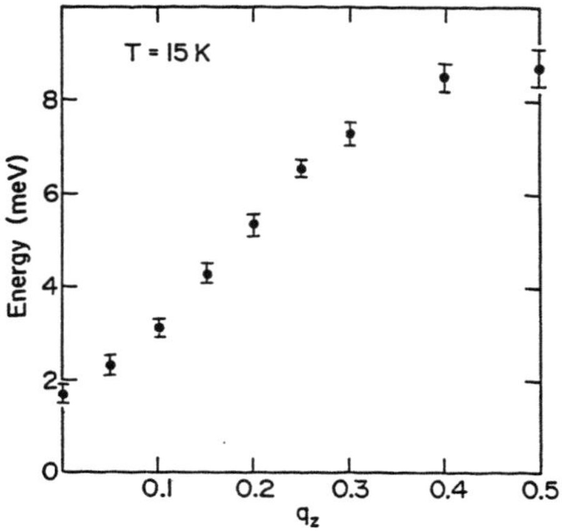

Figure 1. Spin wave dispersion relation in paramagnetic $CsMnBr_3$ at 15K

There are several theoretical approaches to the description of excitations in the Heisenberg antiferromagnetic chain. Unfortunately the numerical methods ([9] and reference therein; [10]) cannot easily be generalized to the specific parameters (temperature, exchange, wavevector) used in an experiment, so that the most useful comparison with experiments is afforded by MORI'S continued fraction formulation of the generalized Langevin equation [11-16]. The theory requires the frequency moments of the scattering function to be input; the first three of these are known exactly at all temperatures [14], which suffices to calculate the first four levels of the continued fraction. In addition a termination procedure for the continued fraction must be used. LOVESEY [13] and DE RAEDT [16] have each suggested such a procedure.

The first calculation of the dynamics using these methods (LOVESEY [13]) involved a termination at the third level, producing a scattering function with three poles. This does not fit the experimental data for TMMC well near the zone centre since it calculates the damping incorrectly and predicts a central mode which is not observed experimentally. DE RAEDT and DE RAEDT [15] showed that an even pole expansion does not suffer from these difficulties. At larger wavevectors, where both TMMC and $CsMnBr_3$ are more Heisenberg-like in their behaviour, the two terminations have not been compared with experiments except at relatively low temperatures in TMMC [17]. Figs. 2 and 3 show such comparisons in $CsMnBr_3$ at wavevectors of (0,0,1.2) and (0,0,1.3) for several temperatures.

In figs. 2 and 3, the data are compared to the predictions of the three-pole and the four-pole expansions with the De Raedt termination and to the three pole expansion with the Lovesey termination. In all cases the theoretical line shapes have been convoluted with the

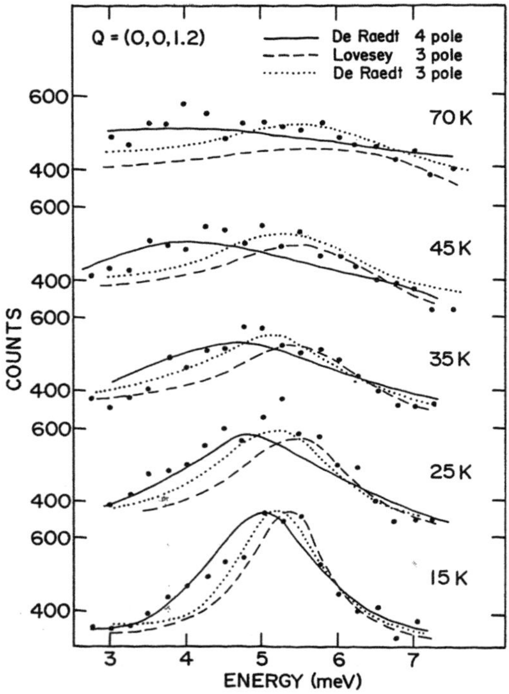

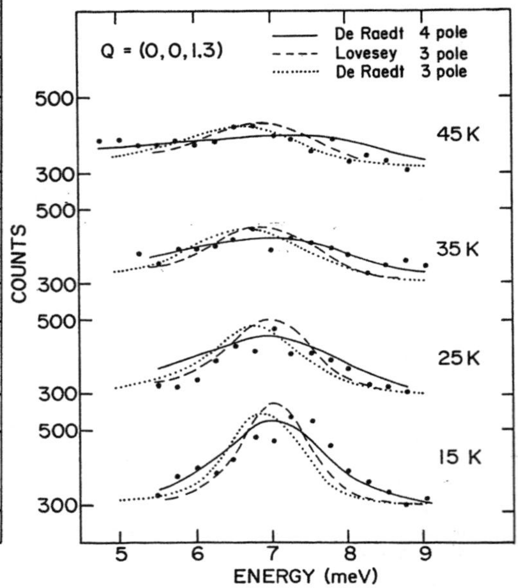

Figure 2. Inelastic scattering at Q = (0,0,1.2) with temperatures of 15, 25, 35, 45 and 70K compared with the predictions of the Lovesey three-pole model and the De Raedt three and four-pole models

Figure 3. Inelastic scattering at Q = (0,0,1.3) with temperatures of 15, 25, 35 and 45K compared with predictions of the Lovesey three-pole model and the De Raedt three- and four-pole model

spectrometer resolution function. There is one intensity parameter to be fitted and this fitting has been carried out at 15K and (0,0,1.2); all other parameters are fixed.

These figures show that the method of termination is as much a factor as the number of poles in the fit (though at smaller reduced wavevectors this ceases to be true). None of the predictions are quantitatively correct, though the qualitative picture is correct. The fitting with all models is better at the larger wavevector (0,0,1.3) despite the fact that the arbitrary intensity parameter was fitted at the smaller wavevector.

There is not a lot to choose between the two De Raedt models in goodness of fit, though the Lovesey model is less good. We conclude that the considerable superiority of the four - pole termination over the three pole at small reduced wavevectors does not extend to mid zone.

We would like to thank Mr. J. Couper and Mr. J. Garrett for technical assistance and the Chalk River Nuclear Laboratories for providing facilities. The financial support of the National Science and Engineering Research Council of Canada is also gratefully acknowledged.

1. M. Steiner, J. Villain and C.G. Windsor: Adv. Phys. 25, 87 (1976).

2. J. Goodyear and D.J. Kennedy: Acta Cryst., B28, 1640 (1972).

3. M. Eibschutz, R.C. Sherwood, F.S.L. Hsu and D.E. Cox: AIP Conf. Proc. 17, 684 (1972).

4. H. Kalt, E. Siegel, N. Pauli, H. Mosebach, J. Wiese and A. Edgar: J. Phys. C16, 6427 (1983).

5. W.J. Fitzgerald, D. Visser and K.R.A. Ziebeck: J. Phys. 15, 795 (1982).

6. W. Breitling, W. Lehman, R. Weber, N. Lehner and W. Wagner: Magn. Magn. Mater 6, 113 (1977).

7. M.F. Collins and B.D. Gaulin: J. Appl. Phys. 55, 1869 (1984).

8. B.D. Gaulin and M.F.Collins: Can.J.Phys. (to appear).

9. M. Takahashi: J. Phys. Soc. Japan 52, 3592 (1983).

10. F.B. McLean and M. Blume: Phys. Rev. B7, 1149 (1973).

11. H. Mori: Prog. Theor. Phys. 34, 399 (1965).

12. S.W. Lovesey and R.A. Meserve: J. Phys. C6, 79 (1973).

13. S.W. Lovesey: J. Phys. C7, 2008 (1974).

14. H. Tomita and H. Mashiyama: Prog. Theor. Phys. 48, 1133 (1972); ibid 51, 1312 (1974).

15. H. De Raedt and B. De Raedt: Phys. Rev. B15, 5379 (1977).

16. H. De Raedt: Phys. Rev. B19, 2585 (1979).

17. M.T. Hutchings and C.G. Windsor: J. Phys. C10, 313 (1978).

Solitons in a *XY (S=1/2)* Quantum System

W.J.M. de Jonge, A.M.C. Tinus, and K. Kopinga

Department of Physics, Eindhoven University of Technology, P.O. Box 513
NL-5600 MB Eindhoven, The Netherlands

In this paper we will report experimental evidence on the existence of solitons in ferromagnetic quantum (S=1/2) XY chains, as observed in the recently discovered system $(C_6H_{11}NH_3)CuBr_3$. We have measured the excess specific heat ΔC, the nuclear spin-lattice relaxation (NSLR) time T_1 in the field range $0 < B < 7$ kOe and the temperature range 1.5 K $< T < 10$ K, although some data are also obtained outside this region. These data will show that both the dynamic behavior of the electronspin system, as monitored by T_1, and an equilibrium property like C can be understood by adopting solitons as dominant excitations in an appropriate temperature and field regime. Parts of these results have been published recently [1].

Several investigations on $(C_6H_{11}NH_3)CuBr_3$, (CHAB), as well as on its Cl isomorph (CHAC) have shown that these compounds can be considered as excellent quasi one-dimensional ferromagnets [2,3]. Detailed electronspin resonance and specific heat experiments on CHAB established that the present compound can be characterized by the Hamiltonian [3]:

$$\mathcal{H} = -2J \sum_i [\bar{S}_i \cdot \bar{S}_{i+1} + (J_{zz} - J)/J \; S_i^z S_{i+1}^z] - \sum_i g\mu_B \bar{B} \cdot \bar{S}_i \qquad (1)$$

which can formally be mapped on a sG Hamiltonian [4]. J/k equals 55 K and $(J_{zz} - J)/J = -0.05$, whereas the interchain interaction $J' \simeq 10^{-3}J$, which is not included in (1), induces an 3d ordered state below $T_c = 1.50$ K. The non-uniaxial part of the anisotropy in the intrachain exchange, $(J_{xx} - J_{yy})/J$, within the easy XY plane is about 7×10^{-4} and is therefore neglected. From these values of J_{xx}, J_{yy} and J_{zz} one may conclude that a cross-over in behavior of the electronspin system from Heisenberg to XY can be anticipated at a temperature of $T < T_{CR} = 0.44JS(S+1)/k = 16$ K [5]. The measurements reported in this paper are restricted to temperatures $T < 10$ K and thus to the XY region.

Although the individual ferromagnetic chains running in the crystallographic c direction are crystallographically equivalent, the local anisotropy axes of the chains do not coincide. The y axis of all chains is parallel to the c axis whereas the x axis is located in the a-b plane at an angle $\varphi = \mp 25^o$ from the b axis, yielding two types of (practically independent) symmetry related chains. The insert in Fig. 1 illustrates this situation.

Fig. 1 shows the excess heat capacity ΔC obtained by subtracting the zero field results from the data obtained in a field B// c. Although ΔC certainly includes the effect of B on the linear spin wave excitations, this contribution in itself would yield a *decrease* of the specific heat with increasing field in the present region. Furthermore, a contribution arising from isolated magnetic ions (or impurities) can be ruled out [6]. Since, as will appear below, the NSLR measurements on this system also suggest a soliton contribution, we will now interpret ΔC in that light.

The soliton contribution to ΔC is [7]

$$\Delta C_s \simeq \frac{k}{a} \left(\frac{8}{\pi}\right)^{1/2} (JS^2)^{-1} kT (E_s^o/kT)^{3/2} . [(E_s^o/kT - 1/2)^2 - 1/2] . \exp(-E_s^o/kT) \qquad (2)$$

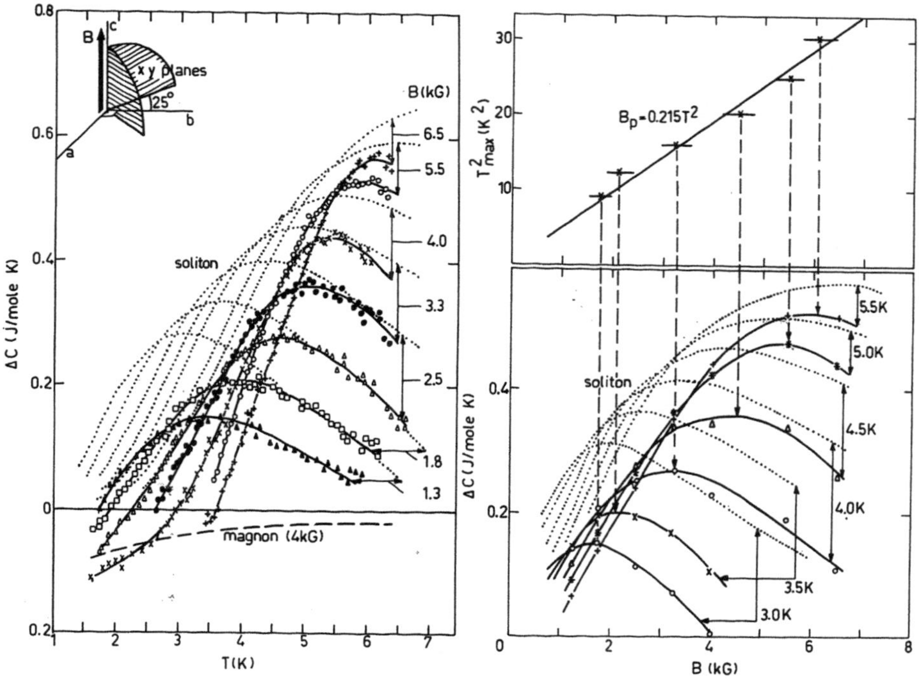

Fig. 1. Temperature dependence (a) and field dependence (c) of ΔC, as well as the location of the maximum of ΔC(B) versus T^2(b). Drawn lines are guides to the eye. Dotted curves represent classical soliton theory with $E_s^o/k=8.4\sqrt{B}$. The calculated magnon contribution is shown by a dashed curve. The inset shows the orientation of the easy XY plane.

where $E_s^o = 8S(Jg\mu_B BS)^{1/2}$, the classical soliton rest energy. Inspection of this expression shows that $\Delta C_s/kT$ is uniquely determined by E_s^o/kT and has a maximum for $E_s^o/kT \simeq 3.9$. If the data obey the relation $E_s^o = \alpha\sqrt{B}$ predicted for a pure XY system, the maximum in ΔC(B) determines α through $B_{peak} = (3.9kT/\alpha)^2$. The data shown in Fig. 1 clearly reveal a quadratic dependence of B_{peak} on T and result in an experimental soliton energy $E^o_{s,exp}/k = 8.4\sqrt{B}$, indicating a reduction of ~20% compared with the classical value $E_s^o/k=10.85\sqrt{B}$, obtained from the general expression inserting the appropriate values for CHAB. This $E^o_{s,exp}$ may be compared with the prediction of MAKI [8] based on a semi-classical approach, which yields $E_s^o/k = 9.4\sqrt{B}$ for this case. The dotted curves in Fig. 1 are calculated with $E^o_{s,exp}$ and show perfect agreement as far as the field and temperature dependence is concerned.

With respect to the overall *magnitude* of ΔC we recall that the data include the contribution of linear magnetic excitations. To illustrate the effect of this contribution we have included in Fig. 1 some results of a linear spin wave calculation based on the field dependence of the energy gap observed by FMR [3]. The corrected magnitude of ΔC, however, is still too low, although the discrepancy decreases for higher fields and temperatures. This feature, which was also observed in $CsNiF_3$ [6] and TMMC [9], may be related to finite temperature corrections [10] or to the effect of out of plane components of the spin [4,11].

Measurements of ΔC were also performed with B in other directions. Some representative examples are shown in Fig. 2. In this figure the results

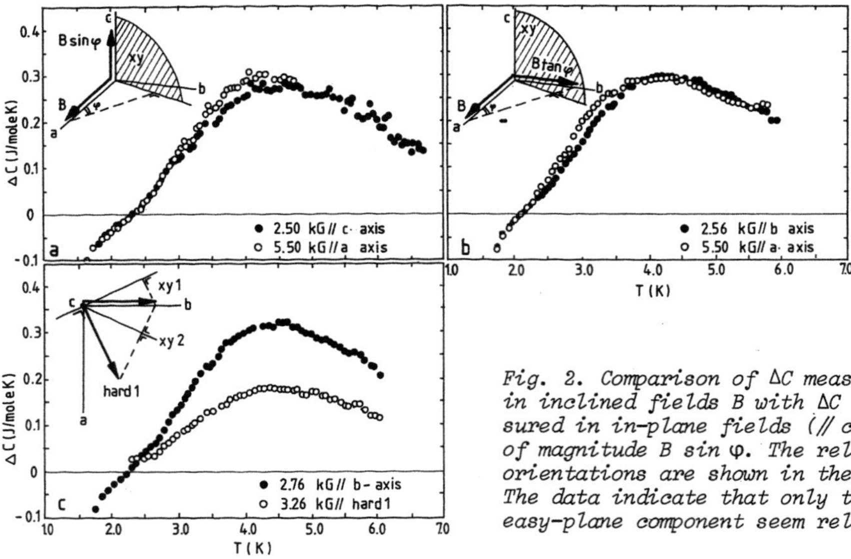

Fig. 2. Comparison of ΔC measured in inclined fields B with ΔC measured in in-plane fields (// c or b) of magnitude B sin φ. The relative orientations are shown in the insert. The data indicate that only the easy-plane component seem relevant.

obtained with a field B in the a-b plane are compared with a field along b or c yielding the *same component* of B in the XY easy plane, which is shown in the insert. This comparison as for instance shown in Fig. 2a and Fig. 2b reveals that the data can almost completely be understood from the effect of the XY plane component of the field. Fig. 2c refers to the situation where B is directed perpendicular to the XY plane of *one* of the two equivalent chains, compared with the configuration B// b yielding the same field component in *both* types of chains. The resulting field effect is shown to be approximately twice as large in this case. These observations corroborate the soliton interpretation since, as long as the hard component of the field is smaller than the out of plane anisotropy, only the in-plane component induces the symmetry breaking required for soliton excitations. Our data can be considered as the first experimental observation of this effect. The out of plane component of the field may alter the stability range of the solitons [12], but the data, which have been collected at fields well below the in-plane critical field above which kink solitons seem to be unstable (15 kG for CHAB) [4], do not indicate such an effect. We will return to this point when discussing the T_1 data. Furthermore, we like to note that although the data should still be corrected for the magnon contribution, this will not exceed $\Delta C \simeq -2.10^{-2}$ J/mole K in this field and temperature range and hence will not affect our conclusion.

The NSLR time T_1 has been measured in the paramagnetic state at the nuclear hydrogen positions. The data were obtained on single crystals, using conventional pulsed NMR, with B applied along the crystallographic axes. The results for B// c are shown in Fig. 3 in reduced form. The majority of the data result from measurements in the temperature range $2\ K \leq T \lesssim 6\ K$ and with external fields up till 60 kOe. The field dependence at four temperatures is also shown in the insert. Although the qualitative behavior of the T_1^{-1} based on the contribution of linear spin-wave excitations, especially at low T [1], is similar to what is actually observed, deviations appear at higher temperatures. Therefore we will try to interpret the data on basis of the soliton concept. When NSLR is due to collisions with a one-dimensional soliton gas, the relaxation rate T_1^{-1} is [13]:

$$T_1^{-1} \simeq \frac{1}{T} \exp(-E_s^o/kT) \simeq \frac{1}{T} \exp\ (-\alpha\sqrt{B}/T) \qquad (3)$$

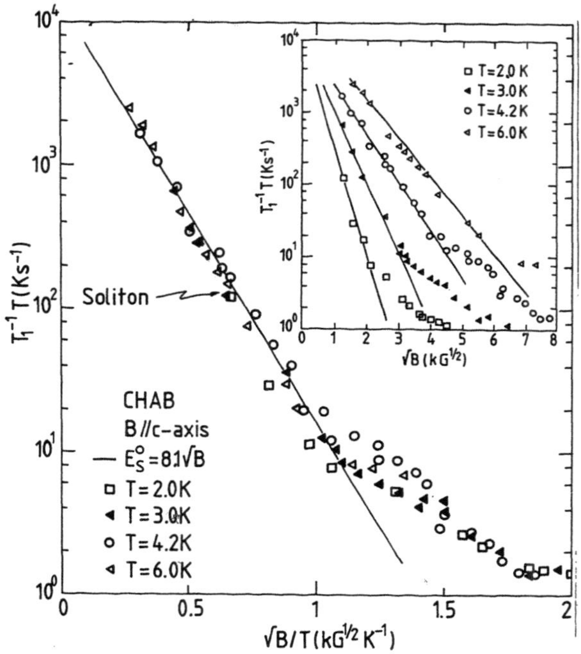

Fig. 3. Experimental behavior of $\ln(TT_1^{-1})$ as a function of $\sqrt{B}/T$. The solid line represents the classical soliton model with $E_s^o/k = 8.1\sqrt{B}$. The insert shows the field dependence more explicitly.

where $E_s^o/k = \alpha\sqrt{B}$ is the soliton rest energy.
Eq. 3 results in a linear relationship of $\ln(TT_1^{-1})$ and $\sqrt{B}/T$ for all fields and temperatures. As can be observed in Fig. 3, the data indeed are located on a straight line for $0.2 < \sqrt{B}.T^{-1} < 1$. At lower temperatures, where a spin-wave description might be adequate, deviations occur. The data result in $E_s^o/k = 8.1\sqrt{B}$, which agrees surprisingly well with the value $8.4\sqrt{B}$ obtained from the ΔC measurements.

Measurements of T_1^{-1} with B along the a and b axes were also performed. If only the in-plane component of B is considered, the data for B// b can also be described by Eq. 3. The data for B// a, however, cannot simply be scaled by the same procedure.

As was stated before, some of the data in Fig. 3 were obtained in fields outside the theoretical stability region which in CHAB extends up till B_c = 15 kOe. In particular the data as function of field plotted in the insert of Fig. 3 do not indicate any specific feature at or near this field value. Instead it appears that the breakdown of the soliton behavior occurs at a fixed value of the parameter $\sqrt{B}/T \simeq 1$, suggesting that, since $\sqrt{B}/T$ is roughly proportional to the density n_s, the deviations from the soliton description originate from the lack of statistical weight of the solitons.

Concluding,we state that the measurements of both ΔC and T_1 reported here yield strong evidence of kink solitons in a ferromagnetic quantum (S=1/2) XY system. The field and temperature dependence of both T_1 and ΔC can very well be described by *classical* soliton theory, both for in-plane and inclined magnetic fields. This is surprising, since in the present quantum limit the validity of the mapping of the spin hamiltonian to the sG hamiltonian is not clear [14]. Although this question triggered part of the extensive theoretical effort [15] which has been devoted to the properties of S=1/2 systems, general results for the infinite system in the presence of a symmetry-breaking field are not yet available.

The authors wish to acknowledge the contributions of Dr. H. Nishihara, J.L. Postulart, P.H.A. Mutsaers, R.A.C.M. Eberson and H. Hadders, and stimulating discussions with Dr. M. Steiner.

References

1. K. Kopinga, A.M.C. Tinus and W.J.M. de Jonge: Phys. Rev. B29, 2868 (1984)
2. C.P. Landee and R.D. Willett: Phys. Rev. Lett. 43, 463 (1979)
3. K. Kopinga, A.M.C. Tinus and W.J.M. de Jonge: Phys. Rev. B25, 4685 (1982)
   A.C. Phaff, C.H.W. Swüste and W.J.M. de Jonge: J. Phys. C17, 2595 (1984)
4. E. Magyari and H. Thomas: J. Phys. C15, L333 (1982)
5. A.C. Faria and A.S.T. Pires: J. Phys. C12, 2637 (1979);
   F. Boersma, Ph.D. Thesis, Eindhoven
6. A.P. Ramirez and W.P. Wolf: Phys. Rev. Lett. 49, 227 (1982)
7. J.F. Currie, J.A. Krumhansl, A.R. Bishop and S.E. Trullinger: Phys. Rev. B22, 477 (1980)
8. K. Maki: Phys. Rev. B24, 3991 (1981)
   H.J. Mikeska: Phys. Rev. B26, 5213 (1982)
9. F. Borsa: Phys. Rev. B25, 3430 (1982)
10. K. Sasaki: Progr. of Theor. Phys. 68, 411 (1982)
11. P. Kumar: Phys. Rev. B25, 483 (1982)
12. D. Hackenbracht and H.G. Schuster: Phys. Rev. B27, 572 (1983)
13. T. Goto and Y. Yamaguchi: J. Phys. Soc. Japan 50, 2133 (1981)
14. J.M. Loveluck, T. Schneider, E. Stoll and H. Jauslin: Phys. Rev. Lett. 45, 1505 (1981)
15. J.D. Johnson and J.C. Bonner: Phys. Rev. Lett. 44, 616 (1980)
    T. Schneider and E. Stoll: Phys. Rev. Lett. 47, 377 (1981)
    M.W. Puga and H. Beck, J. Phys. C15, 2441 (1982)

Mössbauer Relaxation Studies of π-Domain-Wall Dynamics in Pure and Impure Ising-Type Magnetic Chains

H.J.M. de Groot, R.C. Thiel, and L.J. de Jongh

Kamerlingh Onnes Laboratorium der Rijksuniversiteit te Leiden, Nieuwsteeg 18, NL-2311 SB Leiden, The Netherlands

1. Introduction

The subject of domain-wall dynamics in Ising-type ferro- and antiferromagnetic chains has recently become of great interest, since they provide a simple and readily accessible possibility to study nonlinear excitations [1,2]. By Ising-type is meant that there is a preferential direction for the moments, either through crystal field effects or dipolar anisotropy. Consequently, π-domain-walls can be present even in zero magnetic field and their density will determine the intrachain correlation length at low temperatures. In experimental quasi one-dimensional (1-d) magnetic systems, such walls are thermally excited because of entropy considerations. In the paramagnetic regime $T>T_c$, where T_c is the transition temperature to 3-d long-range order (due to the weak interchain coupling), the walls may propagate more or less freely along each of the chains. However, as T_c is approached more closely (e.g. within 10%), the interchain coupling provides a progressive damping mechanism for the wall-motion, and finally below T_c the walls become blocked and form a static 3-d domain structure. Thus, in a certain range above T_c the density of walls and their motion effectively determine the spin correlation function $\langle \vec{S}(o,o)\vec{S}(z,t)\rangle$, which may be studied by neutron scattering and such microscopic probes as NMR and Mössbauer effect experiments [1,2]. In the present work we briefly review a number of Mössbauer studies recently performed at our laboratory, as well as the relevant theoretical descriptions.

2. Theoretical concepts

The line splittings appearing in the Mössbauer spectrum reflect the hyperfine interactions between the nucleus and the solid state and can be separated in two contributions. First, the electric quadrupole interactions may give rise to a partially split spectrum in the paramagnetic regime. Second, when the atom carries a magnetic moment, this can lift the remaining degeneracy of the nuclear energy levels, and may be treated as an effective hyperfine field $\vec{H}_{hf}$ acting on the nucleus. Since $\vec{H}_{hf}$ contains a contribution proportional to the electron spin $\vec{S}$ the fluctuations in $\vec{H}_{hf}$ can be related to the spin-autocorrelation function $\langle \vec{S}(o,o)\vec{S}(o,t)\rangle$. In 3-d magnets, the fluctuations of $\vec{S}$ in the paramagnetic regime are very rapid compared to the nuclear Larmor frequencies ω_n, except for a small region in the immediate neighbourhood of T_c (i.e. $\Delta T/T_c \simeq 10^{-2} - 10^{-3}$, cf. fig. 1b). This implies that a magnetic splitting of the spectrum only occurs below T_c, and $\vec{H}_{hf}$ becomes proportional to the sublattice magnetization (fig. 1a). In quasi 1-d magnets however, the intrachain correlation lengths can be already substantial at temperatures $T \simeq 2T_c$, explaining why the observed magnetic splitting may persist far into the paramagnetic regime (fig. 1a), provided that the spin fluctuations are slower than ω_n^{-1}.

Evidently, spin fluctuations of the order of ω_n^{-1} will broaden the Mössbauer absorption lines. Since $\vec{H}_{hf} \propto \vec{S}(t)$, one may write for the in-

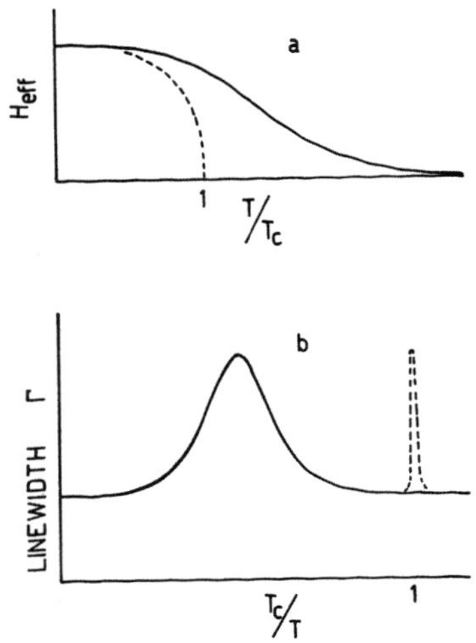

Fig. 1

Effective hyperfine field versus T/T_c (a) and Mössbauer linewidth versus T_c/T (b). Solid lines: Ising type chains, dashed lines: 3-d magnets. (Schematic)

crease in the linewidth:

$$\Delta\Gamma \propto \int_0^\infty dt \langle \vec{s}(o,o)\vec{s}(o,t) \rangle \exp(i\omega_n t) \ . \qquad (1)$$

In terms of the dynamical structure factors $S_\parallel(k,\omega)$ and $S_\perp(k,\omega)$, which are the space-time Fourier transforms of the parallel and perpendicular correlation functions, we have:

$$\Delta\Gamma \propto \sum_k [A_k S_\parallel(k,\omega_n) + B_k S_\perp(k,\omega_n)], \qquad (2)$$

where A_k and B_k are constants related to the nature of the hyperfine coupling. In most cases the term $S_\perp(k,\omega_n)$ can be neglected with respect to $S_\parallel(k,\omega_n)$. This can be understood as follows. Since the passage of a domain-wall turns the spins over by an angle π, only the spins inside the walls (with wall width d_s) contribute to $S_\perp(k,\omega_n)$. Therefore $S_\perp(k,\omega_n)$ will be proportional to the density of walls, n_s, and can be ignored for low n_s, to which case we shall restrict ourselves in this work. Since the walls move with velocities v_s comparable to the magnon velocity, i.e. $v_s \approx 10^{10}$-10^{12} l.u./s, the frequencies corresponding to the perpendicular fluctuations are much higher than $\omega_n \approx 10^7$-10^8 s^{-1}, and have no effect on the linewidth. Only for high values of n_s and/or very broad walls can the perpendicular fluctuations cause line broadening. Otherwise, the passage of a wall is seen as an instantaneous flip in the Mössbauer experiment, and the internal structure of the wall is unimportant.

As regards the parallel fluctuations contributing to $S_\parallel(k,\omega)$ we need to consider the average time between successive passages of domain walls. In the Ising-type magnets the walls separate degenerate groundstate regions differing by a rotation over an angle π of the (sublattice) magnetization, and these regions will each grow or shrink according to the wall propagations. Therefore the average rate of spin-reversals at a given site, i.e. the fliprate Γ_ω, will be proportional to the product of wall density n_s and mean wall velocity v_s. For low values of n_s we may assume the walls to be noninteracting and to propagate independently along the chain. Then the sequence of spin reversals at a given site will be stochastic, and one may take $S_\parallel(k,\omega)$ to a reasonable approximation to be the product of two lorentzian shapes in k and ω:

$$S_\parallel(k,\omega) \propto \left(\frac{\Gamma_\omega}{\omega^2+\Gamma_\omega^2}\right)\left(\frac{\Gamma_k}{k^2+\Gamma_k^2}\right). \tag{3}$$

Since the Mössbauer effect is a local probe one has to integrate over all k-values. For the above expression for $S_\parallel(k,\omega)$ the integration is easily performed as the widths in k and ω separate. For a linear chain the result of the integration is just a multiplicative constant (π), so that:

$$\Delta\Gamma \propto \sum_k S_\parallel(k,\omega_n) \propto \Gamma_\omega/(\omega_n^2+\Gamma_\omega^2). \tag{4}$$

Thus the line-broadening due to the wall propagation appears as an activated process, since Γ_ω depends exponentially on temperature through the relation $\Gamma_\omega \propto n_s v_s$ (see below). At high temperatures the flip-rate Γ_ω is very fast compared to ω_n, so that $\Delta\Gamma \propto 1/\Gamma_\omega$ (since $\Gamma_\omega >> \omega_n$). As the temperature is lowered the broadening will appear as soon as the parallel fluctuations become of the order of ω_n. Similarly, at very low temperatures the wall-density will become too low and the fluctuations too slow compared to ω_n to cause line-broadening. In this limit we may put $\Delta\Gamma \propto \Gamma_\omega$ (since $\Gamma_\omega << \omega_n$). Clearly, the total linewidth $\Gamma = \Gamma_0 + \Delta\Gamma$, with Γ_0 the instrumental resolution, will go through a maximum for $\Gamma_\omega \simeq \omega_n$ (cf. fig. 1b). Thus the presence of solitons may be detected by extracting linewidths from spectra taken at different temperatures, as was shown previously [5]. In practice, however, this procedure has several disadvantages. Indeed, for the ^{57}Fe Mössbauer nucleus there are actually four values of ω_n, differing by about one order of magnitude. Furthermore, for $\Gamma_\omega \to \omega_n$ the lineshape becomes non-lorentzian and the number of lines changes from 2 to 6 or 8. This makes it impossible to extract reliable linewidths close to ω_n, which is just the region where the most crucial information about the flipping process could be obtained. Also the frequency dependence of the line positions and intensities is not taken into account, and thus a lot of potential information is lost.

Therefore we have recently followed another approach, where we actually calculate the resonance conditions for the Mössbauer nucleus in the presence of the nonlinear excitations. The emission probability of the Mössbauer photon may be written as:

$$W_{e,g}(\omega) = 2\Gamma_0^{-1}\ \mathrm{Re}\int_0^\infty dt\ \exp[(i\omega - \tfrac{1}{2}\Gamma_0)t]\langle e|H^-|g\rangle\langle g|U^+H^+U|e\rangle. \tag{5}$$

Here $U(t) = \exp(-iHt)$ is the time evolution operator of the total solid state with hamiltonian H. The latter may be greatly simplified within the approximation in which the magnetic hyperfine interaction (the time-dependent part) is replaced by a fluctuating hyperfine field $\vec{H}_{hf}(t)$. For π-domain walls we may write: $\vec{H}_{hf}(t) = \vec{H}^0_{hf} f(t)$, with f(t) a stochastic step variable jumping between $\pm$ 1. This approach is known in Mössbauer literature as the Blume-Tjon relaxation model [6]. Applied to our present problem, the function f(t) corresponds with the fliprate Γ_ω, whereas the temperature variation of $\vec{H}_{hf}(t)$ corresponds to that of the autocorrelation function in fig. 1a. In principle therefore, $\vec{H}^0_{hf}$ should be constant, corresponding to the value of $\langle\vec{S}(o,o).\vec{S}(o,t)\rangle$ at T=0, and this leaves Γ_ω as the only adjustable parameter to fit the spectra taken at different temperatures. In practice though, we had to allow for a slight temperature dependent decrease of $\vec{H}^0_{hf}$, which however never exceeded 10% in the investigated temperature range ($T \lesssim 3\ T_c$). Such a gradual decrease of $\vec{H}^0_{hf}$ with T may be understood from the fact that the dynamics is not completely determined by π-walls only; for instance magnon processes will have to be considered in addition. However, their effects are small in the strongly anisotropic Ising systems that we shall consider, at least in the temperature range of interest. Thus we may safely state that the fliprate Γ_ω is the only temperature dependent parameter needed to fit the spectra in the relaxation regime, and the so-obtained Γ_ω as a function of temperature may then be directly compared with theoretical predictions for the domain-wall dynamics.

To this end the nature of the π-domain-walls has to be further specified. For classical spins and weak anisotropy, the walls may be considered as Sine-Gordon solitons. The interaction hamiltonian is:

$$\mathcal{H} = \sum_{i=1}^{N-1} \{-2J\, \vec{S}_i \cdot \vec{S}_{i+1} - D_x S_{ix}^2 + D_z S_{iz}^2\}, \tag{6}$$

with $J>0$ or $J<0$ for ferro- or antiferromagnetic chains. The anisotropy terms represent a main planar anisotropy term ($D_z>0$) establishing an easy (XY) plane, with a weaker inplane anisotropy ($D_x>0$) favouring the x-axis. For both $J>0$ or $J<0$ the hamiltonian may be transformed into its Sine-Gordon form:

$$\mathcal{H} = \tfrac{1}{2}E_0 \int_{-\infty}^{\infty} dz(\Psi_z^2 + \frac{1}{c_0^2}\Psi_t^2 + 2m_s^2 \sin^2 \tfrac{1}{2}\Psi) \quad , \tag{7}$$

and the resulting formulae for the various parameters are listed in table I. It is seen that for $J<0$ the soliton motion is driven by the exchange energy ($c_0 = 4|J|S$), whereas for $J>0$ it is driven by the orthorhombic anisotropy ($c_0 = 2\sqrt{D_z J}\, S$). In the usual low-density limit the soliton density is given by [4]:

$$n_s = (2/\pi)^{\frac{1}{2}}\, m_s\, (E_s/k_BT)^{\frac{1}{2}} \exp(-E_s/k_BT) \tag{8}$$

where m_s and E_s are the mass and the excitation energy of the soliton and the average velocity $v_s = c_0(E_s/2k_BT)^{-\frac{1}{2}}$ is determined by the thermal distribution.

Table I Sine-Gordon parameters entering into the hamiltonian for ferromagnetic and antiferromagnetic Ising-type chains

	E_0	m^2	c_0	$E_s = 8mE_0$
$J>0$	$\frac{1}{2}JS^2$	D_x/J	$2\sqrt{D_zJ}\,S$	$4S^2\sqrt{D_xJ}$
$J<0$	$\frac{1}{2}JS^2$	$D_x/\lvert J\rvert$	$4\lvert J\rvert S$	$4S^2\sqrt{\lvert D_xJ\rvert}$

However, the applicability of the Sine-Gordon model for our Fe^{2+} chains with strong anisotropy is questionable, since in that case the wall-width $d_s \propto 1/m_s$ becomes very small and the continuum approximation, essential for Sine-Gordon solitons, is no longer allowed. In addition, for such Fe^{2+} ions the low-temperature magnetic properties often have to be described within a low effective spin formalism, e.g. $S=\frac{1}{2}$, which is far from the classical spin $S=\infty$. We should therefore look for the analogues of Sine-Gordon solitons in a discrete quantum mechanical Ising-type chain, described by the hamiltonian:

$$\mathcal{H} = \sum_{i=1}^{N-1} \{-2J_x S_i^x S_{i+1}^x - 2J_y S_i^y S_{i+1}^y - 2J_z S_i^z S_{i+1}^z\} \tag{9}$$

where $J_z \gg J_x, J_y$ and $S=\frac{1}{2}$. For the antiferromagnetic chain with $J_x=J_y$ the problem has been treated by Villain [7] and others [8]. The nonlinear excitations are domain-wall pair-states, where a wall corresponds with a defect +-+-+ ⋮ +-+-+ in the normal +-+-+- sequence. Thus a possible wall pair-state would be +-+ ⋮ +-+-+- ⋮ -+-+. For the ferromagnetic case it has been

pointed out by Schneider et al. [10] that the analogous nonlinear excitations are the n-magnon bound states, e.g. +++ ⋮ ---- ⋮ ++++ , which again are seen to be wall pairs. For both $J<0$ and $J>0$ the basic excitation energy is approximately equal to $2J_z$. In the extreme Ising limit ($J_x=J_y=0$) these pair states form a complete set of (degenerate) orthogonal eigenfunctions of the system. In a perturbation approach one may add the small transverse (J_x and J_y) coupling terms, by which the degeneracy is lifted and the pair-states now form a small band around the Ising energy $E_I=2J_z$.
Between pair states of n and n±2 spins there is now a finite transition probability, which will result in local spin flips and may be seen as the quantum mechanical analogue of the wall propagation in the classical Sine-Gordon chain. Similarly as for the Sine-Gordon system, one finds for the quantum chain that for the ferromagnetic case a biaxial anisotropy ($J_x \neq J_y$) is needed to move the wall, whereas for the antiferromagnetic chain uniaxial transverse terms ($J_x=J_y$) are sufficient. Also for the quantum chains one may apply the stochastic approximation, as in the above, where the pair-states may be treated as a gas of noninteracting boson quasiparticles. One then finds $n_s \propto \exp(-E_I/k_BT)$, where now the Ising energy $E_I=2J_z$ replaces the classical soliton excitation energy. The average propagation velocity v_s in the quantum case may be derived from the dispersion by taking the expectation value $v_s = \langle \partial E(k)/\partial k \rangle$. In this stochastic approximation it therefore follows that for both Sine-Gordon solitons and the analogous excitations in the anisotropic quantum chains, the fliprate Γ_ω should show an exponential dependence, albeit with different excitation energies.

3. Experiments

Since a brief report on our experiments on antiferromagnetic chains has already appeared elsewhere [5], we shall restrict ourselves to the ferromagnetic case. As an example of an Ising-type chain we show data for the compound $FeCl_2(NC_5H_5)_2$. Below 30 K and down to $T_c = 6.7 \pm 0.2$ K, the system may be considered as an effective spin $S=\frac{1}{2}$ Ising-type ferromagnetic chain. From susceptibility measurements we determined $J_z/k_B = 25 \pm 2$ K. We also show data taken on two diluted versions of this compound, where c = 0.47% and c = 0.94% of the Fe^{2+} ions were replaced by nonmagnetic Cd^{2+}. This effectively reduces the intrachain correlation length and therefore the 3-d ordering temperature is lowered by at least a factor of two ($T_c \simeq 3.0$ K for c = 0.47%).

As typical Mössbauer spectra we show in fig. 2 data for the c = 0.47%

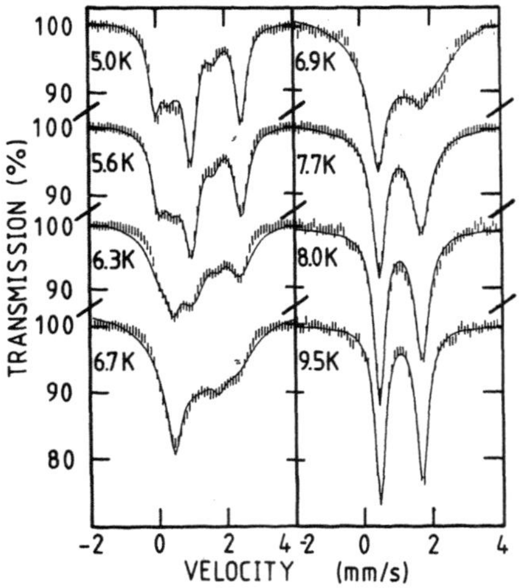

Fig.2

Mössbauer absorption spectra for the c = 0.47% $FeCl_2(NC_5H_5)_2$ sample at different temperatures ($T_c \simeq 3.0$ K). Solid curves represent fits to the model of stochastic domain-wall propagation.

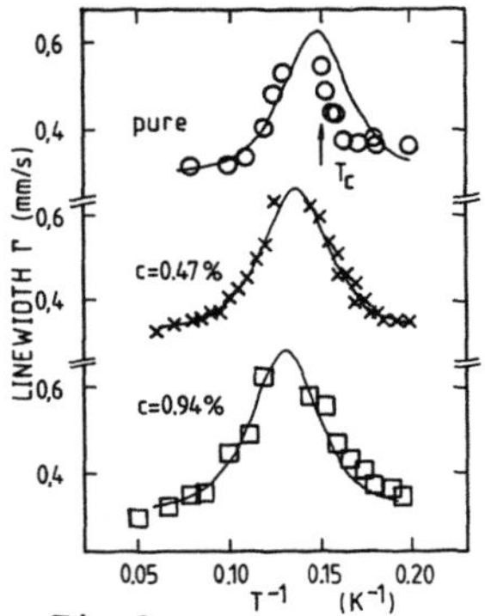

Fig.3
Experimental linewidths versus inverse temperature. Solid curves represent the calculated behaviour according to eq.4.

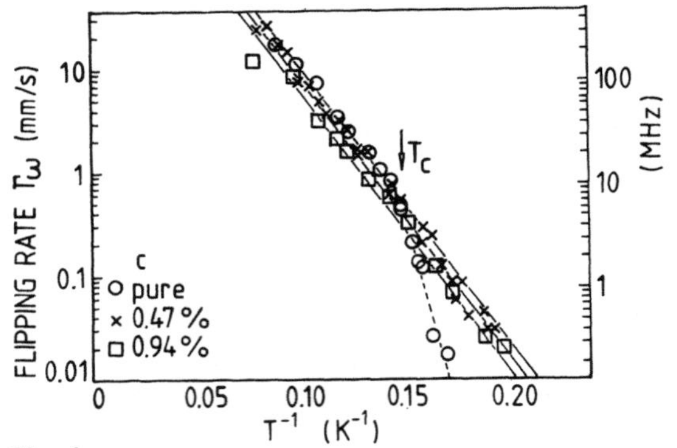

Fig.4

Electronic spins fliprate versus inverse temperature for the three $FeCl_2(NC_5H_5)_2$ samples. The dashed line is a guide to the eye. The parallel solid curves show the exponential dependences predicted by the soliton model.

compound. We emphasize the observation of a fully split spectrum up to T = 6.5 K, whereas $T_c \simeq 3.0$ K! The experimental linewidths extracted from the spectra by a fitting process with 8 or 2 Lorentzian lines are plotted in fig. 3 and are compared with the theoretical expression eq.(4). We note the sudden decrease of the linewidth at T_c to the rest value Γ_0 for the pure compound as a consequence of the blocking of the walls in the 3-d ordered regime. In figure 4, the values obtained for Γ_ω have been plotted against 1/T for the three compounds. The exponential behaviour yields the excitation energy $E_s = 60 \pm 5$ K which is close to $E_s = 2\,J_z = 50 \pm 4$ K, expected for the $S=\frac{1}{2}$ Ising chain using $J/k_B = 25 \pm 2$ K found from the susceptibility. For the diluted compounds T_c is much lower and the relaxation can be followed to the lower limit (0.3 MHz) of the Fe^{2+} Mössbauer frequency window. Finally we remark that the impurities do not seem to affect the wall dynamics apart from a slight decrease of the average velocity with increasing impurity, as may be concluded from the downward shifts of the parallel straight lines through the data in figure 4 with respect to that for the pure compound.

References
1. J.P. Boucher, L.P. Regnault, J. Rossat-Mignod, J.P. Resiard, J. Bouillot and W.G. Stirling, J. Appl. Phys. 52, 1956 (1981)
2. L.J. de Jongh, J. Appl. Phys. 53, 8018 (1982)
3. F. Borsa, Phys. Letters 80A, 309 (1980)
4. H.J. Mikeska, J. Phys. C13, 2913 (1980)
5. R.C. Thiel, H. de Graaf and L.J. de Jongh, Phys. Rev. Letters 47, 1415 (1981)
6. M. Blume and J.A. Tjon, Phys. Rev. 165, 446 (1968); M. Blume, Phys. Rev. 174, 351 (1968).
7. J. Villain, Physica 79B, 1 (1975)
8. See, e.g. S.E. Nagler, R.L. Armstrong and W.J.L Buyers, J. Appl. Phys. 52, 1971 (1981) and references cited therein
9. J.B. Torrance and M. Tinkham, Phys. Rev. 187, 587 (1968)

$k=0$ Excitations in the Quantum Chain Systems $(C_6H_{11}NH_3)CuCl_3$ and $(C_6H_{11}NH_3)CuBr_3$

K. Kopinga, W.J.M. de Jonge, C.H.W. Swüste, and A.C. Phaff
Department of Physics, Eindhoven University of Technology, P.O. Box 513
NL-5600 MB Eindhoven, The Netherlands

R. Hoogerbeets and A.J. van Duyneveldt
Kamerlingh Onnes Laboratorium der Rijksuniversiteit Leiden, Nieuwsteeg 18
NL-2311 SB Leiden, The Netherlands

Recently, much theoretical effort has been devoted to the properties of ferromagnetic quantum (S=1/2) chains [1,2]. The excitation spectrum of Ising-Heisenberg chains has been shown to consist of magnons, magnon continua and magnon bound-states, of which the latter can be interpreted as the quantum analogue of classical envelope solitons [3,4]. For XY-Heisenberg quantum systems, the understanding of the excitation spectrum is less advanced [2,5], whereas in the classical limit these systems show a rich variety of non-linear phenomena, including soliton-like excitations [6]. The lower energy levels of the S=1/2 XYZ system in zero field have been obtained rigorously by mapping onto an eight-vertex model [7].

A number of the studies mentioned above reveal that even a small amount of anisotropy may have a drastic effect on the excitation spectrum and thermodynamic properties, but the experimental evidence is rather scarce. The recently synthesized compounds $(C_6H_{11}NH_3)CuCl_3$ (CHAC) and $(C_6H_{11}NH_3)CuBr_3$ (CHAB) offer the possibility to compare the behaviour of two almost ideal Heisenberg ferromagnetic quantum chains with about 2% Ising-like and 5% easy-plane anisotropy, respectively [8,9]. In this paper we report FMR measurements on both CHAC and CHAB, which enabled us to study the anisotropy in these compounds in more detail and indicated for the first time the existence of magnon bound-states in a nearly isotropic ferromagnetic chain system.

The crystallographic structure of CHAC is orthorhombic, space group $P2_12_12_1$ (z=4). The structure consists of bibridged chains of $CuCl_3^-$ ions running parallel to c. The distance between the chains along a and b amounts to 10 Å and 8.5 Å, respectively [8]. Preliminary X-ray measurements on CHAB indicate a similar structure. The intrachain interaction amounts to $J/k=45 \pm 5$ K for CHAC and $J/k=55 \pm 5$ K for CHAB [9], whereas the interchain coupling is smaller by three orders of magnitude. Below $T_c = 2.21$ K for CHAC and $T_c = 1.50$ K for CHAB a 3d non-compensated antiferromagnetic ordering occurs. In the ordered state the spins lie in the ab plane and are canted away from the b-axis by an angle of 17° for CHAC [8] and 25° for CHAB [10].

The resonance experiments have been performed using a number of conventional microwave spectrometers. The samples (typically 5-10 mm^3) usually were glued to the bottom or wall of the waveguide, which was inserted in a ^4He cryostat. Various samples with different shapes were used. Since the resonance signals, which were obtained using magnetic field modulation, generally were very strong, care was taken to prevent signal distortion. The magnetic field was applied along the crystallographic a, b or c axis or was rotated in a plane perpendicular to one of these axes.

The experiments on CHAC and CHAB revealed several strong resonance signals in the region 1.2 K < T < 4.2 K, which depend on the orientation of the magnetic field. In CHAC a number of additional satellite lines (SL) can be detected, which are located close to some of the main resonance signals. These SL will be discussed in the last part of the paper. Rotation diagrams

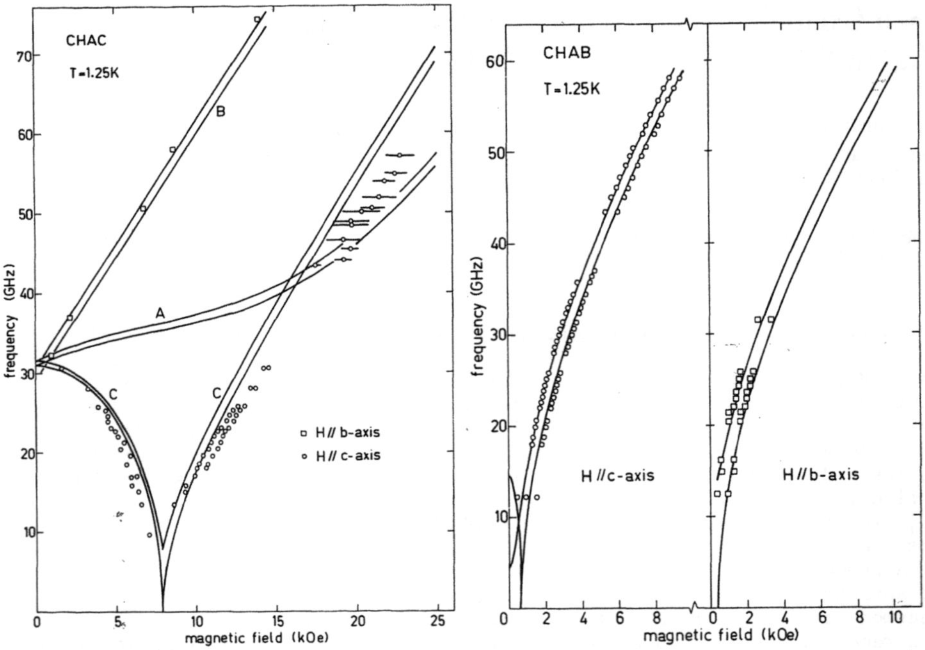

Fig. 1. Frequency-field diagram of the FMR in CHAC and CHAB at T = 1.25 K. Drawn curves are the predictions from linear spin-wave theory.

on CHAC and CHAB revealed two symmetry-related sets of signals. Each set can be associated with one of the two sets of symmetry-related chains, which are present in these compounds [8]. For fields along the crystallographic a, b or c axis the small interchain coupling induces a hybridization of the (degenerate) resonance modes, giving rise to a splitting of the observed line of ~500 Oe. Some representative results are plotted in Figs. 1 and 2.

In order to confront our FMR data with theory we start from the hamiltonian describing a single chain

$$H = -2 \sum_i (J^x S^x_i S^x_{i+1} + J^y S^y_i S^y_{i+1} + J^z S^z_i S^z_{i+1}) - \sum_i g\mu_B \vec{S}_i \cdot \vec{H} \quad . \qquad (1)$$

For a fully isotropic ferromagnetic chain the dynamic structure factor $S_{xx}(q,\omega) = S_{yy}(q,\omega)$ at T=0 (H// z) is exhausted by the magnon δ-function resonance [2]. Since the present compounds contain only a few percent of anisotropy and $kT \ll J$ we will first interpret our data in terms of linear spin-wave theory, where we associate the observed FMR with the k=0 magnon mode. For H along one of the principal axes the dispersion relation reads

$$\varepsilon_k = (A_k^2 - B_k^2)^{1/2}$$

$$A_k = 4SJ^z - 2S(J^x+J^y)\cos ka + g\mu_B H^z \; ; \qquad B_k = -2S(J^x-J^y)\cos ka \; . \qquad (2)$$

In CHAC and CHAB, only the crystallographic c axis corresponds with a principal axis (intermediate). For an arbitrary orientation of H, terms linear in the magnon operators occur, which can be eliminated by a suitable

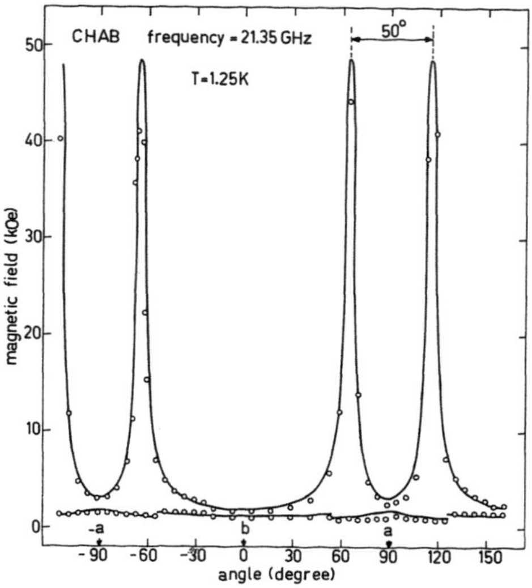

Fig. 2. Rotation diagram of the FMR in CHAB with the magnetic field in the ab plane at ν=21.35 GHz. Drawn curves are the predictions from linear spin-wave theory.

rotation of the xyz coordinate system. The small interchain coupling can be included rather straightforwardly. The results from this approach are represented in Figs. 1 and 2 by drawn curves, whereas the corresponding parameter values are given in table 1. It is obvious, that the experimental data are described very satisfactorily. The values of J^x, J^y and J^z confirm the conjectured [9] complementary anisotropy in CHAC and CHAB. At higher temperatures the observed signals shift to lower frequencies. Although - formally - linear spinwave theory is no longer valid, a fair description of the data can be obtained by normalizing the anisotropies (J^x-J^y) etc. A detailed report of these experiments is published elsewhere [11].

Table 1. Values of the exchange parameters in CHAC and CHAB obtained from a fit of linear spin-wave theory to the FMR data at T = 1.25 K. x, y and z denote easy, intermediate and hard, respectively.

	CHAC	CHAB
J^x/k (K)	45.52	55.92
J^y/k (K)	44.99	55.91
J^z/k (K)	44.49	53.18

As already mentioned above, a number of relatively weak satellite lines (SL) can be detected close to some of the FMR signals in CHAC, especially at higher T. Fig. 3 shows a typical spectrum recorded for H// b, whereas in Fig. 4 the location of the signals collected at T = 2 K is plotted for H// b. The location of the SL was found to be independent of crystal shape and micro-wave inhomogeneity and hence the SL cannot be attributed to magnetostatic modes. Rotation diagrams revealed that the SL always occur at the low frequency side of the FMR signal, whereas both their location with respect to this FMR signal and their spacing shows only minor variations. Hence the SL most likely originate from excitations in single chains. Since, moreover, SL are not observed in CHAB, their occurrence may be connected with

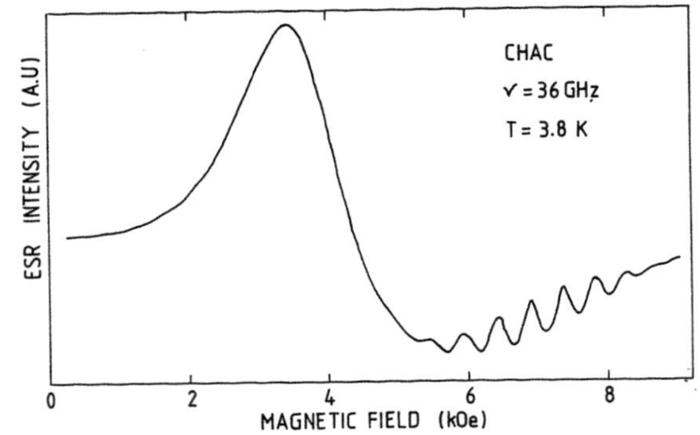

Fig. 3. ESR spectrum of CHAC at ν=36 GHz, H// b and T=3.8 K.

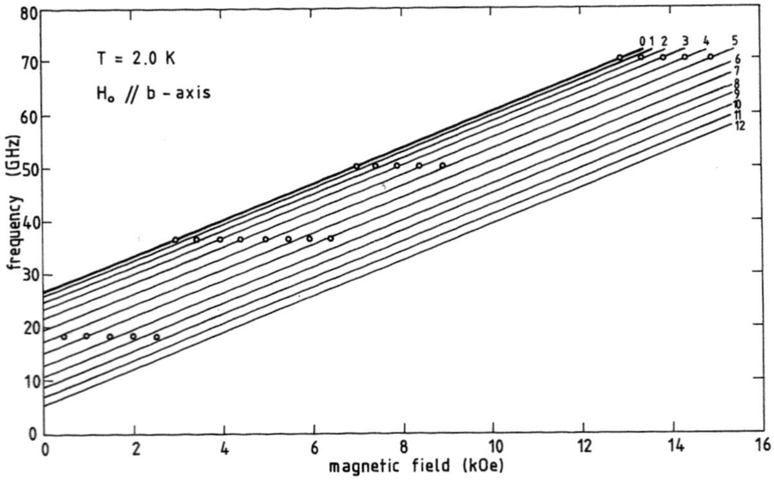

Fig. 4. Field dependence of the ESR frequencies in CHAC at T=2 K and H// b. Circles are the experimental data, drawn lines are the calculated MBR.

the type of exchange anisotropy in CHAC. Since no unique or complete theoretical description of the excitations in the present system exists, we will confront our data with several approximate theoretical models. We will focus our attention to the so-called magnon bound-state resonances (MBR), i.e. transitions between magnon bound-states, which up till now only have been observed in strongly Ising-like systems [12].

To start with, we used the following expression for the energy of the nth bound-state level of a ferromagnetic Ising-Heisenberg S=1/2 chain in a magnetic field along z [4,13]:

$$\varepsilon_n^{IH}(k) = \frac{2J^z}{\gamma}\frac{\sinh\varphi}{\sinh n\varphi}(\cosh n\varphi - \cos k) + nh \qquad (3)$$

where $\gamma = J^z/J^\perp$, $\cosh\varphi = \gamma$, and the reduced field h is parallel to z. Since the microwave field couples only states with the same k and transition probabilities are large for k close to zero, we take k=0.

Alternatively, we used the expression for the bound-state energies of the general XYZ S=1/2 chain in zero field [7]. With the tabulated values of

J^x, J^y and J^z an approximate formula for the k=0 bound-state energies can be obtained:

$$\varepsilon_n^{XYZ} = 16.464 \tanh(0.092n). \qquad (4)$$

For comparison, one might also use the expression for the energy of a classical envelope soliton in a system with uniaxial symmetry [3]:

$$\varepsilon_n^{CL} = \varepsilon_o N_1 \tanh(n/N_1) \;;\; N_1 = 2(J/\varepsilon_o)^{1/2} \qquad (5)$$

where ε_o equals the magnon energy gap (1.51 K in CHAC). If we insert the values given in table 1 we obtain ε_n^{CL}=16.48tanh(0.092n), which agrees surprisingly well with (4). The solution of the XYZ model is formally restricted to H=0. To enable a comparison with our data we assume that the bound-state energies depend - at least in first order - linearly on n and H and simply add a term nh to (4). To a certain extent this is justified by the exact solution for the Ising-Heisenberg model (3) and the experimental fact that the SL show almost the same field dependence as the FMR (cf. Fig. 4, H// b). The SL are now associated with n → n+1 transitions between magnon bound-states. The location of these MBR can be calculated directly from (3) or, alternatively, (4). For both models the calculated splitting of the MBR (i.e. the difference between the n → n+1 and n-1 → n transition in units of field) ranges from about 200 Oe (n=1) to a maximum of 700 Oe (n=7) and goes to zero for n → ∞. Although the experimental data suggest a more or less constant splitting, the overall agreement is rather satisfactory, as can be seen from Fig. 4, where the theoretical prediction is given by drawn lines.

Next, we will turn to the temperature dependence of the SL. In Fig. 5 the resonance fields are plotted for 1.2 K < T < 5 K for H// b and ν=36 GHz. Inspection of this figure shows that the spacing of the SL decreases at higher T. As already mentioned above, the main resonance signals at higher T can be described by a linear spin-wave model if the anisotropies (J_x-J_y) etc. are renormalized. If we calculate the MBR from the XYZ model using the same set of renormalized parameters also a decrease of the splitting is predicted, as is shown in the insert. Secondly, as may be deduced from the figure, SL with a larger spacing from the main signal (MBR with large n)

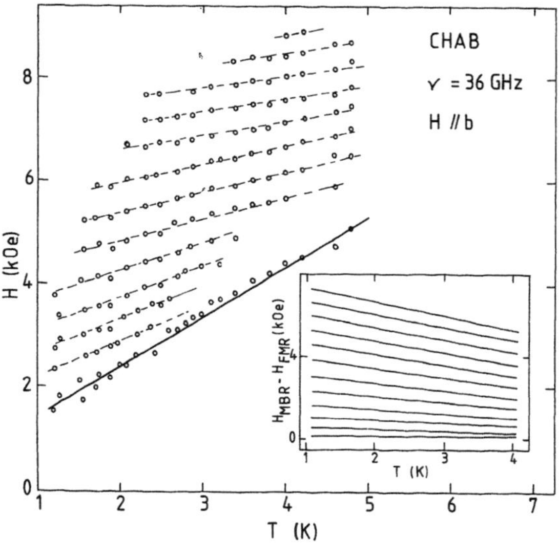

Fig. 5. Temperature dependence of the ESR fields in CHAC at ν=36 GHz and H// b. Circles are the experimental data. Lines through the data are meant as guides to the eye. Dashed lines and the drawn line denote satellite lines and the main FMR signal, respectively. The insert shows the theoretical prediction (see text).

can only be detected at progressively higher temperatures. This is consistent with the small occupation number of the corresponding magnon bound-state levels at low T.

Finally, we like to comment on the fact that SL have not been observed in CHAB, although formally (cf. table 1) this compound also can be described by the XYZ model. This question is elucidated by preliminary calculations, which reveal that the spacing of the SL in CHAB is about a factor 10 smaller than in CHAC, and hence the SL - if present - are most likely obscured by the very strong main FMR signal. Apart from this, the anisotropy within the easy plane in CHAB is very small (cf. table 1), and hence a description in terms of an XXZ model might be more convenient. In this context we mention recent measurements of the heat capacity and nuclear spin-lattice relaxation time in applied fields, which indicate the presence of kink-solitons in this compound [14].

The authors wish to acknowledge L.A. Bosch and P.F.M. v.d. Weyden for their help in collecting the data. We are much indebted to H. Hadders for the preparation of the crystals.

References

1. J.D. Johnson and J.C. Bonner: Phys. Rev. B22, 251 (1980)
2. T. Schneider and E. Stoll: Phys. Rev. B25, 4721 (1982); B26, 3846 (1982)
3. A.M. Kosevich, B.A. Ivanov and A.S. Kovalev: JETP Lett. 25, 486 (1977)
4. H.R. Jauslin and T. Schneider: Phys. Rev. B26, 5153 (1982)
5. T. Schneider, E. Stoll and U. Claus: Phys. Rev. B26, 1321 (1982)
6. See, for instance, H.J. Mikeska: J. Appl. Phys. 52, 1950 (1980)
7. R.J. Baxter: Ann. Phys. 70, 193 (1972); 70, 323 (1970)
   J.D. Johnson, S. Krinsky and B.M. McCoy: Phys. Rev. A8, 2526 (1973)
8. H.A. Groenendijk, H.W.J. Blöte, A.J. van Duyneveldt, F.M. Gaura, C.P. Landee and R.D. Willett: Physica 106B, 47 (1981)
9. K. Kopinga, A.M.C. Tinus and W.J.M. de Jonge: Phys. Rev. B25, 4685 (1982)
10. R. Hoogerbeets and A.J. van Duyneveldt: Physica 121B, 233 (1983)
11. A.C. Phaff, C.H.W. Swüste, W.J.M. de Jonge, R. Hoogerbeets and A.J. van Duyneveldt: J. Phys. C17, 2583, 2595 (1984)
12. J.B. Torrance and M. Tinkham: Phys. Rev. 187, 587 (1969)
    Q.A.G. van Vlimmeren and W.J.M. de Jonge: Phys. Rev. B19, 1503 (1979)
13. I.G. Gochev: JETP 34, 892 (1972)
14. K. Kopinga, A.M.C. Tinus and W.J.M. de Jonge: Phys. Rev. B29, 2868 (1984)

Magnon-Phonon Coupling in Spin-1/2 Quasi 1D Antiferromagnets

D.J. Lockwood

Division of Physics and Division of Microstructural Sciences
National Research Council, Ottawa, K1A 0R6, Canada

Compounds possessing the $CsNiCl_3$ structure [1] have attracted considerable interest recently. The crystal structure (space group Pb_3/mmc) comprises chains of face-sharing $NiCl_6^{4-}$ octahedra lying along the c axis separated by the large Cs^+ ions. For those compounds exhibiting magnetic ordering the exchange interaction between divalent transition-metal ions along the chain direction can be much larger than that between chains, resulting in pronounced one-dimensional (1D) magnetic behaviour [2]. The cobalt salts $CsCoCl_3$, $CsCoBr_3$ and $RbCoCl_3$ have proved to be excellent examples of a spin-½ 1D Ising-like antiferromagnet. All three salts exhibit three-dimensional (3D) magnetic ordering at low temperatures [3-5], and their respective Néel temperatures are listed in Table 1. At low temperatures, below T_N^{II}, the spin arrangements in $CsCoCl_3$ and $CsCoBr_3$ comprise that of a triangular (ab plane) Ising lattice with interchain antiferromagnetic nearest-neighbour coupling, J', very weak ferromagnetic next-nearest-neighbour interchain interactions and strong Ising coupling, J, along the chains. Thus the chains are ordered ferrimagnetically within the ab plane. For temperatures $T_N^{II} < T < T_N^{I}$ one-third of the chains are disordered while the other two-thirds remain antiferromagnetically coupled in the basal plane. Above T_N^I all of the chains are disordered but they retain significant spin correlations along the c axis for temperatures beyond $2T_N^I$. The magnetic structure of $RbCoCl_3$ is not known but it is expected to be similar to that found for the Cs salts [5]. There have been numerous measurements of the spin dynamics in these compounds using both neutron and optical techniques (e.g. [5-8] and references therein). The zone-centre 1D spin wave response consists of a peak superimposed on a continuum extending up to the frequency of the sharp zone-boundary mode. At temperatures below T_N^I the $\vec{q} = 0$ spectrum is remarkable for the appearance of a number of sharp lines whose intensities are strongly temperature dependent. All of these features have been satisfactorily accounted for in terms of an effective Hamiltonian of the form [6]

$$H_{eff} = 2J\sum_i[S_i^z S_{i+1}^z + \varepsilon(S_i^x S_{i+1}^x + S_i^y S_{i+1}^y)] + h\sum_i(-1)^i S_i^z \qquad (1)$$

where $h = h_0 + 2J'\sum_\delta \langle S_{i+\delta}^z \rangle$. Values of J, J', ε and h_0 appropriate to each salt for $T < T_N^{II}$ are given in Table 1. In all three salts a strong spin-phonon coupling has been observed between the magnon and E_{1g} phonon. The experimental evidence for such symmetry-allowed coupling is presented next.

1. $CsCoBr_3$

The E_{1g} phonon at 72.5 cm^{-1} in $CsCoBr_3$ lies well below the magnon band peaked near 100 cm^{-1} at $\vec{q} = 0$. Nevertheless a strong spin-phonon coupling has been observed in both neutron [6] and Raman [7] scattering experiments.

Table 1. Magnetic parameters for various alkali trichloro metalates [3-6]. The E_{1g} phonon frequency is also given [5,7,8]

Compound	T_N^I (K)	T_N^{II} (K)	J (cm^{-1})	ε	J' (cm^{-1})	h_Ω (cm^{-1})	E_{1g} phonon (cm-1)
$CsCoCl_3$	20.8	~ 9	49.9	0.12	1.0	0.6	118
$CsCoBr_3$	28.3	~10	54.1	0.14	3.2	1.1	73
$RbCoCl_3$	~28	~11	~50	~0.1	~1.5	-	115

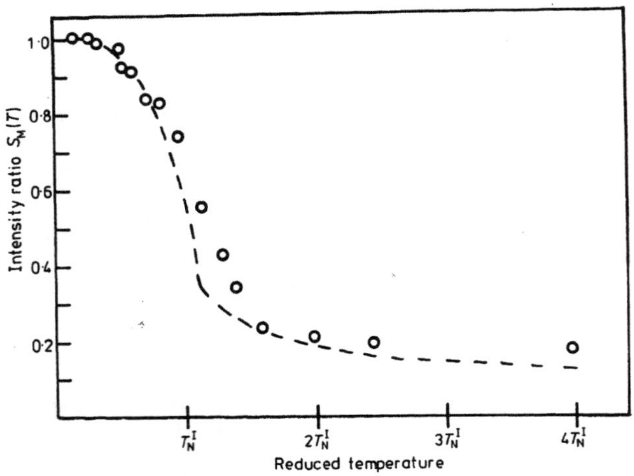

Figure 1. $S_M(T)$ for the E_{1g} phonon in (ZX) polarization in $CsCoBr_3$ (o). The broken curve is the calculated result for $CsCr_2S_4$ [9]

For example, the integrated Raman intensity of the E_{1g} phonon decreases rapidly with increasing temperature up to T_N^I, as is shown in Fig. 1. No other Raman active phonon exhibits such a marked temperature dependent behaviour, even though an E_{2g} phonon at 118 cm^{-1} overlaps the magnon band and could, from symmetry arguments, be coupled to it. The temperature dependence of the E_{1g} phonon intensity has been explained in terms of a spin-dependent mechanism [7]. The Raman scattering intensity is given by

$$I(T) = (\bar{n}+1)|R + M\langle\vec{S}_i \cdot \vec{S}_j\rangle/S^2|^2 \quad (2)$$

where $\bar{n}$ is the Bose population factor and R and M are the spin-independent and spin-dependent scattering amplitudes, respectively. The 3D spin-correlation function may be determined experimentally by plotting the function

$$S_M(T) = \langle\vec{S}_i \cdot \vec{S}_j\rangle/S^2 \simeq [I(T)^{\frac{1}{2}}-I(\infty)^{\frac{1}{2}}]/[I(0)^{\frac{1}{2}}-I(\infty)^{\frac{1}{2}}] \quad (3)$$

as is done in Fig. 1. The close similarity between the $CsCoBr_3$ results and the broken curve ($S_M(T)$ as calculated for the 3D antiferromagnet $CdCr_2S_4$ [9]), demonstrates that the temperature dependence is governed by the 3D ordering. JOHNSTONE and DUBICKI [7] further showed that since the E_{1g} phonon scattering is non-diagonal, only the interchain Co^{2+} ion pairs contribute to the Raman scattering and thus $S_M(T)$ is a direct measure of the spin-correlation function for these pairs alone.

2. $RbCoCl_3$

In $RbCoCl_3$ the E_{1g} phonon is at a higher frequency (see Table 1), because of the smaller anion mass, and now lies within the band of magnon frequencies. The temperature dependence of this magnon-phonon combination as seen in Raman scattering [5] is shown in Fig. 2. A coupled-mode analysis of this scattering at temperatures from 20 to 50 K showed that in the 1D phase the magnon-phonon coupling constant has a weak temperature dependence and is approximately 10% of the phonon energy [5]. The onset of 3D ordering has a marked effect on this coupling, as can be seen in Fig. 2. A peak at 110 cm^{-1} emerges beside the phonon, but is not strongly coupled to it. The origin of this additional feature is not understood. Furthermore, the E_{1g} phonon integrated intensity is temperature independent in the 3D-ordered régime contrary to what was observed in $CsCoBr_3$.

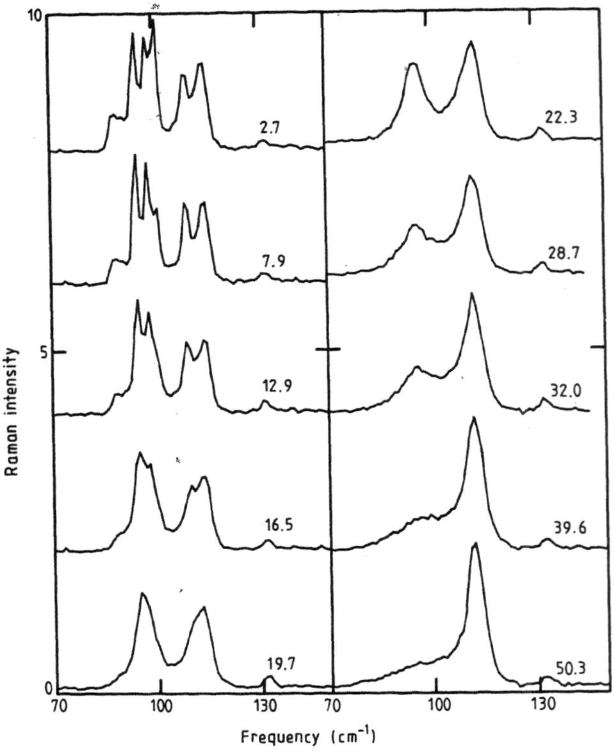

Figure 2. Temperature dependence of the X(ZX)Y Raman spectrum of $RbCoCl_3$ in the 100 cm^{-1} region. Values on the spectra are the temperatures in K

3. $CsCoCl_3$ and $CsCo_{1-x}M_xCl_3$

Pure $CsCoCl_3$ was the first of the cobalt compounds to be studied by inelastic scattering techniques and results obtained for the magnon were at first considerably confused by the presence of the E_{1g} phonon. In both neutron and Raman scattering the $\vec{q} = 0$ E_{1g} mode was found to lie near the end of the magnon continuum, and values for its frequency ranged from 118 to 127 cm^{-1} (see the discussion in [6]). As in the other cobalt compounds, strong magnon-phonon coupling was observed. In the Raman spectrum at low temperature the phonon intensity is extremely small, which lead to some disagreement as to its frequency (see [10] and references therein). However, the most recent work [8] performed at temperatures $T > T_N^I$ clearly shows that the E_{1g} phonon is at 118 cm^{-1}. The E_{1g} phonon intensity increases rapidly with increasing temperature and for $T >> T_N^I$, the magnon-phonon spectrum is similar to that of $RbCoCl_3$.

Dilution of Co^{2+} ions in $CsCoCl_3$ with Mg^{2+} ions results in a most unusual concentration dependence of the E_{1g}-phonon Raman scattering [10]. Although the phonon frequency in $CsCo_{1-x}Mg_xCl_3$ is concentration independent (at least for $0.2 < x < 1.0$), the lineshape, polarization and intensity behave anomalously. The E_{1g} mode appears with considerable intensity in (YX) polarization, where it is forbidden; the mode loses intensity with decreasing x such that it becomes very weak in pure $CsCoCl_3$; a lower-frequency sideband, which is absent in pure $CsMgCl_3$, grows rapidly in intensity relative to the E_{1g} phonon with increasing cobalt concentration. Recent Raman studies of $CsCo_{1-x}Ni_xCl_3$ [11] confirm the presence of the $\vec{q} = 0$ E_{1g} phonon near 118 cm^{-1}, but the introduction of another magnetic ion considerably modifies the magnon-phonon spectrum, as is shown in Fig. 3. A strong magnon-phonon coupling is evident and, as for pure $CsCoCl_3$, the E_{1g} phonon becomes visible only at higher temperatures. The frequencies of the magnon and the phonon, and the magnon-phonon coupling, is strongly influenced by the Ni^{2+} ion concentration. For example, the magnon and E_{1g}-phonon peak frequencies at 15 K are 86.7 and 115.6 cm^{-1} respectively for x = 0.02, and 88.6 and 117.4 cm^{-1} respectively for

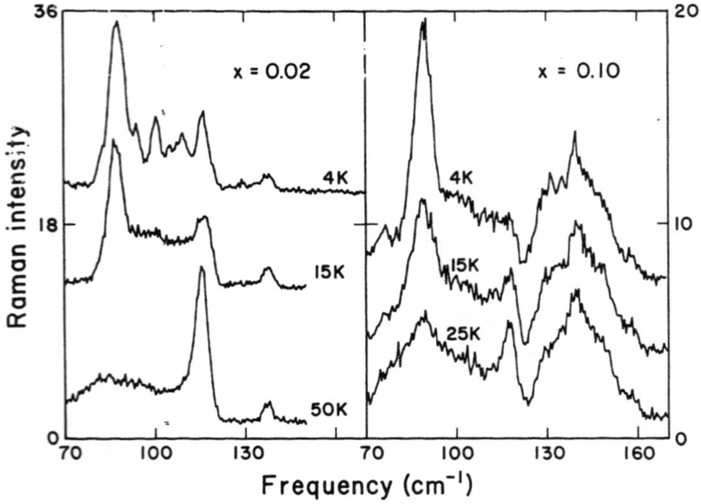

Figure 3. Temperature dependence of the X(ZX)Y Raman spectrum of $CsCo_{1-x}Ni_xCl_3$ in the 100 cm^{-1} region

x = 0.10. Similar results were obtained from studies of other mixed antiferromagnets $CsCo_{1-x}M_xCl_3$, with M = Mn, Cu and Fe [11].

4. Conclusion

This review of the experimental evidence for magnon-phonon coupling in the above cobalt compounds shows that the phenomenon is sensitive to the Co^{2+} ion environment. Extremely variable results regarding the intensity of the E_{1g} phonon and its temperature dependence are obtained depending on the nature of the alkali and halogen ions, and the presence of other metal ions. It should be noted that Raman results for other AMX_3 compounds such as $RbNiCl_3$ or $CsMnBr_3$ show no anomalies in the E_{1g} phonon scattering [5]. From this it is apparent that the E_{1g} mode is particularly sensitive to the chemical nature of the divalent metal ion and its interaction with surrounding ligands. The reason for this sensitivity lies in the nature of the c-axis normal mode displacements of the halogens [10], which modulate the superexchange pathways between interchain and intrachain pairs of transition metal ions. However, as noted above, it appears that the interchain Co^{2+}-ion pairs dominate the Raman polarizability at low temperatures, and this is postulated to be a consequence of the very efficient super-exchange pathway provided by low lying ligand-to-metal charge transfer states coupled with "near resonance" experimental conditions [7].

No definitive theoretical interpretation of the coupling has been established at present. BOSE and CHATTERJEE [12] have explained the experimental observation that the intensity of the magnon Raman scattering in $CsCoCl_3$ falls off drastically as soon as the E_{1g} phonon line position is crossed [8]. The presence of the phonon in the spectrum and its coupling to the magnon was accounted for by including in the Hamiltonian (1) a term

$$H_{ph} = E_{ph}b^+b + (2/N)^{\frac{1}{2}}A \sum_{i=1}^{N} (S_i^+b + HC) \ . \qquad (4)$$

The first term in H_{ph} generates the unperturbed phonon energy in the Bose representation, while the second term represents the magnon-phonon interaction. Assuming a magnon-phonon coupling parameter, A, of the order of 2 cm^{-1} they predicted that for $T < T_N^I$ the intensity of the magnon scattering just below and just above the E_{1g} phonon varies by a factor of 6 because of the coupling. More theoretical work is required to explain the strong spin-phonon coupling in these compounds, particularly with regard to the magnon-phonon lineshape and its temperature dependence, and it is hoped that this paper will stimulate such work.

References

1. H.F. McMurdie, J. DeGroot, M. Morris and H.E. Swanson: J. Res. N.B.S. A 73, 621 (1969)
2. N. Achiwa: J. Phys. Soc. Japan 27, 561 (1969)
3. M. Melamud, H. Pinto, J. Makovsky and H. Shaked: Phys. Status Solidi(b) 63, 699 (1974)
4. W.B. Yelon, D.E. Cox and M. Eibschütz: Phys. Rev. B 12, 5007 (1975)
5. D.J. Lockwood, I.W. Johnstone, H.J. Labbe and B. Briat: J. Phys. C 16, 6451 (1983)
6. S.E. Nagler, W.J.L. Buyers, R.L. Armstrong and B. Briat: Phys. Rev. B 27, 1784 (1983)

7. I.W. Johnstone and L. Dubicki: J. Phys. C 13, 4531 (1980)
8. W.P. Lehmann, W. Breitling and R. Weber: J. Phys. C 14, 4655 (1981)
9. E. Callen: Phys. Rev. Lett. 20, 1045 (1968)
10. I.W. Johnstone, G.D. Jones and D.J. Lockwood: Solid State Commun. 39, 395 (1981)
11. R.W.G. Syme and D.J. Lockwood: to be published
12. I. Bose and S. Chatterjee: J. Phys. C 16, 6681 (1983)

Spin Dynamics in $CsNiF_3$: Spinwaves, 2-Spinwaves, Solitons

M. Steiner
Hahn-Meitner-Institut, Glienicker Straße 100 D-1000 Berlin 39

Abstract:
A survey is given about the present knowledge of the spindynamics in the 1-D ferromagnet $CsNiF_3$ with and without external field. The identification of the different contributions to the spindynamics by neutron scattering expériments is discussed as is the use of the different theoretical approaches. It is shown that despite the wealth of experimental data, basic questions about the system are still open.

$CsNiF_3$ has a hexagonal crystal structure consisting of NiF_6-octahedra, which by sharing common faces build up infinite chains along the c-axis. Susceptibility measurements indicated the dominant interaction to be ferromagnetic, g = 2.4, and the occurrence of threedimensional long range order was found for $T < T_N = 2.66$ K. For a large temperature range $T > T_N$ deviations from a Curie-Weiss-law were found, which could be attributed to ferromagnetic short range order along the chains. One can thus set up the following Hamiltonian to describe the behaviour of the Ni^{2+}-chains (S = 1):

$$(1) \qquad H = -2J \sum_i \vec{S}_i \vec{S}_{i+1} + A \sum_i (S_i^z)^2 - g\mu_B H^y \sum_i S_i^y .$$

The anisotropy parameter A ensures that the spins lie in the x-y plane, if $A > 0$, as found from magnetization measurements [1]. Although (1) is a reasonable starting point, the possiblity of having anisotropic exchange can not be excluded. In what follows we are mainly interested in the spindynamics which are contained in pair correlation functions $\langle S^\alpha S^\alpha \rangle$, α = x, y, z. These correlation functions need not be equal: for H = 0 the ones for x and y are equal, while for $H \neq 0$, all three are different. Once we have measured $\langle S^\alpha S^\alpha \rangle$ we want to compare these results to the results calculated on the basis of (1), which should describe q, ω, H, T-dependences for all α. In addition we want that, in order to obtain a satisfactory theoretical description of the experimental findings, not only the dynamics but also the thermodynamic properties are described by the same type of calculations. The experimental technique used to explore the spindynamics in $CsNiF_3$ was inelastic neutron scattering. The quantity which describes the cross-section is the dynamic structure factor $S(\vec{Q}, \omega)$ which is the space and time dependent Fourier transform of the pair correlation function $\langle \vec{S}_r(o)\, \vec{S}_{r'}(t) \rangle_\perp$, where the subscript $\perp$ denotes spin components perpendicular to the momentum transfer $\vec{Q}$ [1]. Thus, by making a suitable choice of $\vec{Q}$, it is possible to measure $\langle S^\alpha S^\alpha \rangle + \langle S^\beta S^\beta \rangle$, α = x, y, z; β = x, y, z; $\alpha \neq \beta$. By measuring the three possible combinations, in principle the determination of the individual correlation functions should be possible. In practice, however, this turns out to be difficult. There is another possibilty to obtain further selection rules for the scattering and thus to study one specific part of the spindynamics: if polarized neutrons are used and if the spin state of the neutron (+, -) after the scattering is analysed then one can distinguish between spin-flip scattering from transverse spinfluctuations and non-spin-flip scattering from longitudinal fluctuations (with respect to an applied field) [2]. In discussing the result for $CsNiF_3$ we consider three cases: ① $H = 0$; ② $H \neq 0$, T low; ③ $H \neq 0$, T high.

① H = 0 [3].

If T > 4.2 K one can neglect the effect of interchain coupling and the system can be considered to be onedimensional. For T > 4.2 K spinwaves were observed for q_c > 0.1 r.l.u., whose dispersion could be very well described by

$$\hbar\omega = 2S\{(2J - 2J\cos\pi q_c)(2J - 2J\cos\pi q_c + A)\}^{1/2} \tag{2}$$

with J/k = 11.8 K, A/k = 4.5 K confirming the conclusion from susceptibility and magnetization measurements, that $CsNiF_3$ is a ferromagnetic chain system with planar anisotropy. It is this planar anisotropy which yields a gapless dispersion and different line shapes for the excitations in the in-plane fluctuations (IPC) and the out-of-plane fluctuations (OPC). While in the IPC the lines have a large width, which increases linearly with T, the OPC lines are very narrow. This is in agreement with a semiclassical theory [4], which gives the correct relative strength of the two contributions as well. These spinwaves should disappear if the wavelength becomes larger than the correlationlength. The observed breakdown of the spinwaves could be very well described by the above theory. A recent quantum calculation is in excellent agreement with the data too, but no statement can be made as to which theory is more precise [5].

No experimental data do exist for the q-dependence of the linewidth. Since there is some discrepancy in the theoretical predictions such an experiment should be done to clear that point. In conclusion, the spindynamics in $CsNiF_3$ with H = 0 are well described by considering spinwaves as the only collective excitations. No differences between semi-classical and quantum mechanical calculations for the Hamiltonian (1) could be found. This might be surprising, because the fluctuations are very large, even suppressing long range order along the chain.

(2) $H \neq 0$, T low.

As soon as a magnetic field is applied in the easy plane perpendicular to the chain axis, the spinwave dispersion shows a gap, E_o, at the zone center with $E_o \sim \sqrt{H}$. Since the field breaks the rotational invariance in the easy plane, the spindynamics is different in the three different correlation functions: the two transverse correlation functions $\langle S^x S^x \rangle$ and $\langle S^z S^z \rangle$ show spinwaves, whereas $\langle S^y S^y \rangle$, the longitudinal correlation function does not. The spinwave dispersion and the strength in the two transverse correlation functions can be consistently described using a linear quantum spinwave calculation with J/k = 11.8 K and A/k = 9 K (S = 1) for different H and q [6] for low temperatures, where low means $T < E_o/k$. There is, however, another correlation function, whose dynamic can be looked at: $\langle S^y S^y \rangle$. One can observe longitudinal spin fluctuations there only, the simplest of which are two-spinwave processes: two-spinwave sum process at twice the spinwave energy and two-spinwave difference process around zero energy transfer. The sum process can clearly be identified and was observed at rather large fields and low temperatures so that $T \ll E_o/k$. The experimental results could be described by quantum linear spinwave theory [7]. Thus the spindynamics in a field at low enough temperatures are well understood as showing 1- and 2-spinwave processes only. Because of this rather simple situation it was attempted to find out whether (1) is correct or whether exchange anisotropy has to be considered too. This was done by studying selection rules for spin wave creation by polarized neutrons. In contrast to a Heisenberg ferromagnet in a field, down-neutrons can with a certain probability excite spinwaves in a planar-anisotropy system. The data obtained in the accessible H, T and q-range could not discriminate between the behaviour of the two different types of anisotropy. They did show, however, that the planar character of the system is retained up to rather high temperatures, $T > A/k$ [8]. Before we go on to the most difficult regime in H and T, it should be said that linear spinwave theory yields a very good and coherent description of the spindynamics in $CsNiF_3$. The differences between semiclassical and quantum theories do not appear to be serious and can be easily understood as a renormalization of the anisotropy due to the spin value S = 1. It was, however, not possible up to now to make a final statement about the type of anisotropy in the system.

(3) $H \neq 0$; high T.

From what has been discussed above, the high T-regime can be defined by $T > E_0/k$. In this regime one would expect the spinwaves to be still present though they might be damped. The 2-spinwave difference process should be important in this T-regime, because it is a temperature-activated process (simultanous creation and annihilation of 2 spinwaves). In addition it has been predicted that 2π solitons should contribute to the spindynamics [9] yielding an additional cross section around zero energy transfer. In experiment one observes beside the spinwave peaks a strongly T, H and q-dependent scattering around zero energy transfer. In order to discriminate between the two contributions it is necessary to know the differences in the cross sections for the two dynamic processes: 2-spinwave processes are longitudinal fluctuations and can be observed in $\langle S^y S^y \rangle$ only, solitons on the other hand appear in all three correlation functions [10]. Thus the experimental task is to measure the $\langle S^\alpha S^\alpha \rangle$, α = x, y, z, separately, in particular to measure the q-dependence of the central peak intensity, because this would yield information on the shape of the solitons as reflected in $\langle S^x S^x \rangle$.

A careful experimental study of the different correlation functions, in particular of $\langle S^x S^x \rangle + \langle S^z S^z \rangle$, yields clear evidence for a soliton contribution to the cross section. Assuming $\langle S^x S^x \rangle \gg \langle S^z S^z \rangle$ we can obtain the strength of the soliton scattering with respect to the 2-spinwave scattering. We find I_{2sw}/I_{sol} = 2 while sine-Gordon theory yields 0.3 for this ratio. With the exception of this ratio, all experimental data are well described by the sine-Gordon theory [10], which is a purely classical theory. It should be mentioned that the spinwave scattering is described satisfactorily by a corresponding classical theory as well [10]. The assumption $\langle S^z S^z \rangle \ll \langle S^x S^x \rangle$ was crucial for our analysis, but $\langle S^z S^z \rangle$ is believed to be a sensitive measure for soliton instabilities expected in such a system [11]. Therefore an attempt was made using polarized neutrons to get a direct mea-

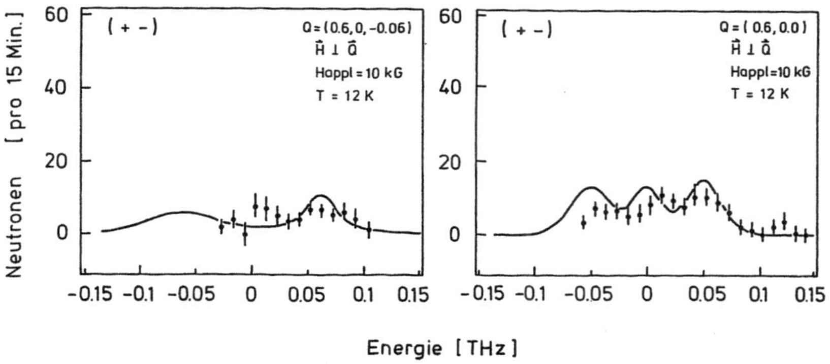

Fig. 1: $\langle S^zS^z\rangle$ as measured by polarized neutrons; solid line is simple sine-Gordon theory.

sure of $\langle S^zS^z\rangle$. The first indirect measure was the observation of a polarization dependence of the central peak in $\langle S^xS^x\rangle$ [8]. But, using all the possibilities of inelastic scattering of polarized neutrons, it was possible to measure $\langle S^zS^z\rangle$ alone. First results are shown in Fig. 1. The solid line is a calculation based upon the sine-Gordon equation using the strength of the soliton scattering observed in $\langle S^xS^x\rangle$. Although, preliminary, these results show that $\langle S^zS^z\rangle$ is small indeed [12]. The conclusion for this $H \neq 0$, large T-regime is, thus, that the spindynamics consist of single spinwaves, two-spinwave processes and sine-Gordon-like solitons. A classical theory including these processes can describe the experimental results consistentely (except for the relative strength of 2-spinwave and soliton scattering). At this moment one should ask whether this agreement is surprising or not, because there is a series of important assumptions to be made to use the above mentioned theory, which are not met by the real system: Finite anisotropy; Discrete system; Soliton-soliton-interaction; Soliton-magnon-interaction; Dilute soliton-gas limit; Soliton stability; Higher order spinwave processes. In addition thermodynamic quantities like specific heat cannot be described by classical models [13] and quantum corrections are apparently serious (see these proceedings).

Thus in general the spindynamics in $CsNiF_3$ are rather well understood at least on a phenomenological basis. One has to distinguish two different regimes: the linear regime for $H = 0$; $H \neq 0$, $T \ll E_0/k$ and a nonlinear soliton regime for $H \neq 0$, $T \gg E_0/k$ (the upper limits of H and T are not yet known). In both regimes classical theories work very well for the spindynamics, while thermodynamics cannot be described in this frame. This raises a fundamental problem in the application of theories to experimental results: A theoretical description of a real system may be called satisfactory only if the theory describes all experiments. And this point has not been reached for $CsNiF_3$.

Acknowledgement: Thanks are due to all colleagues, who contributed to this work: B. Dorner, K. Hirakawa, K. Kakurai, J.K. Kjiems, R. Pynn, G. Shirane.

References:

1 see M. Steiner, J. Villain, C.G. Windsor, Adv. Phys. 25, 87, 1976
2 R. Moon, T. Riste, C. Koehler, Phys. Rev. 181, 920, 1969
3 M. Steiner, B. Dorner, J. Villain, J. Phys. C 8, 165, 1975
4 J. Villain, J. Physique 35, 27, 1974
5 J.K. Kjems, K. Kakurai, M. Steiner, J. Magn. Magn. Mat 31-34, 1133, 1983
6 M. Steiner, J.K. Kjems, J. Phys. C 10, 2665, 1977
7 M. Steiner, K. Hirakawa, G. Reiter, G. Shirane, Sol. State Comm. 40, 65, 1981
8 K. Kakurai, R. Pynn, B. Dorner, M. Steiner, J. Phys. C 17, L 123, 1984
9 H.J. Mikeska, J. Phys. C 11, L 29, 1978
10 M. Steiner, K. Kakurai, J.K. Kjems, Z. Phys. B 53, 117, 1983
11 E. Magyari, H. Thomas, Phys. Rev. B 25, 531, 1982
12 K. Kakurai, R. Pynn, M. Steiner, to be published
13 A.P. Ramirez, W.P. Wolf, Phys. Rev. Lett. 49, 227, 1982

Part II

Spin Wave Dynamics in One Dimension

Classical Ferromagnetic Chain in a Field; Spin-Waves Versus Non-Perturbative Interactions

U. Balucani and M.G. Pini
Istitudo di Elettronica Quantistica CNR, I-50127 Firenze, Italy
A. Rettori, V. Tognetti, and R. Vaia
Dipartimento di Fisica dell'Università, I-50125 Firenze, Italy

1. Introduction

One of the more interesting features of magnetic chains is the lack of spontaneous magnetisation at any finite temperature. In one dimension (1-D) this result is quite general and holds irrespective of the quantum or classical nature of the system, as well as of the presence of exchange or single-ion anisotropies [1]. In the classical Heisenberg case, which we shall explicitly consider in the following, all the thermodynamic quantities of the system can be evaluated analytically at any temperature in zero external field [2]. Unfortunately, in the presence of a magnetic field H, the statistical properties of the model can be determined only in a numerical way, e.g. by transfer matrix techniques [3]. At very low temperatures and for $H \neq 0$, non-interacting spin-wave theory predicts results in good agreement with the numerical data [4, 5]. Larger and larger discrepancies, however, appear as the field is decreased, until for H=0 spin-wave calculations predict a magnetisation $\langle S^z \rangle \to -\infty$, in sharp contrast with the exact result $\langle S^z \rangle = 0$. This dramatic failure of an approach which is so successful in the usual 3-D magnets clearly indicates a breakdown in the conventional many-body techniques based on perturbation theory, and requires a new treatment.

It is the main purpose of this contribution to present a novel approach, valid at not too high temperatures, in which both the limiting situations (namely, the spin wave and the H=0 cases) are obtained analytically within a unified framework, in such a way that the underlying physical implications can be clearly recognised. Moreover, the theory can be systematically extended to give a quantitative description of the crossover between the two regimes, yielding results which are in good agreement with the transfer matrix data for the magnetisation and the specific heat at constant field.

2. Theoretical Framework

The Hamiltonian of a ferromagnetic chain in an external field reads

$$\mathcal{H}_N = - J \sum_{j=1}^{N-1} \vec{S}_j \cdot \vec{S}_{j+1} - g \mu_B H \sum_{j=1}^{N} S_j^z \tag{1}$$

where the N spins of the chain are coupled through the nearest-neighbour exchange integral $J > 0$.

In the following we shall consider the classical case in which the spins are ordinary three-dimensional vectors of length S. In a spherical frame with polar angles ϑ_j, φ_j we may write

$$S_j^{\pm} = S\sin\vartheta_j \exp(\pm i\varphi_j) = \sqrt{2S}\sqrt{n_j}\{1-(n_j/2S)\}^{\frac{1}{2}} \exp(\pm i\varphi_j)$$

$$S_j^z = S\cos\vartheta_j = S - n_j \tag{2}$$

where the variable $n_j = S(1-\cos\vartheta_j)$ has been introduced in such a way that (2) formally resemble a Holstein- Primakoff transformation, with the quantities $a_j = \sqrt{n_j}\exp(i\varphi_j)$ as classical analogues of the Bose operators.

The statistical mechanics of the system is determined by the knowledge of the classical partition function $Z_N^* = (1/S)^N \int d\vec{S}_1 \, .. \int d\vec{S}_N \exp(-\beta^* \mathcal{H}_N)$. Introducing the reduced variables

$$\nu_j = n_j/2S \qquad h = g\mu_B H/JS \qquad T^* = \beta^{*-1} = k_B T/JS^2 \tag{3}$$

the spin length S enters all the thermodynamic properties of the system only as a scaling factor. This well-known feature is typical of classical systems, and ultimately is connected with the commutativity of the classical variables [6].

The integrals over the set $\{\varphi_j\}$ can straightforwardly be performed, giving

$$Z_N^* = (4\pi)^N e^{\beta^*[(N-1)+hN]} \int_0^1 d\nu_1 \ldots \int_0^1 d\nu_N \, e^{-2\beta^*[(h+1)\nu_1 - \nu_N]} \prod_{j=1}^{N-1} A(\nu_j, \nu_{j+1}) \tag{4}$$

where

$$A(\nu_j, \nu_{j+1}) = e^{-2\beta^*(h+2-2\varepsilon\nu_j)\nu_{j+1}} \, I_0\left\{ 4\beta^*\left[\nu_j(1-\varepsilon\nu_j)\nu_{j+1}(1-\varepsilon\nu_{j+1})\right]^{\frac{1}{2}}\right\} \tag{5}$$

and I_0 is the n=0 Bessel function of imaginary argument. In (4), (5) it is convenient to introduce a parameter ε in the anharmonic parts of the Hamiltonian. At the end, ε will of course be set equal to 1 as in the full Heisenberg Hamiltonian.

Moreover, it is found that

$$\int_0^1 d\nu_N \, A(\nu_{N-1}, \nu_N) \, e^{2\beta^*\nu_N} = \beta^*(h+1)^{-1} \exp\{-(\beta^*/\varepsilon)(h+1-2\varepsilon\nu_{N-1})\}$$

$$\times \sinh\left\{(\beta^*/\varepsilon)(h+1)\left[1-\frac{4\varepsilon h}{(h+1)^2}\nu_{N-1}\right]^{\frac{1}{2}}\right\}\left(1-\frac{4\varepsilon h}{(h+1)^2}\nu_{N-1}\right)^{-\frac{1}{2}} . \tag{6}$$

In the case h=0 the procedure can be iterated, eventually giving for $\varepsilon = 1$ $Z_N^*(h=0) = (4\pi)^N(\sinh\beta^*/\beta^*)^{N-1}$, i.e. the exact result of [2] .

Another limiting behaviour which is easily found from (5),(6) is the $\varepsilon \to 0$ case, in which non-linear terms in the quantities ν_j can be neglected. We do not report the explicit calculations (see [7] for details), but merely state that the corresponding magnetisation is found to be $\langle S^z \rangle / S = 1-(T^*/K)$ where $K^2 = h(h+4)$. Not surprisingly, this is identical with the non-interacting spin-wave result, and shows the afore-mentioned unphysical divergence as $h \to 0$.

In the general case two approximations can be made in (6):

(i) Low temperatures, such that $\beta^*(h+1) \gg 1$. This amounts to neglecting the negative exponential present in sinh x. It is important to note that this approximation is valid even for h=0, where it corresponds to the neglect of terms of the order exp $(-1/T^*)$. In contrast, the spin-wave result (implying small deviations from perfect alignment) at low fields corresponds to the much stronger condition $\beta^* \sqrt{h} \gg 1$.

(ii) The square roots in (6) can be expanded. This is possible both in the slightly anharmonic regime (where the effects are probed by ε) as well as for $h \to 0$. As discussed in the following, the main physical effects in these extreme situations are accounted for if the expansion is carried up to terms proportional to ν_{N-1}^2. All subsequent integrals can straightforwardly be performed according to similar recipes.

The approach eventually yields an analytical approximation for the free energy valid at not too high temperatures and for <u>all</u> fields. Even if the leading physical effects are already accounted for by expansions up to ν_j^2, the method can be systematically improved. It is possible to show [8] that a theory for the free energy correct up to order ε^n requires expansions up to ν^{2n}, whereas a correct free energy up to the order K^n implies expansions up to ν^n.

In any case, in the thermodynamic limit $N \gg 1$ the reduced free energy per spin $F^*/N = - T^*(\ln Z_N^*)/N$ follows from

$$N^{-1} \ln Z_N^* = \ln 4\pi + \beta^*(h+1) - \ln 2\beta^* + \ln c + \ln \lambda^+ \tag{7}$$

where $c = 1 - (K-h)/2$ and λ^+ is the largest root associated with the system of the coefficients of the powers of ν_j. In the simplest case we find [7]

$$\lambda^+ = 1-(\beta^*/8\varepsilon)\left(M+ \{ M^2 + 16(\varepsilon/\beta^*)^2 (1-c)[c^2(1-c)(1+c^2)+3(\varepsilon/\beta^*)] \}^{\frac{1}{2}}\right) \tag{8}$$

where $M = (1-c)^2(1+c)^2(1+c^2) + (\varepsilon/\beta^*)(1-c) + 3(\varepsilon/\beta^*)^2$. Noting that $c \to 1$ as the field $h \to 0$, two limiting situations can be recognised:

(i) Field ordering effects greater than thermal disorder, i.e. $(1-c) \gg \varepsilon/\beta^*$. Then (8) can be approximated by

$$\lambda^+ = 1 + \varepsilon\,(T^*/4)\left[1 - (h/K)\right]^2 \tag{9}$$

i.e. an expansion in ε in which anharmonic effects are taken into account in a perturbative way.

(ii) $(1-c) \ll \varepsilon/\beta^*$. This is a physical situation in which temperature effects, albeit small, are sufficient to overcome the ordering action of a magnetic field. Obviously, the zero-field case satisfies this criterion at any finite temperature. In (8) the sequence of large and small terms is now completely reversed and

$$\lambda^+ = 1 + (K/2) - (1/\varepsilon)\,(h/T^*)\,. \tag{10}$$

In this case, any approach base on perturbative many-body techniques is bound to fail. In other words, the physics of the system is dominated by anharmonic interactions (note the factor $1/\varepsilon$). The results (7), (8) thus give at the lowest order an analytical description of the crossover between two physically opposite regimes.

3. Results and discussion

The main equilibrium properties of the system can now easily be evaluated from the free energy and (7). In particular, the reduced magnetisation follows from

$$\langle S^z\rangle/S = 1 - (T^*/K) + (T^*/\lambda^+)\,(\partial\lambda^+/\partial h)\,. \tag{11}$$

Using the simplest result (8), in the limit $(1-c) \gg \varepsilon/\beta^*$ we find

$$\langle S^z\rangle/S = 1 - (T^*/K)\left\{1 + \varepsilon\,(T^*/4)\,[1-(h/K)][1-(h/K)^2]\right\} \tag{12}$$

which is the perturbative "Hartree-Fock" result of interacting spin-wave theory. On the other hand, for $\varepsilon/\beta^* \gg 1-c$ we obtain

$$\langle S^z\rangle/S = 1 - (T^*/2\sqrt{h}) + \{(T^*/2\sqrt{h}) - 1\} = 0 \tag{13}$$

i.e. the correct result at finite temperatures and zero field. The unphysical divergence is now eliminated by the correct treatment of the anharmonic contributions.

Similar evaluations can be made for the magnetic specific heat at constant field C(H). In the weakly anharmonic spin-wave regime using the result (9) we find a linear increase of C(H) with temperature.[7]. Again, for H=0 the result is not correct (even if in this case the breakdown is not connected with a divergence): simply by using (10) the temperature dependence of the free energy is qualitatively changed, eventually leading to a constant specific heat $C(H=0) = k_B$. This is indeed the correct classical result in zero field, if exponentially small terms in $\exp(-1/T^*)$ are neglected, consistent with our previous low temperature approximation.

The most interesting physical situations, i.e. the two extreme cases, can therefore be accounted for in a rather simple analytical way within a unique

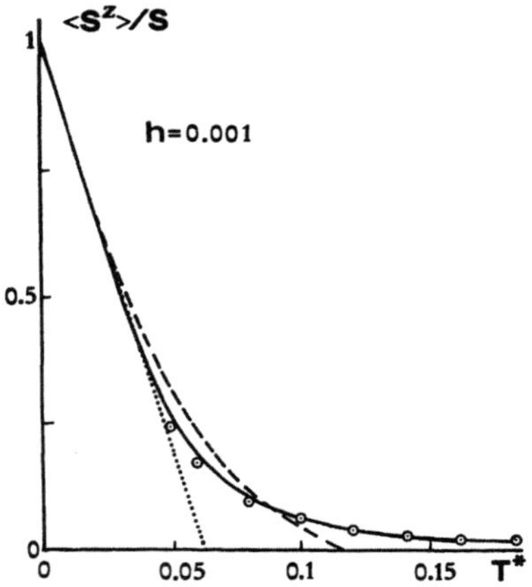

Fig.1 Magnetisation versus temperature with a field h = 0.001. Dotted line: perturbative Hartree-Fock result. Dashed line: full theory up to ν^2 - terms. Solid line: full theory up to ν^4 terms. Open circles: exact transfer matrix data.

theoretical framework. Moreover, in the intermediate region the theory can be systematically improved [8] to give a quantitative description of the crossover between the extreme regimes. The result of such an approach for the magnetisation is reported in Fig.1, which shows an excellent agreement with the exact transfer-matrix data.

Two main points are worthwhile to be finally stressed as far as the physical interpretation is concerned. First, at low temperatures, exponentially small effects in exp $(-1/T^*)$ play no role in establishing the value of $\langle S^z \rangle$ at any field (and, in particular, the absence of long range order in zero field). These effects are related to Dyson's "kinematic" interaction, and their irrelevance at low temperatures in 1-D parallels an analogous result found in 3-D ferromagnets. In both cases, rapid variations of the spin ordering between neighbouring sites are statistically unlikely. On the other hand, in sharp contrast with 3-D, a non-perturbative consideration of the effects of "dynamic" interactions is absolutely essential in 1-D for very low fields or H=0, where anharmonic effects dominate the statistical properties of the system. In other words, neighbouring spins still remain more or less parallel, but over large distances considerable spin deviations play such an important role that the very concept of nearly independent spin wave excitations loses its validity in this regime even at quite low temperatures.

References

1. N.D.Mermin, H.Wagner: Phys.Rev.Lett. 17, 1133 (1966)
2. M.E.Fisher: Am.J.Phys. 32, 343 (1967)
3. M.Blume, P.Heller, N.A.Lurie: Phys. Rev. B 11, 4483 (1975)
4. S.W.Lovesey, J.M.Loveluck: J.Phys.C 9, 3639 (1976)
5. U.Balucani, M.G.Pini, V.Tognetti, A.Rettori: Phys.Rev.B 26, 4974 (1982)
6. U.Balucani, M.G.Pini, V.Tognetti : J.Phys.C (to appear)
7. U.Balucani, M.G.Pini, A.Rettori, V.Tognetti: Phys.Rev.B 29, 1517 (1984)
8. U.Balucani, V.Tognetti, R.Vaia: sent to Phys.Lett.A

The Life Time of the Spin Waves of a Classical Ferromagnetic Chain in an Applied Magnetic Field

U. Balucani and M.G. Pini

Istituto di Elettronica Quantistica del CNR, I-50127 Firenze, Italy

V. Tognetti

Dipartimento di Fisica dell'Università, I-50125 Firenze, Italy
Gruppo Nazionale di Struttura della Materia del CNR

1. Introduction

The dynamics of the one-dimensional Heisenberg ferromagnet in an applied field with Hamiltonian:

$$H = -J \sum_j \vec{S}_j \cdot \vec{S}_{j+1} - g \mu_B H \sum_j S_j^z \tag{1}$$

presents several interesting features whose explanation often requires the use of non-perturbative approaches [1] , [2] .The concept of non-linear excitations (solitons) was also sometimes invoked [3] .However,when the ordering effect of the magnetic field overcomes the thermally induced disorder, the spin -wave approach can be successfully used and the non-perturbative treatment of the magnon interactions is found to remove typical one-dimensional divergences.In the classical case,computer simulations furnish "experimental data",thus permitting us to test the theory. As matter of fact the interesting features of the computed magnetisation [1] and exchange-energy [4] fluctuation correlation functions were well described in the framework of an interacting spin-wave theory.

A non-perturbative calculation of the self-energy associated with one-magnon propagators is found to be crucial for obtaining a physical behaviour of the spin-wave life time. It consists of taking into account of the effect of the magnon bound states and it was performed before for *quantum* magnetic chain including all diagrams with one backward line[5];this is equivalent to breaking the hierarchy of the Green function,extracting only one thermal occupation number.The *classical* case we present here requires more refined decoupling and shows interesting different behaviour .

2. Green-function Formalism

Since the applied magnetic field H orders the spin along the z axis,the standard Dyson-Maleev boson representation can be used and we are dealing with the Fourier transform of the retarded Zubarev Green function (GF) of one magnon,whose equation of motion reads:

$$(E - \omega_k - g \mu_B H) \langle\langle a_k ; a_k^+ \rangle\rangle = (1/2\pi) - (2J/N) \sum_{q,Q,r} \delta_{k+q-2Q} (\gamma_{k-Q-r} - \gamma_{Q+r}) \langle\langle a_q^+ a_{Q+r} a_{Q-r} ; a_k^+ \rangle\rangle \tag{2}$$

where $\gamma_k = \cos k$ and $\omega_k = 2JS(1-\gamma_k)$ is the magnon frequency in the absence of the magnetic field. At this stage the decoupling of the higher-order GFs at the r.h.s. gives the Hartree-Fock renormalisation of the magnon energy:

$$\bar{\omega}_k = \alpha(T)\,\omega_k\,, \qquad \alpha(T) = 1-(1/NS)\sum_q (1-\gamma_q)\langle a_q^+ a_q\rangle\,. \tag{3}$$

In order to obtain the damping effect we have to write down the equation of motion for $\langle\langle a_q^+ a_{Q+r} a_{Q-r}; a_k^+\rangle\rangle$ which involves at r.h.s. also six-point GFs. These can be decoupled extracting one occupation number $n_q = \langle a_q^+ a_q\rangle$, and an equation involving only four-point GFs is obtained. As discussed in detail in [6], the four-boson GF terms at r.h.s. of this equation can lead to damping effects only if they cause a finite momentum exchange among the magnon modes. The other ones cause only renormalisations of the magnon frequencies and at this level can be again decoupled, extracting another occupation number. All this procedure eventually leads to the following equation:

$$\begin{aligned}
(E+\bar{\omega}_q-\bar{\omega}_{Q+r}-\bar{\omega}_{Q-r}-g\,\mu_B H)\,&\langle\langle a_q^+ a_{Q+r} a_{Q-r}; a_k^+\rangle\rangle = \\
&= n_q(\delta_{Q-r-q}\,\delta_{Q+r-k}+\delta_{Q+r-q}\,\delta_{Q-r-k})(E-\bar{\omega}_k-g\mu_B H)\langle\langle a_k; a_k^+\rangle\rangle \\
&+(2J/N)\,\delta_{k+q-2Q}(\gamma_{k-Q-r}+\gamma_{k-Q+r}-\gamma_q-\gamma_k)\,n_{Q-r}n_{Q+r}\langle\langle a_k; a_k^+\rangle\rangle \\
&-(2J/N)(n_{Q-r}+n_{Q+r}+1)\sum_{p,s}\delta_{Q-p}(\gamma_{Q-p+r-s}-\gamma_{p+s})\langle\langle a_q^+ a_{P+s} a_{P-s}; a_k^+\rangle\rangle\,.
\end{aligned} \tag{4}$$

If the four-point GF at r.h.s. were again decoupled, we would obtain the perturbative second-order self-energy. Otherwise, from this integral equation by means of the usual T-matrix technique [7], we arrive at a non-perturbative self-energy which takes into account the two magnon bound states. In the *quantum* case [5], at lowest temperatures, we can neglect the occupation numbers in the factor $(n_{Q-r}+n_{Q+r}+1)$ and we obtain a damping contribution which takes into account all one backward-line diagrams. The bound states are present also at T=0 because they are caused by a term coming from the normal ordering of the boson operators. Moreover the quantum statistics cuts off the contributions of the high wavevectors and the summation over the Brillouin zone can be approximately evaluated in a simple way. The *classical* limit is obtained letting $S\to\infty$ and $\hbar\to 0$ in such a way that the product $\hbar(S(S+1))^{\frac{1}{2}}$ remains finite (equal to S_c) [8]. In this limit the spin operators tend to classical vectors of length S_c and the term 1 is not present in the factor $(n_{Q-r}+n_{Q+r}+1)$ of (4) owing to the commutability of the classical variables. For this reason the leading temperature effect for the spin wave life time is determined by the product of two occupation numbers and, because in the classical limit $n_q = K_B T/(\bar{\omega}_k + g\,\mu_B H)$, a prefactor T^2 must be present.

3. Damping Calçulation

The magnon damping, obtained by the perturbative second-order self-energy at resonance, turns out to be (in units $2JS_c$; $\alpha \equiv \alpha(T)$)

$$\Gamma_k = (T^*/\alpha)^2(1-\gamma_k)(1/N\alpha)\sum_q \frac{(1-\gamma_q)(1+\gamma_{k-q})}{|\sin k - \sin q|\,(h_T+2-2\gamma_q)^2} \qquad (5)$$

where $T^* = K_BT/JS_c^2$ and $h_T = g\mu_BH/JS_c\alpha$ .This integrand diverges for q=k and $q = \pi - k$ and the perturbative approach breaks down.The bound state effects enter through the integral equation (4)in a thermallyinduced way [9] and remove the unphysical divergences.The damping Γ_K is found to be:

$$\Gamma_k = (T^*/\alpha)^2(1-\gamma_k)(1/N\alpha)\sum_q (1-\gamma_q)(1+\gamma_{k-q})F_q/(G_q+H_q) \qquad (6)$$

with

$$F_q = |\sin k - \sin q|\left\{1+(T^*/\alpha^2)\,\frac{h_T+2-2\gamma_q}{h_T+2-2\gamma_k}\left[\frac{K_T-h_T}{2K_T} - \frac{h_T(h_T+2)}{K_T[K_T^2+2(1-\gamma_{k+q})]}\right]\right\}$$

$$G_q = (h_T+2-2\gamma_q)^2 4\sin^2[(k-q)/2]\left\{\gamma_{(k+q)/2} + (T^*/\alpha^2)\left[\frac{B}{K_T}\gamma_{(k-q)/2} - \frac{K_T-h_T}{2K_T}\gamma_{(k+q)/2}\right]\right\}^2$$

$$H_q = \left[2(T^*/\alpha^2)\,\frac{\sin[k/2]}{h_T+2-2\gamma_k}\right]^2(1-\gamma_q)(1+\gamma_{k-q})(h_T+2-\gamma_q-\gamma_k)^2 \qquad (7)$$

$$B = \frac{K_T^2 - h_T(\gamma_k+\gamma_q) - 2h_T\,\gamma^2_{(k+q)/2}}{K_T^2 - 2h_T(\gamma_k+\gamma_q) + 4(1-\gamma_k)(1-\gamma_q)} \quad ; \quad K_T^2 = h_T(h_T+4) .$$

It is worthwhile to note that in the classical case,the thermal occupation number is not more strongly peaked at lowest wavevectors as in the quantum

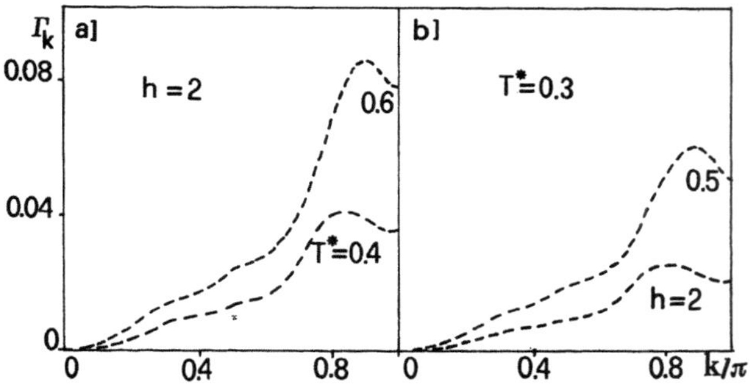

Fig.1:Wavevector dependence of the magnon damping for different fields and temperatures.

case for low temperatures.Thus the summation must be numerically performed over the entire Brillouin zone.Only for low wavevectors and low temperatures can an analytical expression be obtained:

$$\Gamma_k = k^2(\pi \alpha K_T^2)^{-1}(T^*/\alpha)^2 \qquad (k \sim 0) \; . \tag{8}$$

Typical results for Γ_k are shown in fig.1. For these values of the field and temperature the condition $T^* \ll K_T$ is valid so that the ordering effect of the field is predominant and the magnetisation and specific heat can be calculated by the usual spin-wave theory [1] , [4] .

References

1. U.Balucani,M.G.Pini,A.Rettori and V.Tognetti:Phys.Rev.B26,4974(1982).
2. S W.Lovesey:Phys.Letters 86A,43(1981).
3. H.C.Fogedby:J.Phys.A13,1467(1980)--A.R.Bishop:Z.Phys.44,185(1981).
4. V.Tognetti,A.Rettori,M.G.Pini,J.M.Loveluck,U.Balucani and E.Balcar: J.Phys.C16,5641(1983).
5. S.W.Lovesey and M.Hood:Z.Phys.B47,327(1982)..
6. U.Balucani,V.Tognetti and M.G.Pini:J.Phys.C11,L393(1978).
7. R.Silberglitt and A.Brooks Harris:Phys.Rev.Letters 19,30(1967).
8. M.Mason:Phys.Rev.B12,(1975).
9. T.Schneider:Phys.Rev.B24,5327(1981).

Quantum Spin Chains: Phase Behaviour and Dynamic Excitations

Jill C. Bonner[1] and John B. Parkinson[2]

Department of Physics and Astronomy, Michigan State University
East Lansing, MI 48824, USA

Conventional ideas on the phase behavior (at temperature T = 0) of antiferromagnetic (AFM) quantum spin chains have recently been challenged by an interesting conjecture of HALDANE [1]. According to Haldane, half-integer spin chains should behave like the exactly solvable spin-1/2 XXZ model, with a gapless phase for planar anisotropy terminating at an essential singularity at the Heisenberg point, and a phase with long-range order in the ground state together with an excitation energy gap in the case of uniaxial anisotropy. Integer-spin systems, on the other hand, may show an additional phase (subsequently referred to as the "Heisenberg phase") with a non-ordered ground state together with an excitation energy gap, which encompasses the Heisenberg point. This behavior is shown schematically in Fig. 1, where the parameters are defined through the XXZ AFM Hamiltonian:

$$H = J \sum_{i=1}^{N} \left(S_i^x S_{i+1}^x + S_i^y S_{i+1}^y + \lambda S_i^z S_{i+1}^z \right) . \qquad (1)$$

Since the Haldane conjecture has dramatic implications for Kubo-Anderson spin-wave theory, for spin-Peierls theory and for general principles in statistical mechanics which go by the name of smoothness and universality, close scrutiny is warranted. In the absence of a Bethe Ansatz for Heisenberg (XXZ) chains with spin $S > 1/2$, approximate techniques must be used. The first such approximate study was a finite-size scaling calculation, on spin-1 XXZ chains only, by BOTET and JULLIEN [2]. They came to a firm conclusion that the finite-size scaling studies confirmed the Haldane picture for S = 1, identifying the new "Heisenberg phase" and giving values for the critical anisotropies (as defined in Fig. 1) of $\lambda_1 \sim 0$ and $\lambda_2 \sim 1.17$. BONNER and MÜLLER [3] tested the same techniques used by Botet and

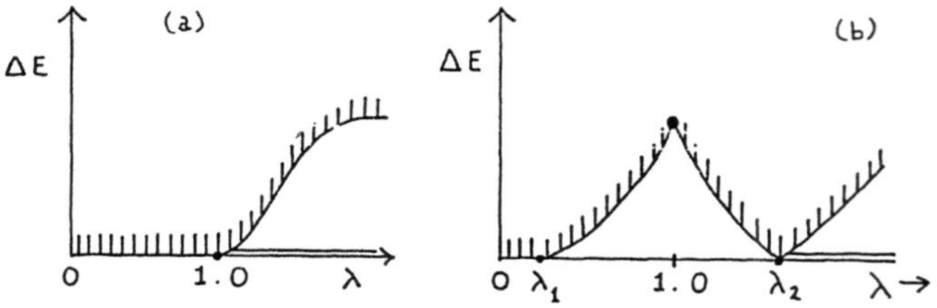

Fig. 1 Excitation energy ΔE versus anisotropy λ according to the Haldane conjecture (a) for half-integer spin chains and (b) for integer spin chains.

[1]On leave from: Physics Department, University of Rhode Island
Kingston, RI 02881, USA
[2]On leave from: Mathematics Department at UMIST, P.O. Box 88
Manchester M60 1QD, United Kingdom

Jullien on the exactly solvable spin-1/2 model, and found qualitatively the same behavior, indicating the presence of a "Heisenberg phase" for $0.5 \lesssim \lambda \lesssim 1.08$ [4]. Since such phase behavior is rigorously known to be incorrect for the case of spin-1/2, Bonner and Müller suggested caution in drawing conclusions from numerical studies especially when the presence of an essential singularity is expected. An essential singularity is well known to slow convergence of numerical approaches, and large finite systems (long chains for this problem) are generally required to obtain reliable estimates of the asymptotic, large N, behavior. Bonner and Müller also emphasized the importance of comparisons of results on chains of various spin values, S = 1/2, 1, 3/2, etc. Subsequent analyses did proceed in this direction [4,5] and, additionally, enlarged the phase diagram parameter space by including a single ion anisotropy term $(D\sum_i (S_i^z)^2)$, a non-trivial parameter for $S \geq 1$ [6,7]. A "consensus" T = 0 phase diagram for S = 1 as a function of D and is shown as Fig. 2[1] and is notable for the extent of the new "Heisenberg phase" which is separated from what may be called a "singlet" phase, which also has a non-ordered singlet ground state and an excitation energy gap, by a line of XY phase! Variational (mean-field) calculations [4] result in a very different phase diagram in that the "Heisenberg phase" is not present. It appears that current evidence for the Haldane conjecture obtained on the basis of (D,λ) phase diagrams can hardly be considered conclusive.

A more direct approach, first noted by Bonner and Müller, is to look at the Heisenberg point, $\lambda = 1$, specifically and extrapolate energy gaps as a function of N^{-1} to $N \to \infty$. Such calculations have now been performed by KOLB, BOTET and JULLIEN [5,8] and the results subsequently checked and extended, and displayed in Fig. 3. For the case of spin-1/2 there is (concave downwards) curvature introducing some uncertainty in the extrapolation, but the trend is clearly consistent with a value zero (no gap) in the $N \to \infty$ limit. For the case of spin-1, however, a dramatic difference occurs. Curvature appears which is concave upwards, and furthermore increases with increasing N (to a maximum value of N = 12), strongly indicating a finite non-zero gap in the $N \to \infty$ limit, consistent with the Haldane conjecture. The T = 0 magnetization isotherm has been calculated as a function of field for several values of N, and the curve for N = 12 is shown in Fig. 4 in comparison with the corresponding curve for spin-1/2. As is well known [9], both curves are step functions. However, in the spin-1/2 case, a smooth curve drawn through the mid-points of the steps is an excellent approximation to the exact analytic curve[2] [10]. A curve drawn through the mid-points of the steps of the spin-1 curve, on the basis of the same assumption, shows interesting features. It is clearly more linear than its spin-1/2 counterpart and over a wider range of magnetic field, much more closely approximating the linear classical $(S = \infty)$ curve also shown in Fig. 3. The distance in field to the foot of the first step is essentially the energy gap as plotted on Fig. 3. Note that the higher excited levels are consistent with the lowest excited level in apparently predicting a non-zero gap in the thermodynamic limit. These numerical results appear to offer the strongest evidence extant for the existence of a "Heisenberg phase" with a gap.

Let us therefore return to Fig. 3 and examine the curve for S = 3/2. The small N values (4,6,8) show the same upward curvature as the parts for S = 1, indicating a non-zero gap in the limit, but the point for largest N

---

1. There is general agreement that the S = 1 phase diagram of Fig. 2 is to be expected on the basis of finite-size scaling studies, but disagreement on its reliability.
2. The exact result is indistinguishable (on the scale of the figure) from the mid-point curve.

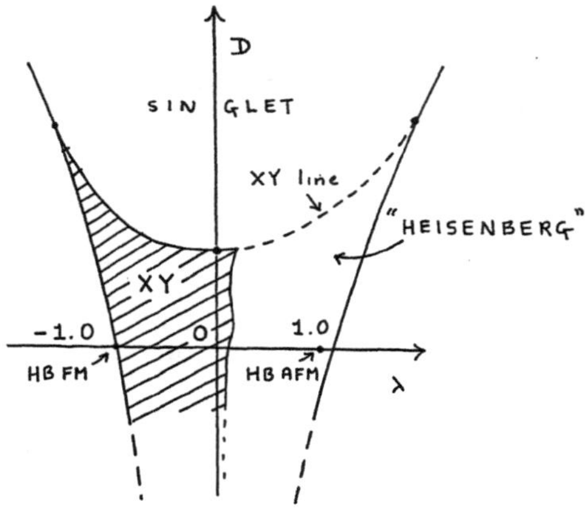

Fig. 2 "Consensus" (D,λ) phase diagram for spin-1 XXZ chains based on finite-size scaling studies.

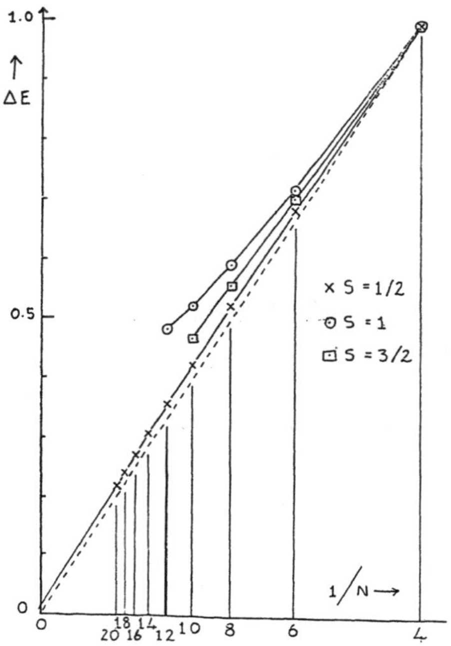

Fig. 3 Excitation energy gaps for Heisenberg spin chains as a function of 1/N.

(=10) reverses this trend. A smooth curve drawn through the four points for S = 3/2 would have an inflection point before a down-turn sets in, indicating the possibility of a zero value in the limit. Hence the available finite Heisenberg chain results are consistent with the Haldane conjecture. But for the possibility of an essential singularity, the extrapolations of Fig. 3 and the form of the T = 0 magnetization curve would provide very convincing support. However, the existence of an "inflection point" for S = 3/2 reflecting a change in trend with increasing N is an indication, presumably, of irregular convergence resulting from the presence of an essential

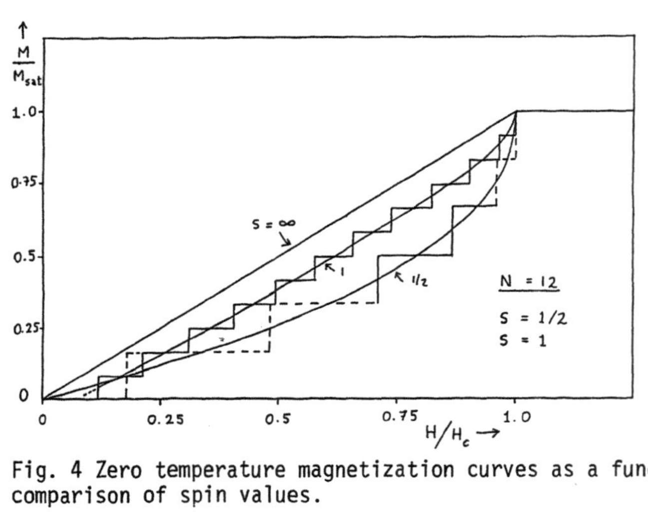

Fig. 4 Zero temperature magnetization curves as a function of field: comparison of spin values.

singularity. BONNER and MULLER [3] estimated that for S = 1/2, chains as long as 25 spins would be required for the asymptotic, large N, behavior to become apparent. The possibility cannot be ruled out that for sufficiently large N (>12), an "inflection point" and ultimate trend of the extrapolation to zero gap may occur for S = 1 as it does for S = 3/2.

If the Haldane conjecture turns out to be correct, quantum spin chains will show a much greater richness of phase behavior than hitherto suspected. This is a stimulus to examine further their dynamic behavior. MÜLLER et al. [11,12] have performed detailed studies of the dynamic behavior, with emphasis on the structure factor (dynamical correlation function) $S(q,\omega)$, for spin-1/2 Heisenberg chains in the presence of a magnetic field. Several unusual phenomena are observed which differ strikingly from classical spin-wave predictions. These quantum effects include (a) the importance of multiple spin-wave double continua (SWDC) in both longitudinal, $S^{zz}(q,\omega)$ and transverse, $S^{xx}(q,\omega)$ dynamical correlation functions, (b) the expected presence of multiple structure in $S^{zz}(q,\omega)$ and $S^{xx}(q,\omega)$ in consequence of (a) and, in particular, the presence of a soft mode[3] in both $S^{zz}(q,\omega)$ and $S^{xx}(q,\omega)$ which tracks across the Brillouin zone as the field increases from zero to the antiferromagnetic critical field H_c (from $q = \pi$ to $q = 0$ in the case of $S^{zz}(q,\omega)$ and vice versa for $S^{xx}(q,\omega)$). The situation for $S^{zz}(q,\omega)$ for $H = 1/2\,H_c$ is sketched in Fig. 5, where two SWDC appear and also the "moving" soft mode. We have performed numerical studies of excitation spectra for Heisenberg chains with spin-1 up to N = 12, and spin - 3/2 up to N = 10. The motivation was two-fold: (a) to examine the various excitation spectra in zero field to look for the possiblity that excited states other than the singlet to triplet combination considered by BOTET et al. [2,8] may extrapolate to zero; and (b) to investigate whether the dynamic excitations of chains for S > 1/2 may show unusual effects which differ from those for S = 1/2. In connection with (a), the lowest-lying states for a given total spin S = 0, 1, 2, 3 . . ., NS-1, NS are all consistent with the singlet-triplet combination in apparently predicting a non-zero gap.[4] This is

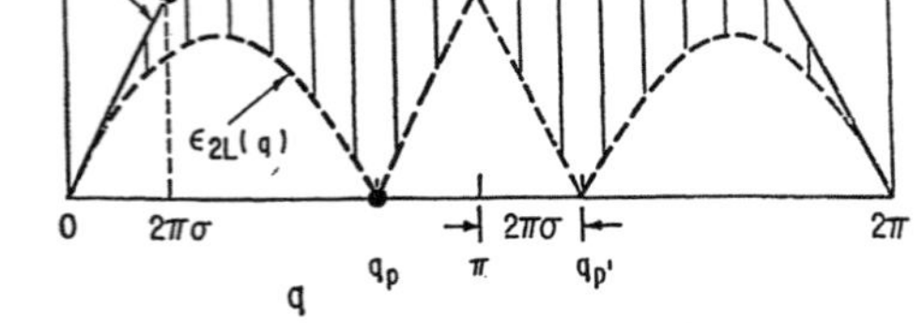

Fig. 5 The longitudinal dynamical correlation function $S^{zz}(q,\omega)$ in a field.

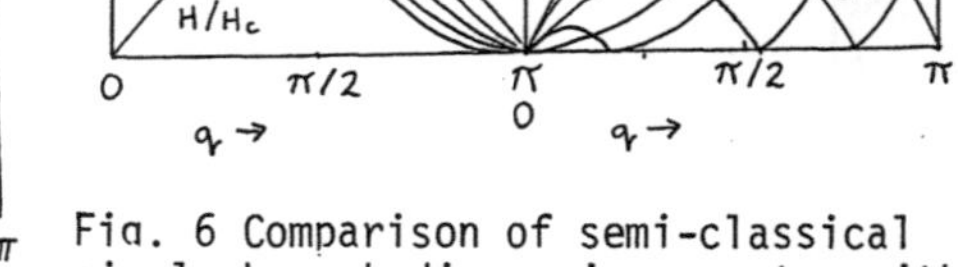

Fig. 6 Comparison of semi-classical single-branch dispersion spectra with the lower boundaries of the spin-1/2 dynamic excitations, as a function of field.

3. An interesting difference appears between "static" and "dynamic" excitations such that a multiplicity of soft modes for $0 \lesssim q \lesssim \pi$ occurs in the static case [13], all but one of which have negligible dynamic spectral weight [12].
4. States close to the maximal spin values were not appropriate for this investigation.

consistent with the form of the spin-1 magnetization curve of Fig. 4. The behavior of low-lying excited singlet states has not yet been exhaustively explored.

The field dependence of the excitations for $S > 1/2$ shows a dramatic difference from the $S = 1/2$ behavior. Interestingly, the differences we shall discuss are not related to the Haldane conjecture, which is irrelevant in the sense that all excitations are measured with respect to the lowest excitation for a given S (S^z). (As is well known, the ground state as a function of field is successively a state of $S^z = 0$ (H = 0), $S^z = 1, 2, 3$, etc., as H increases). The features observed appear rather to be related to the manner in which the field-dependent excitations for the extreme quantum limit of $S = 1/2$ cross over to the semi-classical behavior for $S = \infty$. The situation is illustrated in Fig. 6 based on the work of HEILMANN et al. [14] who were interested in this question from an experimental viewpoint. As is evident from Fig. 6, the manner in which the cross-over might be expected to take place is far from obvious.

The new calculations involve only the lowest-lying excitations for a given $S^z = \Sigma_i S_i^z$ and given wave vector q, since the project involves a Lanczös technique, which can only determine the eigenvalue of greatest magnitude and its associated eigenvector. Hence it has not been possible to look for the SWDC which are a striking feature of spin-1/2 Heisenberg chains. Nevertheless, preliminary results emerge which are very interesting. We found that it is convenient to consider two separate magnetization regimes. The first occupies the range $(1-1/2S)M_{sat} < M \leq M_{sat}$. In this regime there is evidence of a soft mode which tracks across the Brillouin zone with periodicity varying from zero (at $M = M_{sat}$) to π at $M = (1-1/2S)M_{sat}$. For the case of spin-1/2 this comprises the complete range of magnetization. Such behavior is now well established for all S. In the other regime, which exists only for $S \gtrsim 1$, there is little indication of "moving" soft modes, and all dispersion curves (independent of S^z or, equivalently, H) have the form of distorted sine curves. The tempting conjecture presents itself that here may lie the answer to the intriguing question of how the dispersion spectra of quantum spin chains (exemplified by the spin-1/2 curves of Fig. 6) "cross over" with increasing spin to classical behavior. The conjecture is that the quantum behavior characteristic of spin-1/2 persists for all S over a magnetization regime of magnitude $M_{sat}/2S$, thus losing significance as $S \to \infty$. The situation is sketched in Fig. 7, which is a plot of spin value versus magnetization showing the apparent cross-over between the two regimes.

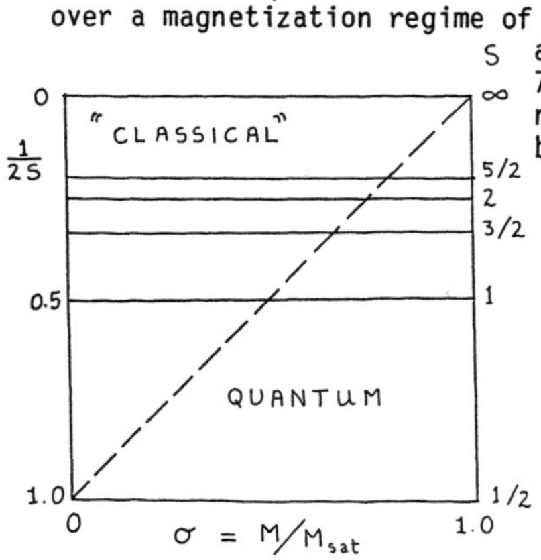

Fig. 7 Sketch of inverse spin value versus T=0 magnetization showing the relative importance of two distinct regimes as $S \to \infty$.

Classical effects, for example, show up in the quasi-linear behavior of the midpoint curve for S = 1 in the T = 0 magnetization curves of Fig. 4 in the field region $0 < H \lesssim H_c/2S$. The space Fourier transform of the T = 0 field-dependent longitudinal spin-spin correlation function (calculated from the eigenvector of the lowest eigenvalue for a given S^z) is the intensity $I^{zz}(q,H)$, an experimentally measurable quantity. For spin-1/2, $I^{zz}(q,H)$ shows cusp anomalies which directly correspond to the dynamic soft mode as it tracks across the Brillouin zone [12]. These anomalies are most pronounced for lower magnetic fields. Curves for $I^{zz}(q,H)$ for S = 1 and S = 3/2 do not show such anomalies, consistent with a more semi-classical picture. It should be emphasized, however, that while the existence of a "quantum" regime, equivalent to the entire excitation regime for S = 1/2, appears to be solidly established for $S > 1/2$, it has only been established that the second regime is very different in nature from the quantum regime. Results for larger chains and higher spin are required to specify the precise nature of the second regime and to determine whether it does, indeed, show semiclassical character.

We are happy to acknowledge M. F. Thorpe for fruitful discussions and useful suggestions. This work was supported through a U.S. NSF Visiting Professorship (JCB), by NSF grant number DMR79-09170 and the ONR and, in part, by NSF grant number DMR80-10819.

References

1. F.D.M. Haldane, Bull. Amer. Phys. Soc. 27, 181 (1982); Phys. Rev. Lett. 50, 1153 (1983); Phys. Lett. 93A, 464 (1983)
2. R. Botet and R. Jullien, Phys. Rev. B 27, 613 (1983)
3. J. C. Bonner and G. Müller, Phys. Rev. B 29, 5216 (1984)
4. J. Sólyom independently arrived at this conclusion and gave the estimate of $\lambda_1 \sim 0.5$ (J. Sólyom and T.A.L. Ziman, preprint) which Bonner and Müller did not have data to estimate quantitatively
5. M. Kolb, R. Botet, and R. Jullien, J. Phys. A 16, L673 (1983)
7. S. T. Chui and K. B. Ma, Phys. Rev. B 29, 1287 (1984)
8. M. Kolb, R. Botet, and R. Jullien, Phys. Rev. B 29, 5222 (1984)
9. J. C. Bonner and M. E. Fisher, Phys. Rev. 135, A640 (1964)
10. R. B. Griffiths, Phys. Rev. 133, A768 (1964)
11. G. Müller, H. Beck, and J. C. Bonner, Phys. Rev. Lett. 43, 75 (1979)
12. G. Müller, H. Thomas, H. Beck, and J. C. Bonner, Phys. Rev. B 24, 1429 (1981)
13. A. Aghahosseini and J. B. Parkinson, J. Phys. C 13, 651 (1980)
14. I. U. Heilmann, G. Shirane, Y. Endoh, R. J. Birgeneau, and S. L. Holt, Phys. Rev. B 18, 3530 (1978)

Nonlinear Excitations in the Classical *XY*-Model

R.W. Gerling [1+2] and D.P. Landau [2]

[1]Institut für Theoretische Physik der Universität Erlangen-Nürnberg
Glückstraße 6, D-8520 Erlangen, Fed. Rep. of Germany

[2]Department of Physics and Astronomy, University of Georgia
Athens, GA 30602, USA

1 Introduction

It has been known for some time [1-4] that the classical XY-model in one and two dimensions shows behavior which is governed by nonlinear excitations.

In the one-dimensional XY-model MIKESKA showed [1,5] that the spin dynamics of the anisotropic Heisenberg chain can be mapped on the sine-Gordon-model. The thermodynamics of the sine-Gordon-model has been discussed by several authors [6,7]. A simple picture was developed in which the solitons are regarded as quasiparticles with a finite mass and a thermal distribution of velocity. Solitons in ferromagnetic chains correspond to moving domain walls, seperating regions with a phase difference of 2π.

The two-dimensional XY-model shows a Kosterlitz-Thouless transition [2-4]. According to the Kosterlitz-Thouless theory the phase transition is driven by the unbinding of vortex-antivortex pairs. The vorticity in a given region is determined by travelling around a square of nearest neighbors sites and adding the relative angle difference between spin directions around the loops. A total difference of zero indicates no vorticity but a total difference of 2π and -2π indicates a vortex and antivortex, respectively.

In the present work we study the one- and two-dimensional XY-model. This model has a planar symmetry and is described by the hamiltonian:

$$H = -J \sum_{(i,j)} (S_i^x S_j^x + S_i^y S_j^y) - h \sum_i S_i^x \qquad (1)$$

where the sum (i,j) runs over all nearest neighbor pairs and the $\vec{S}_i$ are three dimensional classical vectors of unit length. J is a ferromagnetic ($J>0$) coupling constant, and h a magnetic field in the x-direction.

We have used a standard importance sampling Monte Carlo Method [8] to obtain equilibrium spin configurations. As described earlier [9] we obtained 1500 MCS (Monte Carlo steps per spin) at each temperature and each field. Typically the first 500 MCS were not retained, allowing the system to reach thermal equilibrium. We simulated spin chains with periodic boundary conditions and with $N=2000$ spins. For the two-dimensional model we studied systems of size $L\times L$ with periodic boundary conditions; the linear dimension varied between $L=30$ and $L=128$.

2 Spin Solitons in the XY-Chain

MIKESKA gave an expression for the density of solitons in the low temperature limit [1]. His formula reads:

$$n = 4m(\beta m/\pi)^{1/2} \exp(-8\beta m). \tag{2}$$

The mass m of soliton is given by $m = \sqrt{h/JS}$ and $\beta = 1/kT$. An equal number of antisolitons is present.

The density can be determined from the computer experiments simply by counting the solitons and antisolitons. Figure 1 shows results for the density from our simulation. (Actually the number of solitons per 2000 spins is plotted.) The slope obtained by fitting a straight line through our results is about 7.0 instead of MIKESKA's predicted value of 8.

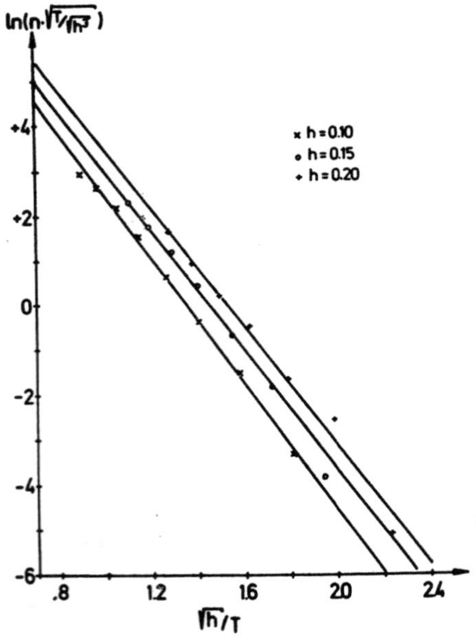

Fig. 1. This figure shows the results for the number of solitons per spin chain n as a function of $\sqrt{h}/T$. The straight lines are best fits. Three different fields as indicated.

There is a large difference in the prefactor of the soliton density. MIKESKA's prediction is $(8/\pi)^{1/2} \approx 1.60$ and we find a value of the order of 9.5. Figure 1 seems to indicate that there is an additional field dependence in the expression for the soliton density. However, up to now the statistical error in our results is too large to determine this field dependence.

We have calculated the phase of $\varphi(\ell)$ of ℓ-th spin by the sum

$$\varphi(\ell) = \Phi_1 + \sum_{i=2}^{\ell} \Delta\Phi_i \tag{3}$$

where $\Delta\Phi_i = \Phi_i - \Phi_{i-1}$ is the relative angle between the neighbors i and i-1. The spin phase $\varphi(\ell)$ in the XY-plane is plotted for states seperated by 200 MCS; a soliton can be seen easily as a phase jump of 2π. Specially selected spin states containing one or two solitons are used as starting configurations for a spin dynamics calculation. The equation of motion for the XY-chain is

$$\frac{d}{dt}\vec{S}_i = \vec{S}_i \times [J(S^x_{i+1}+S^x_{i-1})\,\hat{e}_x + J(S^y_{i+1}+S^y_{i-1})\,\hat{e}_y + h\,\hat{e}_x] \qquad (4)$$

where $\hat{e}_x$, $\hat{e}_y$ and $\hat{e}_z$ are the unit vectors in x, y and z direction. The integration was performed by using a standard integration technique [9].

Figure 2 shows a typical result for the spin phase $\varphi(\ell)$ for two different times between which the soliton travelled about 4 lattice constants. This results in a velocity v=0.3aJ which is small compared to the characteristic velocity c=1.4aJ. Therefore relativistic corrections are not necessary.

The width of the solitons can be determined easily from plots like Fig. 2. The results agree within the statistical error with MIKESKA's predictions.

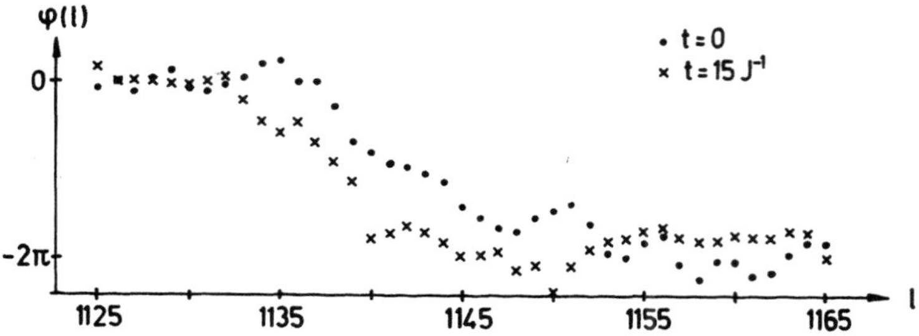

Fig.2. This figure shows the phase $\varphi(\ell)$ of the spins as a function of spin number ℓ. The temperature is 0.2J and the field is 0.1J.

3 The Twodimensional XY-Model

The two-dimensional XY-model does not show long range order for zero field. Nevertheless the zero field specific heat shows a substantial finite peak as a function of temperature. Figure 3

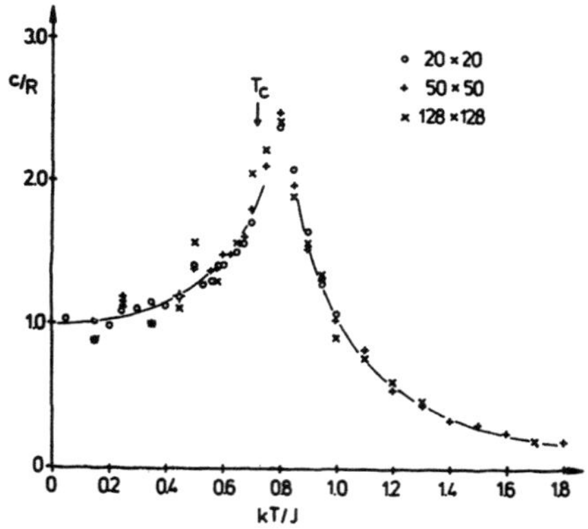

Fig. 3. This figure shows the specific heat c/R as a function of temperature for zero field.

shows the results from our simulation of the two-dimensional system. Despite the fact that the peak in the specific heat occurs well above the Kosterlitz-Thouless temperature the results indicate that the phase transition is at a lower temperature than in the plane rotator model.

Using a method by TOBOCHNIK and CHESTER [10] we calculated the density of the vortex pairs. The Kosterlitz-Thouless theory predicts that the low-temperature density of vortex pairs ρ is given by

$$\rho = R \exp(-2\mu / kT) \tag{5}$$

where 2μ is the creation energy of a vortex-antivortex pair.

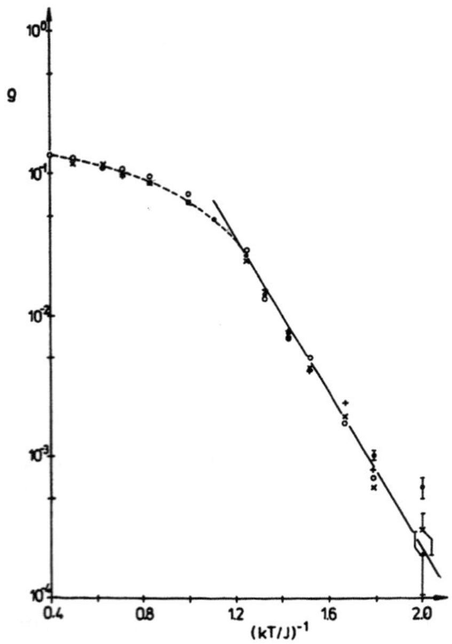

Fig. 4. This figure shows the density of the vortex pairs as a function of inverse temperature for different lattice sizes as indicated. The solid line is a best fit.

Figure 4 shows the results for the vortex pair density from our calculations. For small temperatures (5) describes the temperature dependence quite well. We find a vortex pair creation energy of $2\mu=6.30\pm0.08$. This is in reasonable agreement with the value $2\mu=6.39$ by LANDAU and BINDER [3]. This is also smaller than the value for the plane rotater, which is about 10.

From the unbinding of the vortex pairs one can estimate the phase transition temperature. Therefore we have plotted the vorticity for each square of four nearest neighbors. Figure 5 shows typical results for four different temperatures. The unbinding of the vortex pairs can be seen quite nicely. From this unbinding we estimated the phase transition temperature $kT_c/J=0.725\pm0.010$ (plane rotator: $kT_c/J=0.9$). This is in agreement with findings by LANDAU and BINDER [3].

The formation energy for a vortex-antivortex pair and the phase transition temperature in the XY-model are smaller than

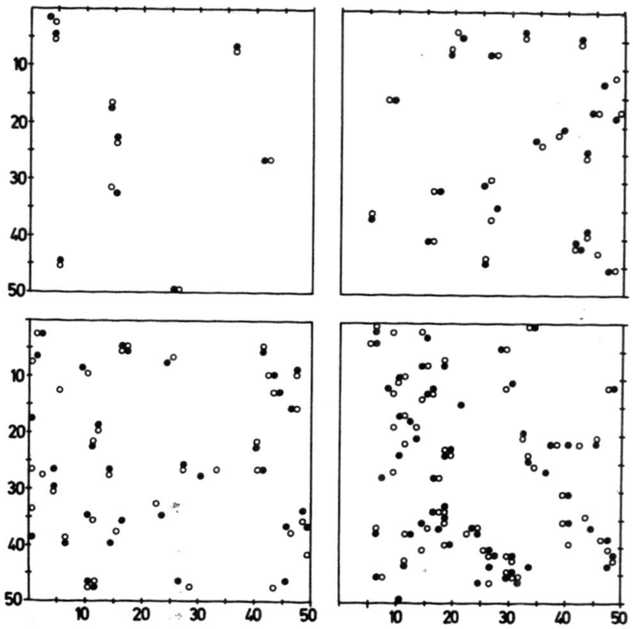

Fig. 5. This figure show the distribution of vortices (o) and antivortices (●) for four different temperatures. Upper left: T/J=0.60; Upper right: T/J=0.70; Lower left: T/J=0.75; Lower right: T/J=0.80. The lattice size is 50x50.

in the plane rotator model. The reason for this is the additional degree of freedom in the XY-model. By tilting spins out of the XY plane new excitations can appear which are not present in the plane rotator model. For example the core energy of the vortices can be reduced by tilting all four spins of a given vortex out of the XY plane.

4 Conclusion

We have demonstrated by the use of computer experiments that in the one- and two-dimensional XY-model nonlinear excitations do exist. For the solitons in the one-dimensional XY-model we could show that our results agree with predictions by MIKESKA [1,5].

In the two-dimensional XY-model further investigations are necessary, to study the behavior of other quantities,e.g. susceptibility at the phase transition temperature. The dynamic behavior of the vortices is also now being studied.

This research was supported by the National Science Foundation and by the Deutsche Forschungsgemeinschaft. We want to thank G.S. Pawley and the Program in Simulational Physics at the University of Georgia for their support.

References

1. H.J. Mikeska, J. Phys. C 11, L29 (1978)
2. J.M. Kosterlitz and D.J. Thouless, J. Phys. C 6, 1181 (1973)
3. D.P. Landau and K. Binder, Phys. Rev. B 24, 1391 (1981)
4. D.P. Landau, J. Magn. & Magn. Mat. 31-34, 1115 (1983)
5. H.J. Mikeska, private communication
6. J.F. Currie, J.A. Krumhansl, A.R. Bishop and S.E. Trullinger, Phys. Rev. B 22, 477 (1980)
7. T. Schneider and E. Stoll, Phys. Rev. B 22, 5317 (1980)
8. K. Binder and D.P. Landau, Phys. Rev. B 13, 1140 (1975)
9. R.W. Gerling and D.P. Landau, Phys. Rev. B 27, 1719 (1983)
10. J. Tobochnik and G.V. Chester, Phys. Rev. B 20, 3761 (1979)

Field Dependent Spin Dynamics of Antiferromagnetic Heisenberg Chains

G.P. Meech

Department of Physics, The University, Southampton SO9 5NH, United Kingdom

S.W. Lovesey

Rutherford Appleton Laboratory, Chilton Oxon OX11 0QX, United Kingdom

1. Introduction

The dynamic properties of simple magnetic chains, in which contiguous pairs of spins are exhange coupled, have been studied with a variety of techniques. These techniques include soliton theories in the presence of weak magnetic fields [1], perturbative studies of a generalized Langevin equation for intermediate and large fields [2,3] and computer simulations [4]. Here we calculate, using a new mode-coupling theory, the response function measured in neutron scattering experiments [5] and computer simulations [4]. Special attention is given to the dynamics at the field induced cross-over from antiferromagnetic to ferromagnetic static correlations.

The appropriate response function is the temporal and spatial Fourier transform of a spin auto-correlation function. If we denote the spin variable at site n by S_n^α, where α = x,y,z, the response function is

$$S^\alpha(k,\omega) = (1/2\pi N) \sum_{n,n'} \exp\{ik(n-n')\}$$
$$\times \int_{-\infty}^{\infty} dt \exp(-i\omega t) \{\langle S_n^\alpha S_{n'}^\alpha(t)\rangle - \langle S_n^\alpha\rangle\langle S_{n'}^\alpha\rangle\} \quad . \tag{1}$$

The angular brackets denote a thermal average of the enclosed variable at a temperature T. For a classical system with inversion symmetry the response function is an even function of the wave vector k, and frequency ω.

2. The Model

The Hamiltonian of an antiferromagnetic Heisenberg chain is

$$H = J \sum_n (\underline{S}_n \cdot \underline{S}_{n+1} - h\, S_n^z)$$

where h is the reduced magnetic field $g\mu_B H/J$. The spin components are expressed in spherical polar coordinates, namely [3]

$$S_n^x = (-1)^n \sin\theta_n \cos\xi_n$$

$$S_n^y = (-1)^n \sin\theta_n \sin\xi_n$$

$$S_n^z = \cos\theta_n \quad .$$

The ground state is given by $\theta_n = \theta = \cos^{-1}(h/4)$, and $\xi_n = \xi$ for all n, where ξ is an arbitrary fixed orientation.

We now change, from θ_n and ξ_n, to new variables ψ_n and ϕ_n representing the deviations of the spins from the ground state configuration. Thus,

$$\theta_n = \theta + \psi_n$$

$$\xi_n = \xi - \phi_n / \sin\theta.$$

The harmonic part of the Hamiltonian is then

$$H_o = (J/2) \sum_k \{a(k)\psi_k\psi_k^+ + b(k)\ \phi_k\phi_k^+\} \quad \text{with} \tag{2}$$

$$a(k) = 2\left[1 - \cos(2\theta)\cos k\right]$$

$$b(k) = 2\left[1 - \cos k\right]$$

and the spin oscillation frequency is determined by

$$\omega_k^2 = a(k)\ b(k).$$

We retain the cubic interaction terms, and neglect quartic and higher-order components. The cubic anharmonic interaction is described by the Hamiltonian

$$H_1 = J \sum_{kpq} \delta_{k+p+q,0} \{\cot\theta(1+\cos k-2\cos p)\ \phi_p\phi_q\psi_k + \tag{3}$$

$$+ \sin(2\theta)\cos k\ \psi_k\psi_p\psi_q\}.$$

In the following section we derive a new mode-coupling equation for the dynamics of the spin system described by the Hamiltonian $(H_o + H_1)$.

3. Mode-coupling theory

The spin equations-of-motion are obtained by commuting the spin components with the Hamiltonian, and they can be written conveniently in matrix form

$$\dot{\Gamma}_k = \alpha\Gamma_k + H_k \quad \text{where} \tag{4}$$

$$\Gamma_k = \begin{pmatrix}\phi_k \\ \psi_k\end{pmatrix}, \quad \alpha = \begin{pmatrix}0 & -a_k \\ b_k & 0\end{pmatrix}, \quad H_k = \begin{pmatrix}F_k \\ G_k\end{pmatrix}.$$

Here, H_k is the nonlinear term derived from the cubic anharmonic interaction, and

$$F_k = -\sum_{p,q} \delta_{p+q,k} \{\cot\theta(1+\cos k-2\cos p)\ \phi_p\phi_q +$$

$$+ \sin(2\theta)(\cos k+2\cos p)\ \psi_p\psi_q\} \quad \text{and}$$

$$G_k = 2\cot\theta \sum_{p,q} \delta_{p+q,k} (1+\cos p-\cos q-\cos k)\ \psi_p\phi_q.$$

The differential equation (4) has the general solution

$$\Gamma_k(t) = e^{\alpha t}\ \Gamma_k(0) + \int_0^t e^{\alpha(t-t')}\ H_k(t')dt'.$$

The key steps in deriving our mode-coupling equation are (a) to make the replacement,

$$e^{\alpha\tau} \to \langle\Gamma_k(\tau)\Gamma_k^+\rangle \langle\Gamma_k\Gamma_k^+\rangle^{-1}$$

and (b) to replace quartic-correlation functions by pair-correlation functions. Arguments to support this scheme for the derivation of mode-coupling equations are reviewed in reference [6].

The Laplace transform

$$\int_0^\infty dt\, e^{-st} \langle\Gamma_k(t)\Gamma_k^+\rangle = \widetilde{\langle\Gamma_k\Gamma_k^+\rangle}$$

$$= \{sI-\alpha + \widetilde{\langle H_k H_k^+\rangle}\langle\Gamma_k\Gamma_k^+\rangle^{-1}\}^{-1} \langle\Gamma_k\Gamma_k^+\rangle \tag{5}$$

where the first equality defines a convenient notation. The correlation functions of interest, involving the ϕ and ψ variables, are derived from (5) by simple matrix algebra.

By setting $s = i\omega$, and noting that the temporal correlation functions are even functions of t, we find, for example,

$$\int_{-\infty}^{\infty} dt\, e^{i\omega t} \langle\psi_k(t)\psi_k^+\rangle = \frac{1}{\pi} \mathrm{Re}\left[\widetilde{\langle\psi_k\psi_k^+\rangle}_{s=i\omega}\right] .$$

Using this relation between temporal Fourier and Laplace transforms we readily obtain expressions for the response functons of interest from the components of the matrix equation (5).

Equation (5) must be solved self-consistently for an arbitrary field using numerical methods. As a first step toward understanding the nature of the mode-coupling equation we have considered a perturbative calculation in which the correlation functions in the self-energy, or memory function, in (5) are replaced by their values for the harmonic theory described by the Hamiltonian H_0. We can perform the calculation in analytic form for the special field $h = 2\sqrt{2}$. The calculation is described in the next section, together with a comparison between the theory and computer simulation data. Note that for the special field $h = 2\sqrt{2}$ the longitudinal wave vector dependent susceptibility exhibits the signature of a cross-over, at low temperatures, from an antiferromagnetic to ferromagnetic static response [4].

4. Results: perturbative calculation and $h = 2\sqrt{2}$

For $h = 2\sqrt{2}$ the oscillation frequency (in units of J) simplifies to

$$\omega_k = 2\sqrt{2}\, |\sin (k/2)| .$$

We find that, in this instance, the Laplace transforms and Brillouin zone integrals involved in the self-energy in (5) can be performed in closed form.

Consider first the results obtained using harmonic correlation functions in the self-energy and setting $\omega = \omega_k$. The latter step is consistent with the assumption that the self-energy is a benign function of k and ω, and that it makes a small correction to the harmonic dynamics. For the longitudinal response we find the result

$$S^z(k,\omega) = \{\omega^2\gamma + \beta(\omega_k^2 - \omega^2)\} \times \{(\omega_k^2 + \Delta - \omega^2)^2 + \gamma^2\}^{-1}$$

with $\gamma = (T/\sqrt{2})\ \{5 + (2\cos(k/2))^4\}$

$$\beta = (T/\sqrt{2})\ \{(2\cos(k/2))^2 + ((2\cos(k/2))^2 - 1)^2\}$$

and $\Delta = -3T\ (\omega_k \cos^2 (k/2))^2$.

From this result it is evident that the longitudinal response displays a distinct peak at $\omega \sim \omega_k$ provided that γ is small compared with ω_k^2; when the latter condition is satisfied the half-width of the collective mode contribution to the response is $\gamma/2\omega_k$ to a good approximation. The results displayed in Fig. 1 are in good agreement with the computer simulation data reported by BALCAR et al. [4].

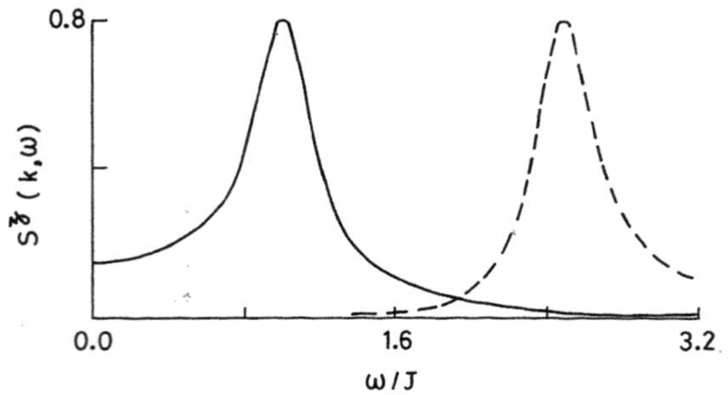

Fig.1 Longitudinal response calculated from (6) for T = 0.1 J. Solid and broken lines are for k = 0.3π and 0.7π, respectively.

When the self-energy is calculated using harmonic correlation function and ω unconstrained we obtain the results displayed in Fig. 2. A zero in the response occurs when $\omega = 4\sqrt{2}\sin(k/4)$. This feature was obtained also by DE RAEDT et al. [3], although other features are not common in the two calculations. The computer simulation data [4] does not show strong evidence for two peaks in the longitudinal response.

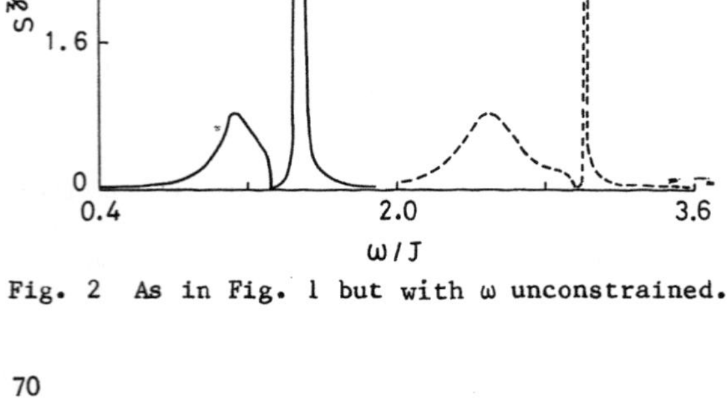

Fig. 2 As in Fig. 1 but with ω unconstrained.

5. Conclusions

The results in Fig. 2, obtained from (5) without repeated iteration, differ from simulation data, and this leads us to believe that further self-consistent iterations of (5) are necessary to obtain meaningful dynamics. Further work for other magnetic fields requires a numerical solution, and it will be reported in a separate paper, together with results for the transverse response.

Acknowledgements

One of us (G.P.M.) gratefully acknowledges the financial support of the Science and Engineering Research Council, U.K. We have benefitted from discussions with Prof. E. Balcar on the interpretation of the computer simulation data reported in [4].

References

1. M. Steiner : J. Magn. Magn. Mat. 31-34, 1277 (1983) and references therein.

2. H. Mori : Prog. Theor. Phys. 34, 399 (1965).

3. B. DeRaedt, H. DeRaedt and J. Fivez : Phys. Rev. B23, 4597 (1980).

4. E. Balcar, S.W. Lovesey and J.M. Loveluck : Z. Phys. B54, 195 (1984).

5. S.W. Lovesey, Theory of Neutron Scattering from Condensed Matter, vol. 2, (Oxford University Press 1984).

6. S.W. Lovesey: Condensed Matter Physics, Dynamic Correlation; Frontiers in Physics vol. 49 (Benjamin/Cummings, Reading 1980).

Is $CsNiF_3$ a Classical System?

M.G. Pini

Istitudo di Elettronica Quantistica del CNR, I-50127 Firenze, Italy

A. Rettori

Dipartimento di Fisica dell'Università ed Unità GNSM, I-50125 Firenze, Italy

Introduction

Recently much interest arose in the non-linear properties of the ferromagnetic chain with Hamiltonian ($J,D>0$)

$$\mathcal{H} = -2J\sum_i \vec{S}_i\cdot\vec{S}_{i+1} + D\sum_i (S_i^z)^2 - g\mu_B H\sum_i S_i^y \tag{1}$$

which in the classical limit can be approximated (in the continuum limit and for $D\to\infty$) by the sine-Gordon (SG) model[1] . However, much experimental data [2] on $CsNiF_3$, a compound well described by (1), addressed to estimating the contribution of non-linear modes to the thermodynamic and scattering properties,have not received a consistent interpretation[3]. Also more recent theories that start with the full Hamiltonian (1) have not yet clarified the problem. Various authors[4] have estimated in the classical case the effect due to out-of-plane fluctuactions on the thermodynamic properties, while Mikeska [5] has calculated the quantum correction to the energy of the static soliton.

In this communication, we show that the universal behavior of the experimental data for the magnetization on the parameter $k_BT/(\tilde{H}J)^{\frac{1}{2}}$ with $\tilde{H}=g\mu_BH$, inexplicable by the classical model (1), is reproduced at low temperatures by a simple *quantum* spin-wave theory. In particular, we find that even at T=0, owing to zero-point motion, the out-of-plane fluctuations are very strong, while their thermal increment is extremely slow. In order to analyze the validity of the classical approximation , we exactly calculate the thermodynamic properties for the classical system with Hamiltonian (1) using a numerical transfer-matrix (TM)method. From the comparison with the experimental data, we conclude that the classical approximation is absolutely inadequate for $CsNiF_3$ in the range of parameters of interest. In all our calculations we have assumed: J=11.8K, S=1, g=2.4 and D=4.5K in the classical case, D=9K in the quantum one.

Thermodynamic Properties

For the magnetization $\langle S^y\rangle$, Cibert pointed out that the experimental data by Rosinski and Elschner[6] are a universal function of $k_BT/(J\tilde{H})^{\frac{1}{2}}$. This universality is shown by the classical planar and Heisenberg models but not by the classical easy-plane model. In order to understand this behavior we have calculated $\langle S^y\rangle$ by TM as well as by a quantum linear spin-wave (LSW) theory.

For the latter case, at the lowest order we obtain

$$H_B = \sum_k \omega_k b_k^+ b_k \tag{2}$$

with b_k^+ and b_k Bose operators and $\omega_k=S\left((J_0-J_k)^2+2(\tilde{D}+\tilde{H})(J_0-J_k)+2\tilde{D}\tilde{H}+\tilde{H}^2\right)^{\frac{1}{2}}$ where $J_k=4J\cos k$, $\tilde{D}=D(1-1/2S)=4.5K$. The magnetization is given by

$$\langle S^y\rangle = S - N^{-1}\sum_k \left[(\alpha_k^2+\beta_k^2)n_k+\beta_k^2\right] \tag{3}$$

where $\alpha_k=\cosh\theta_k$, $\beta_k=\sinh\theta_k$ and $\tanh(2\theta_k)=\tilde{D}(J_0-J_k+\tilde{D}+\tilde{H})^{-1}$, n_k is the Bose factor. In Fig. 1 we report the predictions of quantum LSW and the TM results for $\langle S^y\rangle$ vs $k_BT/(J\tilde{H})^{\frac{1}{2}}$. The quantum results clearly show the existence of a scaling law (and good agreement with experimental data up to $k_BT/(\tilde{J}\tilde{H})^{\frac{1}{2}}\approx 2$) while the classical ones do not present universality. In order to elucidate this different behavior we have calculated the out-of-plane component $\langle (S_i^z)^2\rangle$ which in the quantum case is given by

$$\langle (S_i^z)^2\rangle_q = (S/2)N^{-1}\sum_k (\alpha_k-\beta_k)^2(2n_k+1)\ . \tag{4}$$

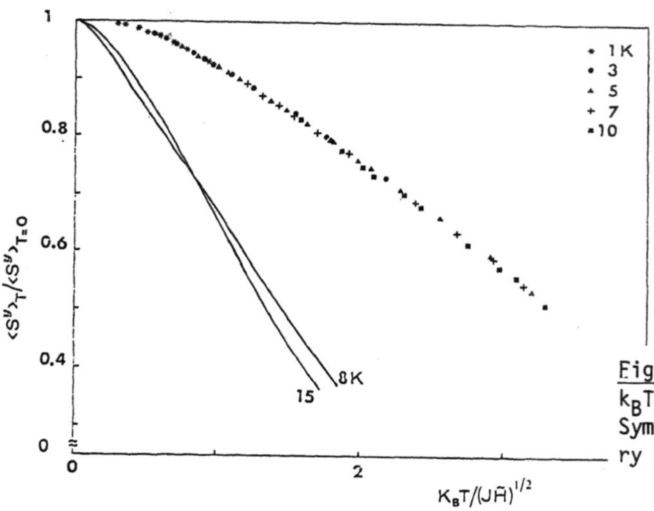

Fig.1 $\langle S^y\rangle_T/\langle S^y\rangle_{T=0}$ vs $k_BT/(\tilde{H}\tilde{J})^{\frac{1}{2}}$. Curves: TM data. Symbols: quantum LSW theory results

This quantity is shown in Fig. 2 together with its classical counterpart $\langle (S_i^z)^2\rangle_{cl}=Sk_BT\left((J_0S+2DS+\tilde{H})^2-(J_0S)^2\right)^{-1}$ and with TM results vs T for H=15kG. $\langle (S_i^z)^2\rangle_{cl}$ vanishes for T→0 and increases almost linearly with T. This strong variation explains the crossover found by TM methods[3] between the planar model at low T and the Heisenberg one at high T for the magnetization. In the quantum case at T=0 the out-of-plane fluctuations for the easy-plane model (1) are large, very near the isotropic Heisenberg model value 0.5. However, contrary to the latter case, their thermal increment is extremely slow in the temperature range examined, thus explaining the universal behavior experimentally found for the magnetization of $CsNiF_3$. Actually, the strength of the anisotropy D is very low to reduce the zero-point out-of-plane fluc-

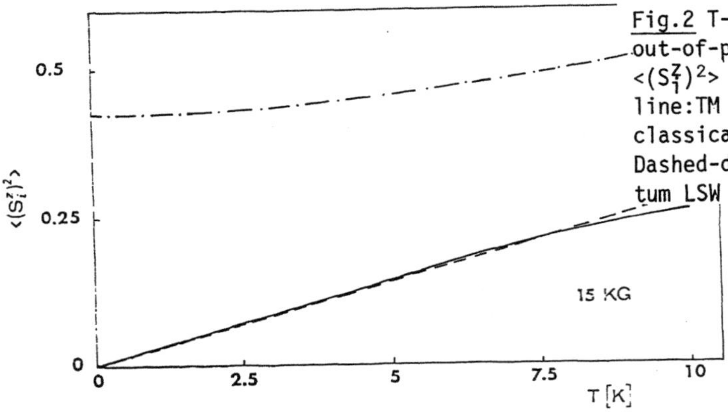

Fig.2 T-dependence of the out-of-plane fluctuations $<(S_i^z)^2>$ for H=15kG. Full line:TM data. Dashed line: classical LSW results. Dashed-dotted line: quantum LSW results

tuations, while, determing the gap of the SW dispersion curve $\omega_{q=0}=S\sqrt{2DH}$, it is very effective to reduce the thermal fluctuations.

In Fig. 3 we report the specific heat difference $\Delta C=C_H-C_{H=0}$ obtained by TM method for the system (1), together with the experimental results by Ramirez and Wolf[7]. A significant comparison is possible if the experimental results are corrected by the addition of the positive quantity $-\Delta C_q^{SW}$. This disagreement is complete both for the amplitude and for the peak position. Also the TM results for the planar model are much higher than the experimental data. Therefore we are led to the conclusion that a classical model is inadequate to describe the specific heat of $CsNiF_3$, at variance with the results obtained for TMMC[8] using a classical model where the peak positions are well reproduced while for the amplitudes there is a discrepancy of only 30%.
A comparison has been made also for the susceptibility and the correlation length which were measured by Kakurai and Steiner[9] using quasi-elastic neutron scattering. In particular in Fig. 4 the experimental results for $\xi_{\parallel}^{-1}$ are shown together with the TM data for the Heisenberg, easy-plane and planar classical models.

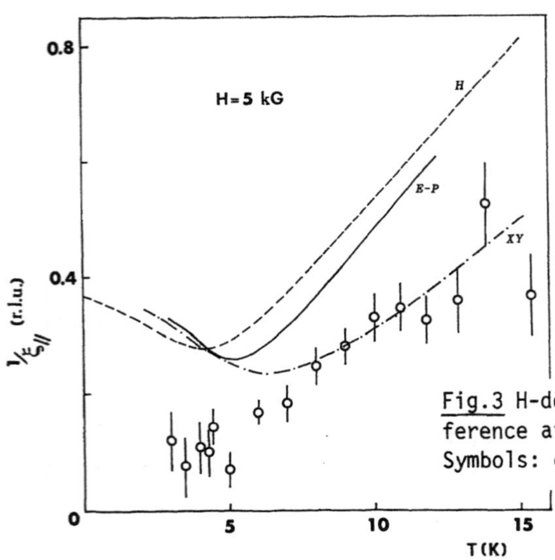

Fig.3 H-dependence of specific heat difference at different T.Full line: TM. Symbols: experimental data(Ref.7)

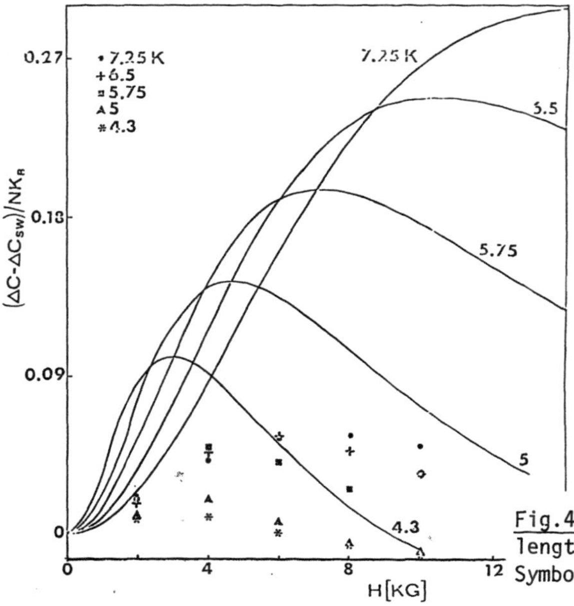

Fig.4 Inverse parallel correlation length vs T for a field of 5 kG. Symbols: exp. data (Ref.9)

It is easily seen that for the easy-plane model (1) there is a complete disagreement in the whole range of T. Similar results were obtained for the susceptibility[10].

Concluding Remarks

We have demonstrated the inadequacy of the classical approximation for $CsNiF_3$ since the exact numerical TM for the classical easy-plane model (1) gives results that strongly disagree with the experimental data for magnetization, susceptibility, correlation length and in particular for the specific heat. These results are very different from those obtained for TMMC where a classical model well reproduced the specific heat[8] and the susceptibility[11]. We think that for $CsNiF_3$ a complete quantum treatment is required. We have shown that the quantum out-of-plane fluctuations are very strong demonstrating the inadequacy of a mapping of Hamiltonian (1) to the quantum SG model. Therefore approaches[5] which include only the leading-order deviations from planar behavior are to be regarded only as starting points for the treatment of the real system $CsNiF_3$.

References

1. H.J. Mikeska, J. Phys. C11, L29 (1978)
2. M. Steiner, K. Kakurai and J.K. Kjems, Z. Phys. B53, 117 (1983), and references therein
3. G. Reiter, Phys. Rev. Lett. 46, 202 (1981); 46, 518(E) (1981)
   J.M. Loveluck, T. Schneider, E. Stoll and H.R. Jauslin, Phys. Rev. Lett. 45, 1505 (1980)

4. V.K. Samalam and P. Kumar, Phys. Rev. 26, 5146 (1982)
H.C. Fogedby, P. Hedegård and A. Svane, Phys. Rev. B28, 2893 (1983); 29, 2861 (1984)
K.M. Leung and A.R. Bishop, J. Phys. C16, 5893 (1983)
K. Osano, preprint (1984)
5. H.J. Mikeska, Phys. Rev. B26, 5213 (1982)
6. J. Cibert, thesis, Grenoble (1981)
7. A.P. Ramirez and W.P. Wolf, Phys. Rev. Lett. 49, 227 (1982)
8. F. Borsa, M.G. Pini, A. Rettori and V. Tognetti, Phys. Rev. B28, 5173 (1983)
9. K. Kakurai and M. Steiner, J. Magn. Magn. Mater. 31-34, 1215 (1983)
10. M.G. Pini and A. Rettori, Phys. Rev. B29, 1 May (1984)
11. H. Harada, S. Sasaki and H. Shiba, Solid State Comm. 40, 29 (1981)

Spin Wave Theory for the Classical Easy Plane Ferromagnet in Zero Field: Exact Results at Low Temperature

G. Reiter

Physics Department, University of Houston, Houston, TX 77004, USA

We present a spin wave calculation of the dynamical response of the classical easy plane ferromagnet

$$H = -J\sum_i \vec{S}_i \cdot \vec{S}_{i+1} + D\sum_i (S_i^z)^2 , \tag{1}$$

$J,D > 0$ for $KT \ll \sqrt{8\,DJ}\,S^2$. One can calculate the dynamical response function

$$R_q^\alpha(z) = \int_0^\infty e^{izt} \langle S_q^\alpha(t) S_{-q}^\alpha \rangle \, dt \tag{2}$$

asymptotically exactly, for all wavevectors, as $T \to 0$. The method used is an extension of that used to treat the fully isotropic system [1] . The response function can be written as [2]

$$R_q^\alpha(z) = i \langle S_q^\alpha S_{-q}^\alpha \rangle \left[z - \omega_{q,\alpha}^2/(z + \gamma_q^\alpha(z) \right]^{-1}. \tag{3}$$

$\omega_{q,\alpha}^2$ is the exact second moment. The equilibrium correlation needed to calculate it may be calculated from spin wave theory. For the in plane correlation, however, it is only possible to calculate this accurately when the separation of spin is much less than the coherence length. One must assume the decay is exponential to obtain the susceptibility at long wavelengths. This has been shown to be correct by the transfer matrix method [3] at low temperatures. We give here only the results for the in plane correlation function R^y. The out of plane correlation has also been obtained. We find, as $T \to 0$

$$\gamma_q^y(t) = \frac{S^2 KT\, N^{-1}}{J_0 - J_q + 2D} \sum_{q_1+q_2=q} \frac{(\Gamma_{12}^0)^2}{\omega_1\omega_2} \cos(\omega_1-\omega_2)t + \frac{4(\Gamma_{12}^1)^2}{\omega_1\omega_2} \cos(\omega_1+\omega_2)t \tag{4}$$

where $\omega_q = S\left[(J_0-J_q)(J_0-J_q+2D)\right]^{\frac{1}{2}}$, and the coefficients Γ_{12}^0, Γ_{12}^1 are obtained from the equation of motion by the method given in Ref. 1. The only difference in the present case is the necessity of introducing a Bogoliubov transformation to diagonalize the Hamiltonian. At long wavelengths and low frequencies, we find that the imaginary part of $\gamma_q(z) = 2CK$, where $C = S\sqrt{2DJ}$ and K, the inverse correlation length is $KT/2JS^2$. The second moment becomes $C^2(q^2 + K^2)$. The real part of γ may be neglected, and we then obtain the result derived by Villain [4] from a gaussian assumption about the dynamical correlations. This assumption is only correct at long wavelengths, as may be seen by comparing the exact second moment with that predicted by Villain's theory. For short wavelengths we can evaluate (4) numerically, and the results for the linewidth as $T\to 0$ are shown in Fig.1 for several values of the anisotropy. Note the crossover to the value for the isotropic case, $2KT \sin q/S$, and the strong deviation from this result for $q = \pi$.

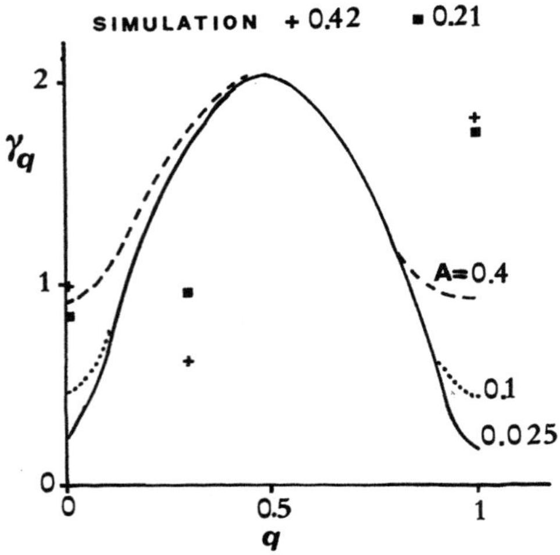

Fig.1 : The fullwidth at half maximum at low temperatures, of the spin waves $\gamma_q = \gamma_q(\omega_q)/KT$. We have compared the results with linewidths shown in the simulations of Ref. 5 at $KT/JS^2 = .1$. The numerical results in Ref. 3 should be related to ours by $\gamma_q = 4\gamma_q''(\omega_q)$, but show large differences. The spin value has been taken to be one. The parameter $A = D/J$.

The method used in Ref. 3 is equivalent to the spin wave theory, and should yield the same results. The numerical results quoted for the linewidths are drastically different, however, indicating a numerical error in the codes used to evaluate the final expression. The simulations of Ref. 5, which allow one to obtain linewidths at several wavevectors and two anisotropy values which should be comparable to the exact result presented here, also show large systematic errors.

References

1 G.Reiter: Phys.Rev.B 21, 5356 (1980)
2 G.Reiter and A.Sjolander: J.Phys.C 13, 3027 (1980)
3 M.Cieplak and A.Sjolander: J.Phys.C 14, 4861 (1981)
4 J.Villain: J. de Physique 35, 27 (1974)
5 J.M.Loveluck, T.Schneider, E.Stoll and J.R.Jauslin: Phys.Rev.Lett. 45, 1505 (1980)

Part III

Soliton Theories

Sine-Gordon Statistical Mechanics

R.K. Bullough and D.J. Pilling

Department of Mathematics, UMIST, P.O. Box 88
Manchester M60 1QD, United Kingdom

J. Timonen

Department of Physics, University of Jyväskylä, Seminaarinkatu 15
40100 Jyväskylä, Finland

The classical partition function

$$Z = \int \mathcal{D}\Pi\, \mathcal{D}\phi\, \exp - \beta H[\phi] \tag{1}$$

in which $\beta^{-1} = k_B T$ and $H[\phi]$ is the sine-Gordon (s-G) Hamiltonian

$$H[\phi] = \gamma_0^{-1} \int [\tfrac{1}{2}\gamma_0^2 \Pi^2 + \tfrac{1}{2}\phi_z^2 + m^2(1-\cos\phi)]\, dz \tag{2}$$

has been evaluated by transfer integral methods [1,2]. The symbols '$\mathcal{D}$' in (1) mean functional integration (inevitable for a *field* $\phi(z,t)$); and Z is evaluated under periodic boundary conditions on a period L after discretization to a lattice of spacing a. The TIM sets $L \to \infty$ in thermodynamic limit, and the problem of evaluating Z and the free energy per unit length $F = -\beta^{-1} L^{-1} \ln Z$ then reduces to finding the lowest eigenvalue ε_0 of the Mathieu equation

$$[-\frac{1}{2m^*}\frac{d^2}{d\phi^2} + (1-\cos\phi) + V_0]\,\psi_\ell(\phi) = \varepsilon_0\,\psi_\ell(\phi) \tag{3}$$

in which $m^* \equiv \beta^2 m^2 \gamma_0^{-2}$ and $V_0 \equiv \gamma_0 a^{-1} m^{-1} \ln(\beta/2\pi a\gamma_0)^{1/2}$. Here and in (2) γ_0 is a dimensionless coupling constant; and ϕ_z means $\partial\phi/\partial z$.

The problem of evaluating (1) is of direct physical interest: a Hamiltonian for the ferromagnetic $CsNiF_3$ in a field is [3,4]

$$H = -J\sum_n \vec{S}_n \cdot \vec{S}_{n+1} + A\sum_n (S_n^z)^2 - g\mu_B B^x \sum_n S_n^x \tag{4}$$

in which $J \simeq 23.6\,K$ and $A \simeq 5\,K$ in classical terms with $A \simeq 9\,K$ in quantal terms. In classical and continuum limits (4) reduces to (2) with $\gamma_0 = (2A/JS^2)^{1/2}$ ($\simeq 0.7$ for $CsNiF_3$ where $S = 1$), $m = (g\mu_B B^x/JSa^2)^{1/2}$ and the velocity $c = aS(2AJ)^{1/2} = 1$: we also set $\hbar = 1$. The Π,ϕ in (2) are canonical with Poisson bracket $\{\Pi,\phi\} = \delta(z-z')$ and the s-G equation

$$\phi_{zz} - \phi_{tt} = m^2 \sin\phi \tag{5}$$

follows from Hamilton's equations. MIKESKA [3] first derived the s-G (5) from (4) and calculated a dynamical form factor $S(q,\omega)$ for its 'kink' and 'antikink' solutions. A similar analysis is available for the staggered field (= continuum limit spin) of the antiferromagnetic TMMC [5,6,7].

The s-G (5) is completely integrable. It has three sorts of soliton solutions, kinks, antikinks and breathers [8,9] and it also has certain weak amplitude harmonic solutions — small amplitude 'phonons' [2]. But complete integrability means there is a canonical transformation from (2) to a new Hamiltonian $H[p]$ in action variables alone, namely [8,9]

$$H[p] = \sum_{i=1}^{N_k}(M^2+p_i^2)^{1/2} + \sum_{j=1}^{N_{\bar{k}}}(M^2+\bar{p}_j^2)^{1/2} + \sum_{\ell=1}^{N_B}(4M^2\sin^2\Theta_\ell+\hat{p}_\ell^2)^{1/2} + \int(m^2+k^2)^{1/2}P(k)dk \tag{6}$$

in which $M \equiv 8m\gamma_0^{-1}$. The set $S_p \equiv \{p_i,\bar{p}_j,\hat{p}_\ell,4\gamma_0^{-1}\Theta_\ell,P(k)\}$ forms a complete set of first integrals in involution, e.g. $\{p_i,P(k)\} = 0$, etc. Complete integrability is then demonstrated from (6) by carrying out the integration explicitly [8,9].

Moreover $H[p]$ (6) is separable: it seems to describe an ideal gas of N_k free relativistic kinks, $N_{\bar{k}}$ antikinks, N_B breathers (with internal momenta $4\gamma_0^{-1}\Theta_\ell$) and a set of phonons labelled by k. Indeed $\int(m^2+k^2)^{1/2}P(k)dk \equiv H_{KG}[p]$ is the Hamiltonian of the free Klein-Gordon (K-G) equation $\phi_{zz} - \phi_{tt} = m^2\phi$ in action variables $P(k)$ alone. Since this relativistic gas is ideal Z should be easy to evaluate by using the set of action variables S_p and their corresponding set of canonical co-ordinates $S_q \equiv \{q_i,\bar{q}_j,\hat{q}_\ell,\Phi_\ell,Q(k)\}$; $\{p_i,q_j\} = \delta_{ij}$, etc. and $\{P(k),Q(k')\} = \delta(k-k')$. All of the phase spaces are determined [8,9]: $-\infty < p_i,q_i < \infty$ etc.; and $0 \le P(k) < \infty$, $0 \le Q(k) < 2\pi$; however $0 < 4\gamma_0^{-1}\Theta_\ell < 2\pi\gamma_0^{-1}$, $0 \le \Phi_\ell < 8\pi$. Note that the continuous infinity of degrees of freedom in (2) is now represented in the K-G modes labelled by k: $-\infty < k < \infty$. We mention this point because, although the complete integrability demonstrated through (6) means the continous infinity of degrees of freedom is correctly counted there, there proves to be a counting problem in evaluating (1) which is not present in either (2) or in its transform (6).

The problem we consider in this paper (necessarily only in sketch) is the evaluation of the partition function (1) by canonical transformation from $H[\phi]$, (2) to $H[p]$, (6). We therefore wish to evaluate

$$Z = \int d\mu \exp -\beta H[p] \tag{7}$$

where $d\mu$ is a measure to be determined. Since the Jacobian of a canonical transformation is unity there seems to be no problem: the obvious measure consists of simple products like $\prod_i(2\pi)^{-1}dp_i dq_i$, etc. and the separability of (6) means that Z factors: $Z = Z_K Z_{\bar{K}} Z_B Z_{KG}$ in obvious notation, and $Z_K = Z_{\bar{K}}$, and Z_B can be expressed in terms of Bessel functions [7]. However we shall see that the separability of (6) is oversimplified.

One problem already is that Z_{KG} is still a functional integral, namely $Z_{KG} = \int \mathcal{D}P\mathcal{D}Q \exp -\beta H_{KG}[p]$: this must be discretized with spacing a whence

$$F_{KG} \equiv -\beta^{-1}L^{-1}\ln Z_{KG} = a^{-1}\beta^{-1}[\ln(\pi\beta/a) - 1 + \tfrac{1}{2}ma]. \tag{8}$$

Thus $\lim L \to \infty$ can be taken but the $a \to 0$ also required cannot. However for this linear problem we have been able to evaluate F_{KG} by various methods [8] (steepest descents, Fourier transform, use of action-angle variables and TIM); and all methods agree except the TIM which finds $F_{KG} - a^{-1}\beta^{-1}(\ln \pi - 1) \equiv F'_{KG}$ instead of F_{KG}.

This oversimplified view that Z factorises would also lead to the result for the whole s-G that

$$F = -\beta^{-1}m\{4(m\beta\pi\gamma_0)^{-1/2} e^{-\beta M}[1 + \tfrac{3}{8}(M\beta)^{-1} + \cdots]$$

$$+ [32\gamma_0^{-2}(M\beta)^{-2} + \cdots]\} + F_{KG} \quad . \tag{9}$$

However the corresponding result from (2) via the TIM and the Mathieu equation (3) is [2,8,10,11]

$$F = -\beta^{-1}m\{8(m\beta/\pi\gamma_0)^{1/2} e^{-\beta M}[1 - \tfrac{7}{8}(M\beta)^{-1} + \cdots]$$

$$+ [\tfrac{1}{4}(M\beta)^{-1} + \tfrac{1}{8}(M\beta)^{-2} + \tfrac{3}{16}(M\beta)^{-3} + \cdots]\} + F'_{KG} \tag{10}$$

In (9) the contributions of kinks, antikinks and breathers are unequivocal: the two former provide the terms in $e^{-\beta M}$ and the breathers the terms in β^{-2}. The corresponding expressions in (10) differ substantially in each case and their origins cannot be so simple. The problem of finding the lowest eigenvalue ε_0 of the Mathieu equation (3) involves methods of matched asymptotic expansions [11] due first of all to Goldstein [12]; and the series in (10) are at best asymptotic [10]. Nevertheless the TIM is relatively easy and likely to be correct. However it has become important to be able to evaluate Z in terms of action-angle variables; for a functional integral like (1) is one example of a quantum theoretical propagator (Green's function) in this case taken in classical limit [8]. In principle we can actually quantise the s-G (5) and find its eigenspectrum and correlation functions by evaluating a functional integral like (1). Moreover we can treat its quantum statistical mechanics.

It is also the particular canonical co-ordinate sets like $S_p \times S_q$ which underlie the so-called quantum inverse methods [13,14] and, as well, they are closely related to the k-space co-ordinates in the quantum Bethe ansatz [15] and in the Bethe ansatz for the quantum statistics of the s-G [16]. Moreover, the quantum s-G at $T = 0$ is equivalent to the massive Thirring model; it relates to the spin-$\frac{1}{2}$ XYZ model [17] and the latter maps to the classical zero-field Ising and 8-vertex models [18]. We should also point out that the same functional integral methods can be used for other integrable spin systems.

It has therefore become profoundly important to understand the reasons for the discrepancies between (9) and (10) which represent a valuable test case of the methods available for solving these systems. We cannot report here all the details of the solution we have now derived for this particular problem. But we can report that we have found how to use (6) to reach (10). We sketch some essential points. First of all to define a functional integral like (1) we must work on finite support L, discretize it so that $L = Na$, work with a finite number of degrees of freedom $2N$ and then set $L, N \to \infty$ followed by $a \to 0$. For a correct thermodynamic limit we must also first impose periodic boundary conditions of period L. Unfortunately usable action-angle variables for the s-G under periodic boundary conditions, even without the discretization, are not available. However we have found how to make systematic corrections to a Hamiltonian like (6) and these are important. For example, for the s-G we replace $H[p]$ by a discretized form

$$H[p,N] = \sum_{i=1}^{N_k} (M^2+p_i^2)^{1/2} + \sum_{j=1}^{N_{\bar{k}}} (M^2+\bar{p}_j^2)^{1/2}$$

$$+ \sum_{\ell=1}^{N_B} (4M^2 \sin^2 \Theta_\ell + \hat{p}_\ell^2)^{1/2} + \sum_{n=-\frac{1}{2}N_{ph}}^{\frac{1}{2}N_{ph}} \omega(\tilde{k}_n) P_n \tag{11}$$

in which $\{P_n, Q_m\} = \delta_{nm}$ and $0 \le P_n < \infty$, $0 \le Q_n < 2\pi$: the P_n, Q_m relate to $P(\tilde{k}_n)$, $Q(\tilde{k}_m)$ by $(2\pi)^{-1} L P_n = P(\tilde{k}_n)$, $Q_m = Q(\tilde{k}_m)$. We then use Floquet theory and the theory of the spectral transform from (2) to (6) to show that the allowed modes $\tilde{k}_n$ satisfy an integral equation of Bethe ansatz form [15]

$$\tilde{k}_n = \frac{2\pi n}{L} + \frac{\Xi(\tilde{k}_n)}{L} , \quad -\tfrac{1}{2}N_{ph} \le n \le \tfrac{1}{2}N_{ph} \quad . \tag{12}$$

Note the number N_{ph} of 'phonon' modes in (12) and (11): there is also an important constraint that $N_k + N_{\bar{k}} + 2N_B + N_{ph} = N$ for each N as $N \to \infty$. Since $\Xi(\tilde{k}_n)$ is bounded as $N, L \to \infty$ it is plain that (11) still reduces to (6) as $N, L \to \infty$. On the other hand the term $L^{-1}\,\Xi(\tilde{k}_n)$ in (12) makes a contribution to (7) which survives as $L \to \infty$. This is the nub of the problem and of its solution.

In order to illustrate the method and to focus attention on the contribution of large amplitude 'phonons' (or 'magnons') to a partition function like (7) and thereby to (1) we shall describe the situation surrounding the sinh-Gordon equation (sinh-G) $\phi_{zz} - \phi_{tt} = m^2 \sinh\phi$ rather than the s-G itself.

The Hamiltonian $H[\phi]$ for the sinh-G is obtained through $\phi \to i\phi$ and $\gamma_o \to -\gamma_o$ in (2). It has no solitons so $H[p]$ (6) becomes $H_{KG}[p]$ simply. However since $N_{ph} \equiv N$ the discretized form corresponding to (11) is

$$H[p,N] = \sum_{-\frac{1}{2}N}^{\frac{1}{2}N} \omega(\tilde{k}_n)\, P_n \tag{13}$$

with $\omega(\tilde{k}_n) \to (m^2+\tilde{k}_n^2)^{1/2}$ as $a \to 0$ while we can show that for $N, L < \infty$ and $a > 0$

$$\Xi(\tilde{k}_n) = \sum_{m \ne n} G(\tilde{k}_n, \tilde{k}_m)\, P_m \tag{14}$$

with

$$G(k,q) = \tfrac{1}{8} m^2 \gamma_o \frac{1}{k\omega(q) - q\omega(k)} \quad . \tag{15}$$

It is obvious that (14) with (15) has an interpretation in terms of phonon-phonon phase shifts with a density of states P_m: however P_m is here the canonical variable and, since there is no upper bound on it, we are concerned with large amplitude interacting phonons. By iterating (12) (with $N_{ph} \equiv N$) once we get $\tilde{k}_n = k_n + L^{-1}\,\Xi(k_n)$. Then to $O(L^{-1})$

$$H[p,N] = \sum_n \omega_n P_n - \frac{1}{L} \sum_{m\neq n} \frac{\partial\omega_n}{\partial k_n} G(n,m) P_n P_m \quad , \tag{16}$$

where $\omega_n = \omega(k_n)$ and $k_n = 2\pi n L^{-1}$: as $N, L \to \infty$ at fixed a, $k_n \to k$ where $-\pi a^{-1} \le k < \pi a^{-1}$ and $\lim a \to 0$ can be taken if possible.

Since

$$Z = \int_0^{2\pi} \int_0^{\infty} \prod_n (2\pi)^{-1} dQ_n dP_n \exp - \beta H[p,N] \tag{17}$$

and to $O(L^{-1})$

$$Z \simeq \int_0^{\infty} \prod_n dP_n [1 + \beta L^{-1} \sum_{m\neq n} \frac{\partial\omega_n}{\partial k_n} G(n,m) P_n P_m] \exp - \beta \sum_n \omega_n P_n \tag{18}$$

we find $Z = \exp - \beta L F$ where

$$\beta L F \equiv \sum_{n=-\frac{1}{2}N}^{\frac{1}{2}N} \ln (\beta\omega_n) - (\beta L)^{-1} \sum_{m\neq n} \omega_m^{-1} G(n,m) \frac{\partial \ln \omega_n}{\partial k_n} \quad . \tag{19}$$

Thus for $N, L \to \infty$

$$F = (2\pi\beta)^{-1} \int_{-\pi/a}^{\pi/a} dk \ln (\beta\omega(k)) - (2\pi\beta)^{-2} \int dk \omega^{-1}(k) \int dq\, G(k,q) \frac{\partial}{\partial q} \ln \omega(q) \tag{20}$$

$$= F'_{KG} + \beta^{-1} m [\tfrac{1}{4}(M\beta)^{-1} + \cdots] \tag{21}$$

In order to reach F'_{KG} not F_{KG} in (21) we use the lattice dispersion relation [8]

$$\omega_n^2 = m^2 + [2a^{-1} \sin (\tfrac{1}{2} k_n a)]^2 \quad . \tag{22}$$

In principle this is also used in the second term in (20) but to $O(a^2)$ $\omega_n = (m^2 + k_n^2)^{1/2}$ can be used there. The sequence of terms (i.e. the $+ \cdots$) in (21) carries both those terms from the further iteration of (12) and the additional terms in the development of $\omega(\tilde{k})$ for the Hamiltonian $H[p,N]$ as in (16): terms $O(L^{-n})$ correcting (16) produce terms $O(M\beta)^{-n})$ correcting (21).

We compare these results (20) and (21) with those obtained for the sinh-G through the TIM via (3): we find

$$F = F'_{KG} + \beta^{-1} m [\tfrac{1}{4}(M\beta)^{-1} - \tfrac{1}{8}(M\beta)^{-2} + \tfrac{3}{16}(M\beta)^{-3} + \cdots] \quad . \tag{23}$$

This result will also follow from (10) by dropping the kink-antikink contribution in $\exp - \beta M$ and replacing γ_0 by $-\gamma_0$. We note that the relationship between sinh-G and s-G parallels that between the repulsive and attractive cases of the non-linear Schrödinger equation (NLS). The repulsive NLS includes the impenetrable Bose gas whose quantum statistical mechanics was pioneered by YANG and YANG [19]. For these systems we expect the classical limit to take the form

$$F = (2\pi\beta)^{-1} \int \ln (\beta\bar{\omega}(k))\, dk \quad , \tag{24a}$$

$$\bar{\omega}(k) = \omega(k) - (\pi\beta)^{-1} \int dq\, G(k,q) \frac{\partial}{\partial q} \ln \bar{\omega}(q) \quad . \tag{24b}$$

By iteration of (24b) we find

$$F = (2\pi\beta)^{-1} \int \ln \beta\omega(k)dk - 2(2\pi\beta)^{-2} \int dk\, \omega^{-1}(k) \int dq\, G(k,q) \frac{\partial}{\partial q} \ln \omega(q) \ , \tag{25}$$

discrepant by a simple factor 2 in the second term when compared with (20). Moreover, by subsequent iterations and using (22) we find

$$F = F'_{KG} + \beta^{-1}m[\tfrac{1}{2}(M\beta)^{-1} - \tfrac{1}{8}(M\beta)^{-2} + \tfrac{3}{16}(M\beta)^{-3} + \cdots] \quad . \tag{26}$$

The TIM result (23) now means of course that for the sinh-G the correct classical limit is not F as defined in (24a) but is $F - m/4M\beta^2$. Our result (21) not only agrees with the TIM at the order $(M\beta)^{-1}$ quoted, but now agrees too with the corrected form of the Bethe ansatz as (20) with (21) show in detail.

Notice that this result for sinh-G apparently shifts the description of the corresponding series in (10) for the s-G from the contribution of s-G breathers to that from large amplitude phonon-phonon interactions. This viewpoint has obvious connections with REITER's [20] interpretation of the $S(q,\omega)$ for $CsNiF_3$ in terms of coupled (i.e. large amplitude) spin waves rather than soliton contributions. But we must analyse this remarkable solution for the s-G itself in all its detail elsewhere.

References

1. R.A. Guyer and M.D. Miller: Phys. Rev. A17, 1205 (1978)
2. J.F. Currie, J.A. Krumhansl, A.R. Bishop and S.E. Trullinger: Phys. Rev. B22, 477 (1980)
3. H.J. Mikeska: J. Phys. C11, L29 (1978)
4. J.K. Kjems and M. Steiner: Phys. Rev. Lett. 41, 1137 (1978)
5. H.J. Mikeska: J. Phys. C13, 2913 (1980)
6. K. Maki: J. Low Temps. Phys. 41, 327 (1980)
7. J. Timonen and R.K. Bullough: Phys. Lett. 82A, 183 (1981)
8. R.K. Bullough, D.J. Pilling and J. Timonen: "Action-Angle Variables in the Statistical Mechanics of the sine-Gordon Field" in Nonlinear Waves, edited by L. Debnath (CUP, Cambridge 1983), pp. 326-355
9. R.K. Bullough and P.J. Caudrey: Solitons (Springer, Heidelberg 1980), and references there
10. M. Stone and J. Reeve: Phys. Rev. D18, 4746 (1978)
11. M. Stirland, A. Dinsmore and R.K. Bullough: to be published
12. S. Goldstein: Proc. Royal Soc. Edinburgh, 49 210 (1929)
13. L.D. Faddeev: "Integrable Models in 1+1 Dimensional Quantum Field Theory", in Proc. of the "Ecole d'Eté de Physique Théoretique", Les Houches 1982, edited by R. Stora and J.B. Zuber (North-Holland, Amsterdam 1983)
14. E.K. Skylanin, L.A. Takhtadzhyan and L.D. Faddeev: Teor. Mat. Fiz. 40, 194 (1979)
15. H.B. Thacker: Rev. Mod. Phys. 53, 253 (1981)
16. M. Fowler and X. Zotos: Phys. Rev. B24, 2634 (1981)
17. A. Luther: Phys. Rev. B14, 2153 (1976)
18. R.J. Baxter: Ann. Phys. (N.Y.) 70, 193 (1972)
19. C.N. Yang and C.P. Yang: J. Math. Phys. 10, 1115 (1969)
20. G. Reiter: Phys. Rev. Lett. 46, 202, 518 (1981)

Solitons in Lattice Systems

H.W. Capel

Institute for Theoretical Physics, University of Amsterdam,
Valckenierstraat 65, NL-1018 XE Amsterdam, The Netherlands

1. Introduction

In the last decade much progress has been made in the theory of integrable nonlinear partial differential equations (PDE's) [1,2]. The inverse scattering formalism provides an exact linearization in the sense that the initial value problem of the PDE is reduced to solving linear problems. In fact, for suitable boundary conditions the time evolution of the scattering data is governed by linear relations, so that the solutions of the PDE can be obtained with the help of a Gel'fand-Levitan equation. Recently, however, FOKAS and ABLOWITZ [3] have proposed a direct method of exact linearization for the Korteweg-de Vries equation, which is based on a linear integral equation having an arbitrary measure and contour. Since then various other PDE's have been treated as well [4,5,6] and it has been shown that Bäcklund transformations can be generated by an appropriate transformation of the measure or the source term in the integral equation [7,8].

In applying the treatment of PDE's to lattice systems it should be kept in mind that in a continuum limit many different lattice equations tend to one and the same integrable PDE and furthermore that most of these lattice equations are not integrable and may give rise to chaotic behaviour. It is therefore of interest to have a systematic method to deal with the integrable lattice versions of a given PDE. In the literature lattice systems have been treated on the basis of discretized spectral problems [9], or on the basis of bilinear identities [10]. In this paper we shall show how the integral equation with arbitrary measure and contour can provide a unified description of a PDE and its discrete analogues. Although the method can be applied to various PDE's with the associated integral equations [11,12], we shall restrict ourselves here to the sine-Gordon equation (SG), which is convenient from a mathematical point of view and which is important in connection with the observed soliton phenomena in one-dimensional magnets [13,14].

2. Direct Method of Linearization

Consider the integral equation

$$u_k + i\,\rho_k \int_C d\lambda(\ell)\, \frac{u_\ell}{k+\ell} = \rho_k \tag{1}$$

in which ρ_k is a source term depending on a spectral parameter k and other, for the moment unspecified, variables. The left-hand side of (1) contains an integration over a contour C in the complex k plane with the measure $d\lambda(k)$. The contour and measure in (1) can be arbitrary apart from the restriction that the homogeneous integral equation without the source term on the right-hand side has only the zero solution. For given ρ_k, C and $d\lambda(k)$ one can then solve the wave function u_k depending on k and the other variables, and by an appropriate integration of u_k over the same contour C with the same measure $d\lambda(k)$ one can obtain a potential v depending only on the other variables. The important feature is now that, for any choice of the contour C and the measure $d\lambda(k)$, the potential v is a solution of one and the same integrable system.

In the continuum case with the position x and the time t as other variables, ρ_k may be chosen to be a plane-wave factor with dispersion $\omega(k)$, i.e. $\rho_k = \exp[i(kx-\omega(k)t)]$. For various choices of $\omega(k)$ one can work out the corresponding integrable PDE's. For instance, if $\omega(k)=k^3$, then the function $v(x,t) \equiv \equiv \int d\lambda(k)\partial_x u_k(x,t)$ obtained by an integration over the contour C satisfies the Korteweg-de Vries equation [3], and in the case $\omega(k) = k^{-1}$, the function $\theta(x,t)$ defined by $v(x,t) \equiv i \int d\lambda(k)u_k(x,t)k^{-1} - 1 \equiv \exp[\tfrac{1}{2}i\theta(x,t)]$ satisfies the sine-Gordon equation $\partial_x\partial_t\,\theta = \sin\theta$ [4].

Choosing various contours and measures one can find a large class of solutions of integrable systems. The N-soliton solutions in particular can be obtained directly taking a measure which is a linear combination of single poles

$$d\lambda(k) = \sum_{\alpha=1}^{N} \frac{c_\alpha}{k-k_\alpha}\,dk \cdot \frac{1}{2\pi i} \tag{2}$$

and a contour C surrounding the points k_α. Using Cauchy's theorem for the contour integration, eq. (1) reduces to a set of N linear algebraic equations for the u_{k_α}

$$u_{k_\alpha} + i\,\rho_{k_\alpha} \sum_{\beta=1}^{N} \frac{c_\beta\, u_{k_\beta}}{k_\alpha + k_\beta} = \rho_{k_\alpha} \tag{3}$$

and the N-soliton solution is obtained as

$$v \equiv i \int_C d\lambda(k) u_k k^{-1} - 1 = i \sum_{\alpha=1}^{N} c_\alpha\, u_{k_\alpha}\, k_\alpha^{-1} - 1 \quad . \tag{4}$$

Starting from the integral equations (1) we shall now derive an integrable difference-difference equation for fields $v_{n,m}$ defined at the sites n,m of a two-dimensional lattice.

This difference-difference equation arises in the form of a Bianchi-identity expressing the commutativity of Bäcklund transformations. Furthermore, applying suitable continuum limits one can obtain an integrable differential-difference equation for time-dependent fields $v_n(t)$ defined at the sites n of a one-dimensional chain as well as a PDE for functions $v(x,t)$, together with their direct linearizations.

3. Bäcklund Transformations

Consider a transformation of the measure in the integral equation, i.e. $d\lambda(\ell) \to d\bar{\lambda}(\ell) = d\lambda(\ell)(p+\ell)/(p-\ell)$. Let $\bar{u}_k$ be the solution of the integral equation (1) with the new measure $d\bar{\lambda}(\ell)$ and let the new potential $\bar{v}$ be obtained by an integration over the same contour with the new measure, i.e. $\bar{v} \equiv i \int d\bar{\lambda}(k)\bar{u}_k k^{-1} - 1$. Then the transformation $v \to \bar{v}$ can be considered as a Bäcklund transformation (BT) relating different solutions of an integrable system. From the explicit expression for $d\bar{\lambda}(\ell)$ it is clear that the BT can increase the number of solitons by 1. In fact, if v is an N soliton solution obtained by a measure as in (2), then the new measure $d\bar{\lambda}(k)$ is a linear combination of N+1 poles and the new potential $\bar{v}$ can be an N+1 soliton solution. Alternatively one can generate a BT by a singular transformation of the source term, i.e. $\rho_k \to \hat{\rho}_k = \rho_k(p+k)/(p-k)$. Let $\hat{u}_k$ be the solution of the integral equations with $\hat{\rho}_k$ instead of ρ_k, then it is easy to show that $\bar{u}_k = \hat{u}_k(p-k)/(p+k)$, leading to the equality of the potentials

$$\hat{v} \equiv i \int d\lambda(k)\hat{u}_k k^{-1} - 1 = \bar{v} .$$

To work out the BT we consider a second integral equation with an additional factor 1/k on the right-hand side for a wave function v_k, i.e.

$$v_k + i\rho_k \int_C d\lambda(\ell) \frac{v_\ell}{k+\ell} = \frac{\rho_k}{k} . \tag{5}$$

We apply the transformation $\rho_k \to \hat{\rho}_k = \rho_k(p+k)/(p-k)$ and for the new $\hat{\rho}_k$ eq. (5) has the new solution $\hat{v}_k$. Multiplying (5) with $\hat{v}_k$ and $\hat{\rho}_k$ by (p-k) one can derive an integral equation for $(p-k)\hat{v}_k$ with the old ρ_k. In fact, decomposing the factor $(p+k)\rho_k$ in front of the integral into a term $(p-\ell)\rho_k$ which is needed for the integral equation and a term $(k+\ell)\rho_k$ which cancels the factor in the denominator we arrive at

$$(p-k)\hat{v}_k + i\rho_k \int_C d\lambda(\ell) \frac{(p-\ell)\hat{v}_\ell}{k+\ell} = \frac{p}{k}\rho_k - \hat{v}\rho_k \tag{6}$$

with

$$\hat{v} \equiv i \int_C d\lambda(k)\hat{v}_k - 1 = i \int_C d\lambda(k)\hat{u}_k k^{-1} - 1 \tag{7}$$

in which $\hat{u}_k$ is the solution of (1) with $\hat{\rho}_k$ instead of ρ_k and the equality sign in (7) can be inferred directly from (1) and (5). From the fact that the homogeneous integral equation (1) or (5) has only the zero solution, we immediately have

$$(p-k)\hat{v}_k = p\, v_k - \hat{v}\, u_k . \tag{8}$$

A second relation is obtained via the inverse BT which can be found by the substitution $p \to -p$ and the interchange of hats and no hats. The result is

$$(p+k)v_k = p\hat{v}_k + v\hat{u}_k . \tag{9}$$

Solving $\hat{v}_k$ and $\hat{u}_k$ as linear combinations of v_k and u_k, we have

$$(p-k)\begin{pmatrix}\hat{v}_k\\ \hat{u}_k\end{pmatrix} = \underline{\underline{L}}_k \cdot \begin{pmatrix}v_k\\ u_k\end{pmatrix} , \quad \underline{\underline{L}}_k = \begin{pmatrix} p & -\hat{v} \\ -k^2 v^{-1} & p\hat{v}v^{-1}\end{pmatrix} . \tag{10}$$

4. Difference-Difference Equation

In a similar way one can apply a second BT by changing ρ_k to $\tilde{\rho}_k = \rho_k(q+k^{-1})/(q-k^{-1})$. The integral equations (1) and (5) with $\tilde{\rho}_k$ instead of ρ_k have solutions $\tilde{u}_k$ and $\tilde{v}_k$ and the new potential $\tilde{v}$ is given by (7) with $\tilde{v}_k$ and $\tilde{u}_k$ instead of $\hat{v}_k$ and $\hat{u}_k$. Proceeding in the same way as before one can show that

$$(q-k^{-1})\begin{pmatrix}\tilde{v}_k\\ \tilde{u}_k\end{pmatrix} = \underline{\underline{M}}_k \cdot \begin{pmatrix}v_k\\ u_k\end{pmatrix} , \quad \underline{\underline{M}}_k = \begin{pmatrix} q\tilde{v}v^{-1} & -k^{-2}v^{-1} \\ -\tilde{v} & q\end{pmatrix} . \tag{11}$$

Using (10) and (11) one can evaluate $\hat{\tilde{v}}_k = \tilde{\hat{v}}_k$ and $\hat{\tilde{u}}_k = \tilde{\hat{u}}_k$. The compatibility $\hat{\underline{\underline{M}}}_k \cdot \underline{\underline{L}}_k = \tilde{\underline{\underline{L}}}_k \cdot \underline{\underline{M}}_k$ yields

$$v\,\tilde{v}\,\hat{v}\,\hat{\tilde{v}} - pq\,(v\,\hat{\tilde{v}} - \tilde{v}\,\hat{v}) = 1 , \tag{12}$$

which is a Bianchi identity expressing the commutativity of the BT's.

Let us now write down the integral equation (1) for wave functions $u_{k,n,m}$ at the sites n,m of a two-dimensional lattice

$$u_{k,n,m} + i\,\rho_{k,n,m} \int_C d\lambda(\ell)\,\frac{u_{\ell,n,m}}{k+\ell} = \rho_{k,n,m} \tag{13}$$

and the corresponding potential

$$v_{n,m} = i \int_C d\lambda(k) u_{k,n,m}\,k^{-1} - 1 . \tag{14}$$

Taking

$$\rho_{k,n,m} = \left(\frac{p+k}{p-k}\right)^n \left(\frac{q+k^{-1}}{q-k^{-1}}\right)^m \tag{15}$$

it is clear that the translation over one lattice distance in the n direction corresponds to the BT with hats, i.e. $\rho_{k,n+1,m} = \hat{\rho}_{k,n,m}$, and similarly $\rho_{k,n,m+1} = \tilde{\rho}_{k,n,m}$. Hence, if we identify in the Bianchi identity $v \equiv v_{n,m}$, $\hat{v} \equiv v_{n+1,m}$, $\tilde{v} \equiv v_{n,m+1}$, $\hat{\tilde{v}} \equiv v_{n+1,m+1}$, we obtain an integrable difference-difference equation which reads

$$\sin \tfrac{1}{4}(A + B) - p\,q\,\sin \tfrac{1}{4}(A - B) = 0 , \tag{16}$$

where $v_{n,m} \equiv \exp[\tfrac{1}{2} i\,\theta_{n,m}]$, $A \equiv \theta_{n,m} + \theta_{n+1,m+1}$, $B \equiv \theta_{n+1,m} + \theta_{n,m+1}$. Equation (16) is the lattice SG previously treated in [15]. Its solutions can be found solving $u_{k,n,m}$ from (13) and doing the integration (14) with the same contour and measure. The N-soliton solution is given by (3) and (4) with

(15) for $\rho_{k,n,m}$ and the Lax representation is given by (10) and (11) with $v \equiv v_{n,m}$, $\hat{v} \equiv v_{n+1,m}$, $\tilde{v} \equiv v_{n,m+1}$.

5. Continuum Limit

To have a continuum limit we must introduce a small lattice parameter b , so that we can consider $b \to 0$, $m \to \infty$, $mb = t$, in which t is to be identified as the (continuous) time variable. From (16) it is clear that we can take q^{-1} as a small parameter, provided that $\theta_{n,m+1} - \theta_{n,m}$ is of the order q^{-1} .
Taking $\theta_{n,m} \equiv \theta_n(t)$, $\theta_{n,m+1} \equiv \theta_n + 2iq^{-1}\partial_t\theta_n$, eq. (16) changes into

$$\tfrac{1}{2}\, i\, p\, \partial_t\, (\theta_n - \theta_{n+1}) + \sin \tfrac{1}{2}\, (\theta_n + \theta_{n+1}) = 0 \qquad , \qquad (17)$$

which is the discrete SG for time-dependent fields defined at the sites n of a one-dimensional chain. Equation (17) is integrable, since its solutions can be found from the linear integral equation

$$u_{k,n}(t) + i\,\rho_{k,n}(t) \int_C d\lambda(\ell)\, \frac{u_{\ell,n}(t)}{k+\ell} = \rho_{k,n}(t) \qquad ,$$

$$v_n(t) = i \int_C d\lambda(k) u_{k,n}(t) k^{-1} - 1 = e^{\frac{1}{2} i \theta_n(t)} \qquad , \qquad (18)$$

in which $\rho_{k,n}(t)$ is obtained from (15) by the limiting process

$$\rho_{k,n,m} = \left[\frac{p+k}{p-k}\right]^n \left[1 + \frac{2k^{-1}q^{-1}}{1-k^{-1}q^{-1}}\right]^m \rightarrow \rho_{k,n}(t) = \left[\frac{p+k}{p-k}\right]^n e^{-ik^{-1}t} \; . \qquad (19)$$

In a second continuum limit $a \to 0$, $m \to \infty$, $na = x$, in which x is the (continuous) space variable, we may take a proportional to p^{-1} . Inserting $\theta_n(t) \equiv \theta(x,t)$, $\theta_{n+1} = \theta - 2ip^{-1}\partial_x\theta$ eq. (17) changes into the SG $\partial_x\partial_t\theta = \sin\theta$ and its solutions can be found from (1) with the plane-wave factor $\rho_k = \exp[i(kx-k^{-1}t)]$ which is obtained as the limit of $\rho_{k,n}(t)$.

Summarizing, we have treated the lattice SG, the discrete SG and the SG together with their direct linearization on a unified basis using an integral equation with arbitrary measure and contour. This can be done for other systems as well. In fact, from (1) one can also obtain lattice and discrete analogues of the Korteweg-de Vries equation, including in particular the Toda lattice. From a slightly more complicated integral equation one can derive discrete versions of the nonlinear Schrödinger equation and the Heisenberg spin chain [11,12]. Recently, a generalization of the integral equation (1) has been proposed which connects an arbitrary solution of a Lax representation, rather than the free-particle type of solution ρ_k, to another, new, solution of the Lax representation [16,17]. This type of integral equations can also be used in the investigation of various discrete integrable systems [18].

References

1. M.J. Ablowitz and H. Segur: Solitons and the Inverse Scattering Transform (SIAM, Philadelphia, 1981)
2. F. Calogero and A. Degasperis: Spectral Transform and Solitons, Vol. 1 (North-Holland, Amsterdam, 1982)
3. A.S. Fokas and M.J. Ablowitz: Phys.Rev.Lett. 47,1096(1981)
4. F.W. Nijhoff, G.R.W. Quispel, H.W. Capel and J. van der Linden: Physica 119A,101(1983)
5. G.R.W. Quispel, F.W. Nijhoff and H.W. Capel: Phys.Lett. 91A,143(1982)
6. F.W. Nijhoff, H.W. Capel, G.R.W. Quispel and J. van der Linden: Phys.Lett. 93A,455(1983)
7. F.W. Nijhoff and H.W. Capel: Phys.Lett. 91A,431(1982)
8. G.R.W. Quispel, F.W. Nijhoff, H.W. Capel and J. van der Linden: Physica 123A,319(1984)
9. M.J. Ablowitz and J.F. Ladek: Stud.Appl.Math. 55,213(1976); 57,1(1977)
10. E. Date, M. Jimbo and T. Miwa: J.Phys.Soc.Japan 51,4116, 4125(1982); 52,388,761,766(1983)
11. F.W. Nijhoff, G.R.W. Quispel and H.W. Capel: Phys.Lett. 97A,125(1983)
12. G.R.W. Quispel, F.W. Nijhoff, H.W. Capel and J. van der Linden: Physica A, in press
13. H.J. Mikeska: In Physics in One Dimension, ed. by J. Bernasconi and T. Schneider, Springer Series in Solid State Sciences Vol. 23 (Springer, Berlin, Heidelberg, New York, 1981) p. 153
14. L.P. Regnault, J.P. Boucher, J. Rossat-Mignod, J.P. Renard, J. Bouillot and W.G. Stirling: In Physics in One Dimension, ed. by J. Bernasconi and T. Schneider, Springer Series in Solid State Sciences Vol. 23 (Springer, Berlin, Heidelberg, New York, 1981) p. 161
15. S.J. Orfanidis: Phys.Rev. D21,1507(1980)
16. P.M. Santini, M.J. Ablowitz and A.S. Fokas: preprint I.N.S. Clarkson (1983)
17. J.Y. Pasquier and R. Pasquier: Phys.Lett. 99A,205(1983)
18. F.W. Nijhoff, H.W. Capel, G.L. Wiersma and G.R.W. Quispel: to be published

Thermodynamics and Correlation Functions of Non-Linear Excitations in the Easy-Plane Ferromagnetic Chain

H. Fogedby
Institute of Physics, University of Aarhus, DK-8000 Aarhus C, Denmark

1. Introduction

The easy-plane ferromagnetic chain with an in-plane symmetry-breaking magnetic field defined by the Hamiltonian

$$H = -J\sum_n \vec{S}_n\vec{S}_{n+1} + A\sum_n (S_n^x)^2 - g\mu_B H_0 \sum_n S_n^z \tag{1}$$

is generally believed [1] to describe the excitation spectrum of the magnetic salt $CsNiF_3$. Owing to the fact that the model (1) in the classical long wavelength limit besides the usual linear extended spin wave spectrum also supports non-linear localized finite-energy domain wall or soliton excitations, this system has attracted much attention both from an experimental and a theoretical point of view.

Experimentally the system has especially been investigated by means of neutron scattering techniques [2], most recently also using polarized neutrons [3], in order to identify the domain wall contribution to the central diffusive peak in the dynamic structure factor and the role of out-of-plane fluctuations. Measurements of a maximum in the field dependent specific heat have also been analyzed in terms of thermally excited domain wall excitations [4].

From the theoretical point of view the interesting non-linear aspects of the system was realized by MIKESKA [5], who in the limit of large anisotropy mapped the classical continuum version of the model (1) onto the domain wall bearing Sine-Gordon system [6] and, furthermore, evaluated the domain wall contribution to the dynamic structure factor within the scheme of the ideal gas phenomenology [6,7]. For finite anisotropy the out-of-plane spin fluctuations give rise to corrections to the static and dynamic properties of the Sine-Gordon model [8,9,10]. Here the controlling parameter is the ratio of the field to the anisotropy. At a characteristic anisotropy (or, equivalently, a characteristic field) the static Sine-Gordon domain wall does in fact become unstable to out-of-plane fluctuations and is driven towards spontaneous motion [11,12,13,14]. Recently, the influence of these features on the dynamic structure factor has also been treated within the ideal gas phenomenology [15].

In the present context we shall not be concerned with the dynamic properties but shall summarize some recent work on the thermodynamics [16,17] and the static correlation functions (form factors) [18] for the classical continuum form of the model (1),

$$H = \int dx \left[\tfrac{1}{2} j \left(\frac{d\vec{S}}{dx}\right)^2 + a(S^x)^2 - h(S^z - 1) \right], \tag{2}$$

based on the steepest descent or path integral approach [19,20,21]. This technique, which is essentially an adaptation of Laplace's asymptotic method to functional integrals, provides a systematic way of determining the low temperature continuum and thermally activated contributions to the static properties. The method is complementary to the ideal gas phenomenology and yields directly the static linear and non-linear elementary excitation spectrum (quasi-particles), that is the spin waves and the domain walls.

2. Thermodynamics

The thermodynamics is given by the free energy $F = -kT\log Z$, where Z is the partition function

$$Z = \int \prod_x d^3S \exp(-H/kT). \tag{3}$$

In the low temperature limit the regions of phase space corresponding to the local minima of H yield the physically important contributions to the functional integral (3). The location of the minima is given by $dH/d\vec{S} = 0$, subject to the length constraint $(\vec{S})^2 = 1$. Introducing canonical coordinates $p = \cos\theta$ and $q = \phi$, where θ and ϕ are the polar and azimuthal angles of the spin $\vec{S}$, the minimum conditions become $dH/dp = 0$ and $dH/dq = 0$, that is the static form of the Hamiltonian equations of motion. Consequently, only the ground state and the static domain wall configurations and the associated static spin wave fluctuations contribute in the evaluation of Z. In that sense the steepest descent method, albeit more rigorous, is more limited than the ideal gas phenomenology [6,7,15] which also samples low velocity domain walls.

For large a the minimum of H corresponds to spin configurations in the easy-plane. H reduces to the Sine-Gordon form with θ as the field variable [5,16] and the static Sine-Gordon domain wall solution [6]

$$S^z = 1 - 2/\cosh^2((x - x_1)/\xi_{in}), \tag{4}$$

centred at x_1 and of width $\xi_{in} = (j/h)^{\frac{1}{2}}$ and energy E,

$$E = 8(jh)^{\frac{1}{2}} \tag{5}$$

contributes to Z. Keeping only the contributions Z from the local minima of H and treating the domain wall configurations as a dilute gas of non-overlapping indistinguishable "impurities" positioned on the chain of length L, we obtain by simple combinatorics

$$Z = 1 + Le^{-E/kT} + (L^2/2!)e^{-2E/kT} + \ldots$$

which sums to an exponential function and yields the "unrenormalized" free energy density $F/L = -kT\exp(-E/kT)$. This result is, however, not correct. In the spirit of Laplace's method we must also include fluctuations about the local minima of H. Such a program requires a somewhat technical linear stability analysis of the ground state and domain wall configurations.

For the ground state we have the usual band of in-plane and out-of-plane static spin waves with correlation lengths ξ_{in} and ξ_{out},

$$\xi_{in} = (j/h)^{\frac{1}{2}} \quad \text{and} \quad \xi_{out} = (j/(h+2a))^{\frac{1}{2}}. \tag{6}$$

The stability analysis of the domain wall, on the other hand, leads to in-plane and out-of-plane eigenvalue problems where the domain wall solution (4) enters as an effective potential in the stability matrices. The in-plane fluctuations consist of a band of phase-shifted spin waves and a localized symmetry-restoring translation mode corresponding to a rigid displacement of the domain wall [6,16]. The out-of-plane fluctuations also include a band of phase shifted spin waves but in addition to that a localized rotation mode corresponding to a rigid rotation of the domain wall [22], and a localized meridional mode equivalent to an out-of-plane tilting of the domain wall [11, 12,13,16,17]. For $a > 3h/2$ the totality of modes are stable and the Sine-Gordon domain wall is a local minimum of H. For $a < 3h/2$ the meridional mode becomes unstable and a more detailed analysis [11,12,13,14] shows that the domain wall is driven towards spontaneous motion.

Including the spin wave fluctuation spectrum in the evaluation of Z by diagonalizing the stability matrices, taking into account the corrections to the spin wave density of states due to the domain wall induced phase shifts, integrating over the domain wall positions, and performing the associated Gaussian integrals, we obtain within the low temperature dilute domain wall gas approximation $Z = Z_{sw}Z_{dw}$. Here Z_{sw} arises from the band of thermally excited linear spin waves about the ground state whereas the domain wall part Z_{dw} corresponds to the thermally activated gas of domain walls. The associated free energy densities are

$$F_{sw}/L = kT(-\log T + \tfrac{1}{2}(\xi_{in}^{-1} + \xi_{out}^{-1}) + \text{const.}), \tag{7}$$

$$F_{dw}/L = - kT8\pi^{-\frac{1}{2}}(j/kT)^{\frac{1}{2}}(h/j)^{3/4}g(2a/h)e^{-E/kT}. \tag{8}$$

We note that the spin wave fluctuations give rise to a "dressing" or renormalization [6] of the domain wall free energy by the factor

$$8\pi^{-\frac{1}{2}}(j/kT)^{\frac{1}{2}}(h/j)^{3/4}g(2a/h).$$

The anisotropy function g has the form

$$g(x) = \frac{[1+(1+x)^{\frac{1}{2}}][2+(1+x)^{\frac{1}{2}}]}{(x-3)^{\frac{1}{2}}x^{\frac{1}{2}}}, \tag{9}$$

where g(2a/h) determines the density of domain walls. In the limit of large anisotropy $a >> h$ the out-of-plane fluctuations are suppressed. g(2a/h) approaches unity and we obtain the Sine-Gordon result [6]. For increasing h/2a g(2a/h) grows monotonically corresponding to an increase in the available phase space. At the characteristic field $h_c = 2a/3$, g(2a/h) formally diverges signalling the drift of the static Sine-Gordon domain wall towards spontaneous motion [13]. For the present one-dimensional system this singularity in F_{dw} is clearly spurious. Including "non-Gaussian" contributions in the evaluation of Z, the singularity is rounded off by fluctuations [16]. An analogous situation is encountered for the double anisotropic Heisenberg model where similar singularities appear within the steepest descent scheme [23]. An ideal gas approach which samples low velocity domain walls yields a smooth free energy [24], indicating that the singularities are associated with the limitation to static domain walls configurations in the steepest descent method. From F_{dw} we also infer the domain wall density n, $n = -F_{dw}/LkT$,

$$n = g(2a/h)n_{SG}, \tag{10}$$

where n_{SG} is the corresponding density for the Sine-Gordon model [6]

$$n_{SG} = 8\pi^{-\frac{1}{2}}(j/kT)^{\frac{1}{2}}(h/j)^{3/4}e^{-E/kT} . \tag{11}$$

The specific heat C follows from F, $C = -Td^2(F/L)/dT^2$. We have $C = C_{sw} + C_{dw}$, where $C_{sw} = k$ and

$$C_{dw} = k(8\pi)^{-\frac{1}{2}}g(2a/h)(E/j)(E/kT)^{5/2}e^{-E/kT} . \tag{12}$$

For $CsNiF_3$ the parameters in (2) are $j = 23K$, $a = 4.5K$ and $h = 0.16K$ per kG. In Fig. 1 we have depicted the specific heat as a function of the field at various temperatures both in the easy-plane case and the Sine-Gordon case and compared with the experimental data [4]. We note that the domain wall calculation correctly predicts the observed specific heat maximum but that the data in general lie below the theoretical values indicating that quantum corrections [25,26] come into play rather strongly at the temperatures where measurements

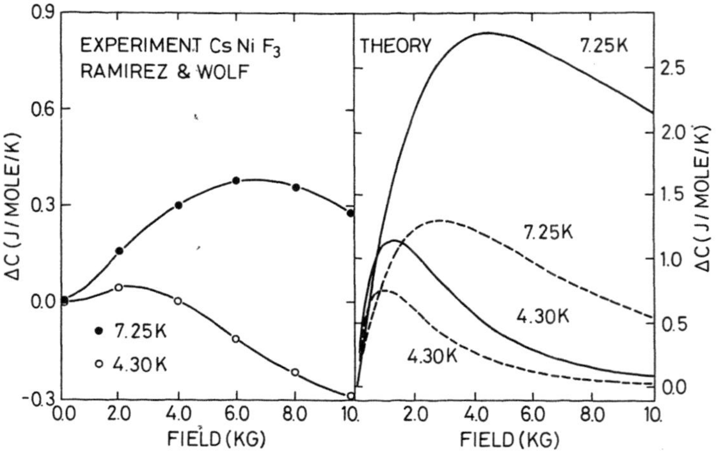

Fig. 1. The specific heat difference $\Delta c = c(H_0) - c(H_0 = 0)$ as a function of H_0 at temperatures 4.3K and 7.25K. In the right panel are shown the theoretical values for Δc. The dashed curves are the Sine-Gordon results and the full-drawn curves the present results for $a = 4.5K$. Note the difference in scale for the two figures.

have been performed. This is also confirmed by a recent Monte Carlo and transfer matrix calculations [27,28] on the model (1). Also, the numerical calculations did not show any enhancement of the specific heat in the vicinity of the characteristic field h_c, which for $CsNiF_3$ is at 18 kG, as tentatively suggested in [16] and [17].

3. Correlation Functions

The static correlation functions for the continuum model (2) can also be evaluated systematically within the steepest descent scheme [18,20,21]. They are of interest in the context of neutron scattering measurements on $CsNiF_3$ [2,3] since they are related to the observed integrated intensities.

We distinguish between the in-plane longitudinal (L) and transverse (T) correlation functions

$$G^L_{in} = \langle S^z(x)S^z(0)\rangle - \langle (S^z)^2\rangle \quad \text{and} \quad G^T_{in} = \langle S^y(x)S^y(0)\rangle,$$

which have the Sine-Gordon form [5,9,20] and the out-of-plane correlation function

$$G_{out} = \langle S^x(x)S^x(0)\rangle$$

which is specific to the easy-plane magnet. Here $\langle .. \rangle$ denotes a thermodynamic average. The correlation functions are now evaluated to order n (the domain wall density) and T (the spin wave density) by inserting the domain wall and phase shifted spin wave configurations in the dilute domain wall gas approximation, diagonalizing the stability matrices, carrying out the associated Gaussian integrals, and integrating over the domain wall positions. These calculations are somewhat lengthy and will not be given here [18].

Generally, the correlation functions consist of a domain wall part and a spin wave part. The longitudinal in-plane correlations G^L_{in} only have a domain wall part. The corresponding form factor is

$$G^L_{in}(k) = 4\pi^2 n(k\xi_{in})^2/\sinh^2(\pi k\xi_{in}/2), \tag{13}$$

exhibiting the characteristic non-Lorentzian line shape [5,9]. We note that the out-of plane fluctuations enter through the domain wall density n given by (10) and (11) as an enhancement of the Sine-Gordon result [15,18]. The transverse in-plane correlations G^T_{in} have both domain wall and spin wave parts. For the form factor we obtain

$$G^T_{in}(k) = 4\pi^2 n(k\xi_{in})^2/\cosh^2(\pi k\xi_{in}/2) + (kT/j)/(k^2 + \xi_{in}^{-2}) \ . \tag{14}$$

The domain wall part has a characteristic two-peak structure [5,9] and is enhanced by the density n. On that structure is superposed the Lorentzian spin wave part of order T.

The out-of-plane correlation function G_{out} is the most interesting one since it directly probes the out-of-plane fluctuations. In the large distance limit

$|x| \gg \xi_{out}$

$$G_{out} \sim \exp(-|x|/\tilde{\xi}_{out}) , \qquad (15)$$

where the renormalized out-of-plane correlation length $\tilde{\xi}_{out}$ depends on the domain wall density n,

$$\tilde{\xi}_{out} = \xi_{out} + 6n\xi_{in}\xi_{out}/(\xi_{in} - \xi_{out})(\xi_{in} - 2\xi_{out}) . \qquad (16)$$

As we approach the characteristic field $h_c = 2a/3$ (18 kG for $CsNiF_3$) the correlation length $\tilde{\xi}_{out}$ grows strongly signalling the softness of the Sine-Gordon domain wall towards spontaneous motion [13,14]. Correspondingly, the associated form factor $G_{out}(k)$ is enhanced as we increase the field. In Fig. 2 we have depicted the out-of-plane form factor for $CsNiF_3$ at T = 10K and for field values H_0 = 1.9 kG, 5.6 kG, and 9.4 kG. This form factor is accessible to polarized neutron scattering measurements as the integrated structure factor for the out-of-plane fluctuations.

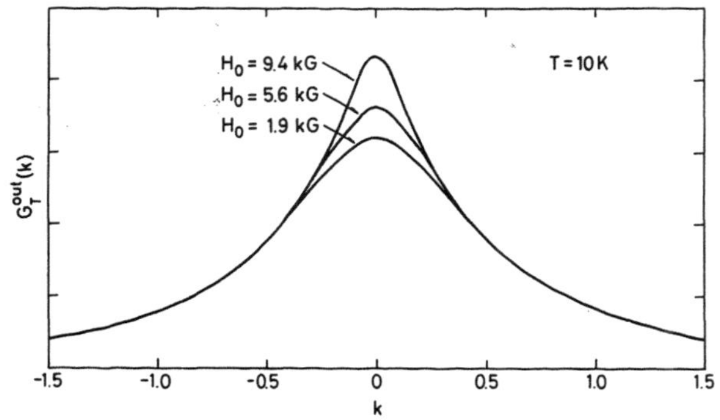

Fig. 2. The out-of-plane form factor $G_{out}(k)$ for T = 10K, a = 4.5K and H_0 = 1.9 kG, 5.6 kG, and 9.4 kG. The k scale is in units of the inverse lattice distance (the lattice distance for $CsNiF_3$ is 2.6Å). The amplitude scale is in arbitrary units.

References

1. M. Steiner, J. Villain, and C.G. Windsor: Adv.Phys. 25, 87 (1976)
2. M. Steiner, K. Kakurai, and J.K. Kjems: Z.Phys. B 53, 117 (1983)
3. K. Kakurai, R. Pynn, B. Dorner, and M. Steiner: J.Phys. C 17, L123 (1984)
4. A.P. Ramirez and W.P. Wolf: Phys.Rev.Lett. 49, 227 (1982)
5. H.J. Mikeska: J.Phys. C 11, L29 (1978)
6. J.F. Currie, J.A. Krumhansl, A.R. Bishop, and S.E. Trullinger: Phys.Rev. B 22, 477 (1980)
7. J.A. Krumhansl and J.R. Schrieffer: Phys.Rev.B 11, 3535 (1975)
8. E. Allroth and H.J. Mikeska: J.Phys. C 13, L725 (1980) and Z.Phys. B 43, 209 (1981)
9. H.J. Mikeska: J.Appl.Phys. 52, 1950 (1981)
10. V.K. Samalam and P. Kumar: Phys.Rev. B 26, 5146 (1982)
11. P. Kumar: Phys.Rev. B 25, 483 (1982) and Physica 5D, 359 (1982)

12. E. Magyari and H. Thomas: Phys.Rev. B 25, 531 (1982)
13. H.J. Mikeska and K. Osano: Z.Phys. B 52, 111 (1983)
14. K. Osano: J.Phys. C 17, 843 (1984)
15. K. Osano: preprint
16. H.C. Fogedby, P. Hedegård, and A. Svane: Phys.Rev. B 28, 2893 (1983)
17. K.M. Leung and A.R. Bishop: J.Phys. C 16, 5893 (1983)
18. H.C. Fogedby, P. Hedegård, and A. Svane: Phys.Rev. B 29, 2861 (1984) and preprint
19. J.S. Langer: Ann.Phys. 14, 108 (1967)
20. K.M. Leung: J.Appl.Phys. 53, 1858 (1982) and Phys.Rev. B 26, 226 (1982)
21. A.M. Polyakov: Nucl.Phys. B 120, 429 (1977)
22. A.R. Bishop, K. Nakamura, and T. Sasada: J.Phys. C 13, L515 (1980)
23. H.C. Fogedby, P. Hedegård, and A. Svane: to appear in J.Phys. C (1984)
24. C. Etrich and H.J. Mikeska: J.Phys. C 16, 4889 (1983)
25. H.J. Mikeska, Phys.Rev. B 26, 5213 (1982)
26. K. Maki: Phys.Rev. B 24, 3991 (1981)
27. O.G. Mouritsen, H. Jensen and H.C. Fogedby: to appear in Phys.Rev. B (1984)
28. M.G. Pini and A. Rettori: Phys.Rev. B 29, 5246 (1984)

Is TMMC a Sine-Gordon System?

M. Fowler, N.F. Wright, and M.D. Johnson
Physics Department, University of Virginia, Charlottesville, VA 22901, USA

1. INTRODUCTION

It is by now widely accepted [1] that the magnetic salt TMMC ($(CH_3)_4NMnCl_3$) is very well described by the one-dimensional Hamiltonian

$$H = -2J\sum_n \vec{S}_n \cdot \vec{S}_{n+1} - 2D\sum_n S_n^z S_{n+1}^z - H\sum_n S_n^x \qquad (1)$$

where $J = -6.7K$, $\delta = D/|J| = 0.016$, $S = 5/2$ and H is the applied external (transverse) field. Mikeska [2] has given a derivation of the intuitively appealing result that the continuum limit of (1) is a sine-Gordon system for sufficiently large easy plane anisotropy parameter δ. For the particular case of TMMC, however, δ is actually rather small. It has been argued that the sine-Gordon picture may nevertheless be valid for TMMC [3], at least for $kT \ll JS^2$ (which is the temperature range of interest) because the strong antiferromagnetic coupling keeps neighboring spins almost antiparallel, so significant out-of-plane deviations necessarily involve many spins and consequently cost substantial amounts of energy. The purpose of this note is to point out that although the sine-Gordon model is valuable in providing a physical understanding of many features of TMMC, its numerical predictions do not accord well with the experimental results, and the main reason for this is the neglect of out-of-plane fluctuations. The two sine-Gordon predictions considered here are the specific heat as a function of temperature, and the magnitude of the (in-plane) magnon mass renormalization at zero temperature. Both can be found very accurately in the sine-Gordon model, and both correspond poorly with what is observed in TMMC. For the magnon mass renormalization, we have shown explicitly to second order in S^{-1} that the out-of-plane fluctuations account for the discrepancy between sine-Gordon predictions and experimental observations for TMMC.

2. THE SPECIFIC HEAT

The quantum sine-Gordon system is completely integrable and thermodynamic quantities can be calculated by standard Bethe Ansatz techniques. The present author and Zotos [4] have calculated the specific heat of the quantum sine-Gordon system as a function of temperature and coupling. Subtracting off the phonon specific heat gives a one parameter family of Schottky-like soliton specific heat curves. It is important to note that the position and height of the soliton specific heat peak cannot be adjusted independently. Zotos [5] compared the sine-Gordon predictions with the TMMC experiments and found a good fit for a particular value of the soliton mass renormalization. These results were presented at the Les Arcs conference in 1982. [5] Unfortunately, the work was in error by a factor of two. One step involved going from the discrete system given by equation (1) above to a discrete sine-Gordon system, with two TMMC sites going to a single sine-Gordon site. [3] This factor of two was omitted, so the actual sine-Gordon specific heat per site is only half that compared with experiment. There is little room left for further adjustment of the sine-Gordon predictions. The quantum sine-Gordon fit is substantially worse than that given by classical sine-Gordon thermodynamics, [6] yet we know from the work of Maki [7] that if TMMC is a sine-Gordon system, it is far from the classical limit. We conclude from this that TMMC thermodynamics is not well described by a sine-Gordon analysis and therefore the out-of-plane fluctuations are important.

3. THE MAGNON MASS RENORMALIZATION

The specific heat calculation covers a temperature range in which the system is highly excited-many solitons are present. One might hope that at lower temperatures out-of-plane effects are less important and the sine-Gordon Hamiltonian would be a more reliable guide. We find this not to be the case for a very basic parameter - the magnon mass. The bare (in-plane) magnon mass is $g\mu_B H$. This is renormalized by the anharmonic magnon-magnon coupling. In the sine-Gordon model, the standard field theoretic renormalization analysis [7] predicts a physical magnon mass 30% below the bare value for TMMC in a 5T field. This renormalization is not observed to take place [8] - the experimental results indicate a physical magnon mass close to the bare value.

Riseborough and Reiter [9] have given a formal graphical expansion for TMMC spin-spin Green's functions using Holstein-Primakoff type operators in our equation (1). This expansion, of course, includes both in-plane and out-of-plane motion for the spins. We have used these methods to calculate the magnon mass renormalization to second order in S^{-1}, a total of about a hundred graphs. We find for TMMC over a range of magnetic fields 3T to 7T very good cancellation between in-plane and out-of-plane renormalization terms, in good agreement with experiment. We find a net renormalization of 7% downwards for 3T, going to 5% downwards at 7T and above. Thus, even at low temperatures where there may be little out-of-plane motion for TMMC, the virtual out-of-plane magnons still have important physical consequences. It must be admitted that a series in S^{-1} with $S = 5/2$ is not necessarily fast converging. We find, however, that the first order renormalization ranges from 16% at 3T to 5% at 10T and the second order terms give a smaller contribution of opposite sign, going from 9% at 3T to below 1% at 10T.

Within each order, we find individual graphs giving large contributions, but these are almost completely cancelled by others. This is partly because the large terms come from logarithmically divergent integrals and the lower cutoffs for in-plane and out-of-plane excitations are close to each other for the magnetic field strengths we have examined. Details of these calculations will be published elsewhere.

4. CONCLUSION

To summarize - the quantum sine-Gordon thermodynamic prediction for the specific heat of TMMC gives about half the experimentally observed value. Presumably the rest of the energy goes into out-of-plane modes. At zero temperature, the sine-Gordon model predicts a substantial magnon mass renormalization. This is not observed, and we find by calculation that to second order in S^{-1} it is almost exactly cancelled by out-of-plane fluctuations. Thus for two very different physical quantities, we find the non-sine-Gordon out-of-plane fluctuations to be as important as the in-plane fluctuations for TMMC.

REFERENCES

[1] M. Steiner: Journ. Magnetism and Magnetic Materials 31-34, 1277 (1983).

[2] H. J. Mikeska: J. Phys. C 13, 2913 (1980).

[3] K. M. Leung et al: Phys. Rev. B 21, 4017 (1980).

[4] M. Fowler and X. Zotos: Phys. Rev. B 25, 5806 (1982).

[5] X. Zotos: Journal de Physique, Colloque C3, 44, 477 (1983).

[6] F. Borsa: Phys. Rev. B 25, 3430 (1982).
F. Borsa et al: Phys. Rev. B 28, 5173 (1983).
K. Sasaki and T. Tsuzuki: Journ. Magnetism and Magnetic Materials 31-34, 1283 (1983).

[7] K. Maki: Phys. Rev. B 24, 3991 (1981).

[8] I. U. Heilmann et al: Phys. Rev. B 24, 3939 (1981).

[9] P. S. Riseborough and F. G. Reiter: Phys. Rev. B 27, 1844 (1983).

Dynamics of Phase Transitions in Ferromagnetic Domain-Wall Structures

E. Magyari and H. Thomas

Institut für Physik, Universität Basel, Klingelbergstraße 82
CH-4056 Basel, Switzerland

Phase transitions (PT's) are announced dynamically by a critical slowing-down phenomenon which manifests itself in general by "softening" of a normal-mode frequency. Thus, bulk PT's both in lattice dynamics and in magnetism are associated with an extended soft mode (soft phonon and soft magnon, respectively), announcing the transition to a new ground state of the system. Similarly, PT's in domain walls (DW's) in lattice-dynamical systems (e.g. in ferroelectrics) are connected with a soft localized mode signalling the transition to a new DW structure. In ferromagnetic systems, however, there occur DW-instabilities which are associated with an essentially different mechanism of critical slowing-down, consisting of a softening of the velocity change induced by a perturbation of the DW structure. In the present lecture, this new dynamical phenomenon is discussed by considering as specific examples the instabilities of 180° and 360° DW's in biaxial [1,2] and in planar [3-6] ferromagnets, respectively. In this approach the spins are considered as classical vectors of constant length, and the DW's are assumed to be long-wavelength nonlinear excitations of the spin field so that the continuum approximation applies.

1 180° DW's in Biaxial Ferromagnets, [1]

We first consider a one-dimensional biaxial Heisenberg ferromagnet described by the Hamiltonian

$$H = \sum_i \{-J\vec{S}_i \cdot \vec{S}_{i+1} + A(S_i^z)^2 - C(S_i^x)^2\} , \tag{1.1}$$

where J and C are positive constants and A is a parameter satisfying A>-C such that the x axis is the easy axis. The hard axis is along y for -C<A<0 and along z for A>0. In the classical continuum approximation one obtains the energy functional

$$E[s] = \frac{1}{2}\int_{-\infty}^{+\infty} (s'^2 + as_3^2 - s_1^2 + 1)dz \tag{1.2}$$

and the equation of motion

$$\dot{\vec{s}} = \vec{s} \times (\vec{s}'' + s_1\vec{e}_1 - as_3\vec{e}_s) , \tag{1.3}$$

where $\vec{s}(z,t) = \vec{S}(z,t)/S$, $\vec{s}^2 = 1$, $a = A/C>-1$, the prime indicates $\partial/\partial z$, the dot denotes $\partial/\partial t$, and the units of z,t and E are $[z] = (J/2C)^{1/2}d$ (d is the lattice constant), $[t] = (2SC)^{-1}$ and $[E] = S^2(2JC)^{1/2}$, respectively.

For $a>-1$, the ground states of the system are the domains $\vec{s} = (\pm 1,0,0)$ while for $a<-1$ this role would be taken over by the uniform solutions $\vec{s} = (0,0,\pm 1)$. The dispersion relation for the small-amplitude oscillations (spin waves or magnons) in the domains $(\pm 1,0,0)$,

$$\omega_q^2 = (1+a+q^2)\ (1+q^2)\ , \tag{1.4}$$

shows that these configurations are indeed linearly stable for $a>-1$, and that the first order PT ("bulk instability") to the new ground states $\vec{s} = (0,0,\pm 1)$ occurring at $a = -1$ is associated with a soft magnon (Fig. 1).

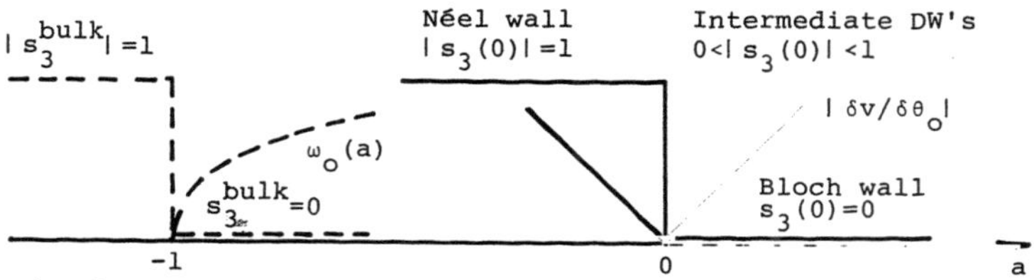

Fig.1 The bulk PT at $a=-1$ is associated with a soft magnon with frequency $\omega_o(a)=(1+a)^{1/2}$ (----); the domain wall PT at $a=0$ is associated with a soft velocity change $|\delta v/\delta\theta_o| = a$ (——)

We now consider solitary waves $\vec{s}(z,t) = \vec{s}(Z)$, $Z = z-vt$, connecting the two ground states $(\pm 1,0,0)$ of the system. These are 180° DW's satisfying the boundary conditions $\vec{s}(\mp\infty) = (\pm 1,0,0)$. There are two static ($v=0$) DW's (plus their trivially symmetric counterparts) satisfying these boundary conditions. One of them is the Bloch wall

$$s_1 = -\tanh z\ ,\ s_2 = \operatorname{sech} z\ ,\ s_3 = 0 \tag{1.5}$$

with energy $E_B = 2$, and the other is the Néel wall

$$s_1 = -\tanh\{(+a)^{1/2}z\}\ ,\ s_2 = 0\ ,\ s_z = \operatorname{sech}\{(1+a)^{1/2}z\} \tag{1.6}$$

with energy $E_N = 2(1+a)^{1/2}$. The widths of these DW's are given (in physical units) by the expressions $\delta_B = (J/2C)^{1/2}d$ and $\delta_N = (J/2C)^{1/2}(1+a)^{-1/2}\ d$, respectively. Therefore, the Bloch and Néel walls are long-wavelength excitations of the spin field satisfying the validity requirement of the continuum approximation ($\delta>>d$) only if $J>>C$.

For $a \neq 0$, any "intermediate" wall between the Bloch and Néel wall is a moving DW. These moving DW's may also be given in simple analytic form [7]. It turns out [7] that for fixed $a \neq 0$ the possible wall velocities are restricted to the interval $|v| < |1-(1+a)^{1/2}|$. For $a = 0$, on the other hand, there exists a family of intermediate static DW's which is described by

$$\vec{s} = (-\tanh z\ ,\ k_2 \operatorname{sech} z\ ,\ k_3 \operatorname{sech} z\ )\ , \tag{1.7}$$

where the constants $k_{2,3}$ are subjected to the condition $k_2^2+k_3^2=1$.

Let us now examine the energetic stability of the Bloch and Néel walls. If we represent the spin field in spherical coordinates $\vec{s} = (\cos\theta\cos\phi\ ,\ \cos\theta\sin\phi\ ,\ \sin\theta)$, where $|\theta|\leqslant\pi/2$ and $0\leqslant\phi\leqslant2\pi$, the Bloch wall (1.5) becomes $\theta_B = 0$, $\sin\phi_B = \mathrm{sech}\, z$. Diagonalizing the second variation of the energy functional (1.2) with respect to small deviations (α,β) from the Bloch wall (θ_B,ϕ_B) we get the following eigenvalues (ε) and eigenfunctions: (1) $\varepsilon=0$, $\alpha=0$, $\beta=\mathrm{sech}\, z$, which is the Goldstone mode reflecting the marginal stability of the Bloch wall against translations, and (2) $\varepsilon=a$, $\alpha=\mathrm{sech}\, z$, $\beta=0$. Therefore, the Bloch wall is energetically stable against θ perturbations for any $a>0$, but at $a=a_c=0$ it becomes unstable with respect to a perturbation connecting it with the family of static 180° DW's given by (1.7). A similar analysis of the Néel wall (1.6) shows that this static spin configuration is energetically stable for $-1<a<0$ and becomes unstable also at $a=0$ with respect to a perturbation connecting it with the same family of static 180° DW's (Fig.1).

Let us now turn to the examination of the domain wall PT at $a=0$ in terms of the dynamics of the system by linearizing the equations of motion around the Bloch wall, e.g., according to

$$\begin{aligned}\theta(z,t) &= \theta_B(z) + \alpha(z)\ \exp(-i\omega t)\\ \phi(z,t) &= \phi_B(z) + \beta(z)\ \exp(-i\omega t)\ .\end{aligned} \tag{1.8}$$

One obtains the system of coupled eigenvalue equations

$$(L-a)\alpha=i\omega\beta\ ,\quad L\beta=-i\omega\alpha \tag{1.9}$$

where $L = -1+2\mathrm{sech}^2 z + d^2/dz^2$.
In the region of energetic stability $a>0$ the Bloch wall must also be dynamically stable. On the grounds of comparable lattice-dynamical situations [8,9], one might expect that the instability at $a=0$ is associated for $a\neq0$ with a localized dynamic mode of the wall $(\alpha(z,a),\beta(z,a))$ corresponding to a frequency $\omega(a)$ whose square is positive for $a>0$ and negative for $a<0$, such that for $a\to0$, $\omega(a)$ goes to zero and $(\alpha(z,a),\ \beta(z,a))$ approaches a linear combination of the static instability mode $(\mathrm{sech}\, z\ ,0)$ with the Goldstone mode $(0,\mathrm{sech}\, z)$. This is, however, not the case. In the present model, the nonexistence of such a soft dynamic localized spin mode can explicitly be proven, since by the ansatz $\alpha=(1-k)W_0 - (1+k)W_1$, $\beta=(1+k)W_0 + (1-k)W_1$, where $k=(a-2i\omega)^{1/2}(a+2i\omega)^{-1/2}$, the system (1.9) may be reduced to the following system of trivially coupled equations:

$$(2L-a+\Omega)W_0=0\ ,\quad (2L-a-\Omega)W_1=-4i\omega W_0\ , \tag{1.10}$$

where $\Omega = (a^2 + 4\omega^2)^{1/2}$.

It is easy to show that the system (1.10) admits localized solutions only for $\omega=0$. Thus, the only localized solutions of Eqs. (1.9) are identical to the static eigenmodes, i.e. the Goldstone mode $(\alpha,\beta) =(0,\mathrm{sech}\, z)$, and at $a=0$ in addition the instability mode $(\alpha,\beta) = (\text{sech}\, z, 0)$. Therefore, the energetic instability of the Bloch wall at $a=0$ is not associated with a localized soft dynamic spin mode. The dynamic stability analysis of the Néel wall leads to a similar result.

How can the absence of a soft mode be reconciled with the existence of an instability ? As we have shown, the family of intermediate static DW's (1.7) bifurcate from the Bloch wall at the critical field $a_c=0$. On the other hand, any infinitesimal θ deviation from the Bloch wall leads for any value of a to a DW moving with infinitesimal velocity [7], except for $a=0$ where it connects to a static DW with infinitesimal k_3 . This result suggests that in the present problem the role of the soft eigenmode is taken over by the θ perturbation which carries the static DW into a moving one, and the role of the soft-mode frequency is taken over by the resulting velocity change. The "softening" of the velocity change δv at the critical field $a_c=0$ may be seen explicitly by calculating δv produced by a perturbation with maximum deviation $\delta\theta_0$ as function of a . The simple and convincing result is $\delta v/\delta\theta_0=-a$ (Fig.1).

This picture is confirmed by a more careful stability analysis. Instead of making the exponential ansatz (1.8) we linearize the equation of motion (1.3) for deviations $\alpha(z,t)$ and $\beta(z,t)$ from the Bloch wall, with arbitrary time dependence. In this way, in addition to the Goldstone mode $(\alpha_1, \beta_1)=(0, \text{sech}\, z)$ we find a nonexponential solution $\alpha_2=-(v/a)\,\text{sech}\, z$, $\beta_2=-vt\,\text{sech}\, z$. For $a\neq0$, $v\neq0$, this solution describes a DW moving with infinitesimal velocity v . For $a\to0$, $v\to0$, v/a finite, it reduces to the static instability mode $(\alpha,\beta)=(\text{sech}\, z, 0)$.

The above considerations also suggest that a soft-relaxation mode picture of critical slowing down may be realized in the presence of spin damping. Indeed, if one includes in (1.3) a phenomenological damping term of the Gilbert form, $\dot{\vec{s}}=\vec{s}\times(\vec{s}''+s_1\vec{e}_1-as_3\vec{e}_3-\eta\dot{\vec{s}})$, $0<\eta<<1$, then the linearized equations of motion around the Bloch wall,

$$(L-a+i\eta\omega)\alpha=i\omega\beta \ , \ (L+i\eta\omega)\beta=-i\omega\alpha, \tag{1.11}$$

do in fact yield a soft relaxation frequency $\omega=-ia\eta$, showing dynamic stability for $a>0$ and dynamic instability for $a<0$. The corresponding soft eigenmode $(\alpha,\beta)=(-\eta\,\text{sech}\, z, \text{sech}\, z)$ describes for $a>0$ the attenuation and for $a<0$ the amplification of the initial velocity produced by the α perturbation. Similar conclusions result also in the case of the Néel wall.

2 360° DW's in Planar Ferromagnet [3,6]

A second class of 1D model ferromagnets includes exchange as well as single-ion anisotropies, whose global effect results in a planar symmetry of the magnet, and a constant external field $\vec{B}$ applied in the easy plane. The corresponding Hamiltonian is

$$H=-J\sum_i\{g(S_i^xS_{i+1}^x+S_i^yS_{i+1}^y)S_i^zS_{i+1}^z\}+A\sum_i(S_i^z)^2-g_M\mu_B B\sum_i S_i^x \ , \tag{2.1}$$

where g is a positive dimensionless parameter which measures the exchange anisotropy, J>0 and the other notation is standard.

In the classical continuum approximation we obtain the energy functional

$$E[s]=\frac{1}{2}\int_{-\infty}^{+\infty}\{g(s_1'^2+s_2'^2)+s_3'^2+as_3^2+2b(1-s_1)\}\,dz \qquad (2.2)$$

and the equation of motion

$$\dot{\vec{s}}=\vec{s}\times\{g(s_1''\vec{e}_1+s_2''\vec{e}_2)+s_3''\vec{e}_3-as_3\vec{e}_s+b\vec{e}_1\} \quad , \qquad (2.3)$$

where $a=g-1+AJ^{-1}$, $b=g_M\mu_B(2JS)^{-1}B$ and the other notations are the same as in Sec.1, but z,t, and E are now expressed in units of $[z]=2^{-1/2}d$, $[t]=(2JS)^{-1}$ and $[E]=2^{1/2}JS^2$, respectively. We assume throughout in this paper that a>0 , i.e. that the interplay of exchange and single-ion anisotropy results in a planar character of the magnet. For convenience, we represent the spin field $\vec{s}$ in the same spherical coordinates as in Sec.1.

We consider solitary wave solutions $\vec{s}=\vec{s}(Z)$, Z=z-vt, of the equation of motion, starting from and returning to the ferromagnetic state, whose ϕ-variation is a 360° kink. They are singled out by the boundary conditions $\theta(\mp\infty)=0$, $\phi(\mp\infty)=0,2\pi$. We restrict our investigation to spin configurations in the easy plane and in its infinitesimal neighborhood.

It is easy to see that for $\theta=0$ the only kink solution of (2.3) satisfying the boundary conditions is the static (v=0) sine-Gordon (sG) , [10], or azimuthal kink

$$\theta=0 \ , \ \sin(\phi/2)=\mathrm{sech}\{(b/g)^{1/2}z\} \ , \qquad (2.4)$$

with energy $E_0=8(gb)^{1/2}$ and width (in physical units) $\delta=(g/2b)^{1/2}d$. It represents a long-wavelength excitation if b<<g .

Let us now consider out-of-plane solutions of the equation of motion (2.3) for infinitesimal θ. To zeroth and first order in θ we obtain: $g\phi''=b\sin\phi$ and $\theta''+(g\phi'^2-a-b\cos\phi)=v\phi'$, respectively. Hence, in this limit the azimuthal structure of the spin field coincides with the sG-configuration (2.4), whereas the meridional (out-of-plane) structure is described by

$$M\theta= 2v(g/b)^{1/2}\mathrm{sech}\,\zeta \ , \qquad (2.5)$$

where M is the hermitian operator $M=d^2/d\zeta^2-g(1+ab^{-1}-6\,\mathrm{sech}^2\zeta)$ with $\zeta=(b/g)^{1/2}Z$.

In the static limit v=0, (2.5) becomes homogeneous. This Schrödinger-type equation admits localized solutions only for a limited number of well-defined values $b_n=ag\{(p-n)^2-g\}^{-1}$ of the applied field [6,11], where $2p=-1+(1+24g)^{1/2}$ and n=0,1,2,..., but n<p . Hence (for the g and n values for which $b_n>0$) , the values b_n represent bifurcation points, where the static in-plane kink solution (2.4) is coexistent with a static out-of-plane kink solution with infinitesimal θ given by the eigenfunction

$$\theta_n^o = A(\mathrm{sech}\zeta)^{n-p} d^n (1-\tanh^2\zeta)^p / d(\tanh\zeta)^n \ , \qquad (2.6)$$

which is of parity $(-1)^n$.

For $v \neq 0$, $b \neq b_n$ one obtains a family of moving kinks, whose meridional component is to $O(v)$ given by the solution of the inhomogeneous equation (2.5) (which is of even parity). At the bifurcation fields b_n one obtains the following results [6]: For $b=b_n$ with n even, there exists no out-of-plane solution other than the static kink bifurcating at this value of b . This static kink is the $b \to b_n$ limit of the family of moving kinks. For $b=b_n$ with n odd, there exists in addition to the bifurcating static kink a family of moving kinks. In this case the static kink represents an isolated odd-parity solution of the equations of motion.

As an illustration, we consider two examples: (1) g=1 and (2) g=2 . In the first case we have a single bifurcation field, $b=b_o=a/2$. The θ-component of the static kink bifurcating at b_o is given by $\theta_o^o \propto \mathrm{sech}\,\zeta$, with $\zeta=(3b_o)^{1/2}z$. The solution of the inhomogeneous equation (2.5) can be given in closed analytic form $\theta=v(3b)^{1/2}(b-b_o)^{-1}\mathrm{sech}\,\zeta$. For $v \to 0$, $b \to b_o$, $v(b-b_o)^{-1}$ finite, this solution becomes the static out-of-plane kink solution. In the second example we obtain two bifurcation fields $b_o=2a/7$ and $b_1=a$ The static out-of-plane configurations bifurcating at these fields vary as $\theta_o^o \propto \mathrm{sech}^3\zeta$, $\theta_1^o \propto \sinh\zeta\,\mathrm{sech}^3\zeta$, respectively, where $\zeta=(b_{o,1}/2)^{1/2}z$. For $b=b_o$ no moving kink solution exists, but for $b=b_1$ the θ-component of the $v \neq 0$ kink is to $O(v)$ given by the expression $\theta=(2v/3)(2/b_1)^{1/2}(1-\tanh^2\zeta)\mathrm{sech}\,\zeta$.

We turn now to the relation of the bifurcation fields b_n to changes in energetic stability of the azimuthal kink, and to the investigation of the dynamic behaviour in the neighborhood of these field strengths. Diagonalizing the second variation of the energy functional (2.2) with respect to small deviations (α,β) from the azimuthal kink (2.4) we get, in addition to the Goldstone mode $\{\varepsilon=0$, $\alpha=0$, $\beta \propto \mathrm{sech}\zeta\}$ a sequence of eigensolutions $\{\varepsilon=\varepsilon_n$, $\alpha=\alpha_n$, $\beta=0\}$, where $\varepsilon_n=ag(b^{-1}-b_n^{-1})$, and α_n is just the θ_n^o given by (2.6). Therefore, the bifurcation fields b_n mark sign-changes of the eigenvalues ε_n , i.e. changes in the energetic stability of the azimuthal kink. We conclude that the azimuthal kink is energetically stable only if $b<b_o$, where b_o is the smallest bifurcation field given by

$$b_o=\frac{1}{2}a(25g-1)^{-1}\{1+10g+(1+24g)^{1/2}\}. \qquad (2.7)$$

(In the isotropic case g=1 one has $b_o=a/3$, in agreement with [12].) At $b=b_o$ the azimuthal kink becomes energetically unstable with respect to a static out-of-plane configuration described by $\theta=\alpha_o \propto \mathrm{sech}^p\zeta$. For increasing b there occurs at each of the bifurcation fields b_1 , b_2 ,... an instability of the in-plane kink with respect to an additional static out-of-plane mode.

In order to examine the dynamics associated with the instabilities at $b=b_n$ we introduce again Gilbert damping term into (2.3) and linearize this equation around the kink (2.4). With an ansatz similar to (1.8), we get the coupled eigenvalue equations

$$(M+igb^{-1}\eta\omega)\alpha=igb^{-1}\omega\beta \quad ,(N+ib^{-1}\eta\omega)\beta=-ib^{-1}\omega\alpha \quad , \tag{2.8}$$

where $N=-1+2\mathrm{sech}^2\zeta+d^2/d\zeta^2$. For the particular case $g=1/3$, (2.8) yields the closed-form solution

$$\{\omega=2i\eta(b-b_o) \quad , \alpha=-\eta\mathrm{sech}\zeta \quad , \beta=\mathrm{sech}\zeta\} \quad , \tag{2.9}$$

where $b_o=a/2$, $a=A/J-2/3>0$ and $\zeta=(3b)^{1/2}z$. For $g\neq 1/3$ no closed-form solutions of (2.8) are available, but a perturbation calculation around the static instability modes $\{\omega=0$, $b=b_n$, $\alpha=\theta_n^o$, $\beta=0\}$ with n even yields a similar result [6]. One sees that (2.9) is a soft-relaxation mode of the kink implying its stability below and its dynamic instability above the critical field $b=b_o$. In the undamped ($\eta=0$) system, however, this mode disappears, and thus the instability of the 360° kink at $b=b_o$ is signalized dynamically again by a "softening of the velocity change". This "soft velocity change" picture holds for any g and $b=b_n$ with n even, [6]. In the neighborhood of the bifurcation fields $b=b_n$ with n odd, on the other hand, the situation resembles the lattice-dynamical results. The static instabilities at b_1 , b_3 ,... are associated even in the undamped system with soft localized modes of the kink. For the particular case g=2 , for instance, there exists a localized dynamic mode whose frequency in the neighborhood of the bifurcation field $b=b_1=a$ is given by $\omega^2=(9a/5)(b_1-b)$ which goes to zero for $b\rightarrow b_1$ [6]. Similar results are obtained in the neighborhood of any b_n with n odd. However, since the azimuthal kink loses its stability already at $b=b_o$, this soft-mode dynamics will not appear in a physical system.

References

1. E. Magyari and H. Thomas: Phys. Rev. Lett. 51, 54 (1983)
2. C. Etrich and H.J. Mikeska: J. Phys. C16, 4889 (1983)
3. E. Magyari and H. Thomas: J. Phys. C16, L535 (1983)
4. H.J.Mikeska and K. Osano: Z. Phys. B52, 111 (1983)
5. H.J. Mikeska: Physica 120B, 235 (1983)
6. E. Magyari, H. Thomas and R. Weber: Z. Phys. B55, (1984)
7. E. Schlömann: Appl. Phys. Lett. 19, 274 (1971)
8. J. Lajzerowicz and J.J. Niez: "Structure and Stability of Domain Walls-Phase Transition" in Solitons in Condensed Matter Physics, ed. by A.R. Bishop and T. Schneider (Springer-Verlag, 1978) pp 195-198
9. E. Magyari and H. Thomas: Phys. Lett. 100A, 11 (1984)
10. H.J. Mikeska: J. Phys. C11, L29 (1978)
11. E. Magyari and H. Thomas: J. Phys. C15, L 333 (1982)
12. E. Magyari and H. Thomas: Phys. Rev. B25, 531 (1982)

Part IV

Mixed and Disordered Magnets

Stress Fluctuations and Local Structural Distortions in Ferromagnetic Metallic Glasses: Their Role on the Relaxation of the Magnetic Permeability

P. Allia and F. Vinai
Istitudo Elettrotecnico Nazionale "Galileo Ferraris", I-10125 Torino, Italy
Gruppo Nazionale Struttura della Materia del CNR, u.r. 24
I-10125 Torino, Italy

1. Introduction

The geometrical nature of the structure of metallic glasses can be described in two complementary ways: by making use of the concept of Bernal holes [1] and by introducing the analysis of Voronoi polyhedra [2][3]. While the description in terms of Bernal holes implies the introduction of a set of elementary polyhedra constituting the glassy structure and composed of adjacent atoms, the Voronoi approach has the clear advantage of describing the atomic configuration around a single atom. The rise of interest for the adoption of entirely local, single-atom criteria to describe the structure of amorphous metals is confirmed by various other attempts, recently made to introduce local order parameters very atomistically defined and randomly fluctuating from site to site. In this way, the geometry of the environment of any atom in a glassy metal can be fully characterized.

Typical examples of such quantities are the atomic-level stresses and the site-symmetry coefficients defined by EGAMI and coworkers [3][4], the bond orientational order parameter used by STEINHARDT et al. [5], the tetrahedron distortion coefficient introduced by KIMURA and YONEZAWA [6]. However, most of these parameters have been used only in the context of the analysis of computer generated structures of glassy metals. Only the second moments of local stress fluctuations have been explicitly related so far to measurable quantities, such as the x-ray structure factor [3], the electrical resistivity [7], and the magnetic aftereffect [8].

In the present paper, a new local order parameter is introduced, which gives detailed information about the distortions of the atomic environments in amorphous structures. The interest in such a parameter lies in the fact that it is not merely derived from structural considerations, but it is the natural output of a theory on the local magneto-elastic energy of ferromagnetic metal-metalloid glasses, explaining the results of magnetic aftereffect measurements [8] in these materials.

2. Atomic-level stresses and magneto-elastic energy of amorphous ferromagnets.

Local stresses and strains are inherently present in metallic glasses. The atomic-level (hydrostatic and shear) stresses are randomly defined at

(1) supported by the Ministry of Education (M.P.I.) and C.N.R. - Progetto Finalizzato Metallurgia.

every site in terms of the derivatives of the interatomic potential [4]. Similarly, local strains can be defined, to picture the degree of structural distortion of the shell of nearest neighbors of any atom in the material [4]. The best way to describe the strain fluctuation at every site is to use a symmetry representation, in which each independent component of the local strain tensor transforms, under rotation, like a spherical harmonic function with even l. In particular, the isotropic strain (compression/dilation of the site) is described by the harmonic with $l = 0$, while the anisotropic strain ε_μ (dipolar distortion) is fully described by the five harmonics with $l = 2$, $-2 \leqslant m \leqslant +2$. When a real representation of the strains is used, the five ε_μ's transform like tesseral harmonics $Z_{2\mu}$ [9].

These structural strains, not related to any macroscopic external stress, are always present in magnetic amorphous metals. On the other hand, the interactions between magnetic atoms of typical metallic glasses containing 3-d transition metals are strongly anisotropic in nature, and rather sensitive to deformation. As a consequence, a typical, site-dependent, magneto-elastic energy develops, representing the effect of the local structural strains on the magnetic anisotropy energy F_a. The latter can be expressed, in a general if phenomenological way, as a pseudo-multipolar series [10]

$$F_a(\bar{r}_{ij}) = f(r_{ij})\, P_2(\cos\theta) + g(r_{ij})\, P_4(\cos\theta) + \dots \tag{1}$$

where $P_{2n}(\cos\theta)$ are Legendre polynomials, $\bar{r}_{ij}$ is the coordination vector between the reference atom i and anyone of its neighbors, θ is the angle between $\bar{r}_{ij}$ and the magnetization direction, and f, g are both decreasing functions of the interatomic distance $|\bar{r}_{ij}|$. The magneto-elastic energy $F_{me} = \delta F_a(\bar{r}_{ij})$ can be written in a simple form by making use of the symmetry representation for stresses and strains. By using only the pseudo-dipolar term of (1), as usual [10], and by taking into account that the shear stress components, σ_μ, are proportional to the ε_μ's through an average shear elastic constant, $\bar{C}_\gamma$ [4], one gets [11]

$$F_{me} = \Sigma_{\mu\mu'} (\lambda_i\, \delta_{\mu\mu'} + \lambda_{\mu\mu'})\, \sigma_{\mu'}\, Z_{2\mu}(\Omega_M) \tag{2}$$

where the double summation is performed over the independent shear components (labelled by μ, μ'); and $\Omega_M \equiv (\theta_M, \phi_M)$ are the angular components of the magnetization vector in spherical coordinates. Moreover,

$$\begin{aligned} \lambda_i &= -\sqrt{\frac{8\pi}{15}}\ \frac{\bar{z}}{2}\ \bar{C}_\gamma^{-1}\ f(a) \\ \lambda_{\mu\mu'} &= -\sqrt{\frac{8\pi}{15}}\ \frac{4\pi}{5}\ \bar{C}_\gamma^{-1}\ [\tfrac{1}{2}\, a\, f'(a) - f(a)]\ \Sigma_j\, Z_{2\mu}(\Omega_{ij})\, Z_{2\mu'}(\Omega_{ij}) \end{aligned} \tag{3}$$

where a is the average n.n. distance, $\bar{z}$ the average coordination number, and the summation is performed over neighbors of atom i. Both λ_i and

$\lambda_{\mu\mu'}$ are local magnetostriction constants. However, λ_i is homogeneous, while $\lambda_{\mu\mu'}$ is site - dependent and contains information about the symmetry of the bond directions Ω_{ij} around the atom i. Note that F_{me} is composed of a diagonal plus an off-diagonal term. In addition, only shear stress components appear in (2), indicating that the hydrostatic pressure gives essentially no contribution to F_{me}.

3. Permeability aftereffect and the order parameter m.

Compelling experimental evidence [8] strongly suggests that the microscopic magneto-elastic energy (2) is the very source of the aftereffect of the magnetic permeability in amorphous ferromagnets. This effect is the decrease with time after demagnetization of the low-field permeability of a ferromagnetic material. For more phenomenological information, the reader is referred to papers [8], [12].

It is possible to show [11] that the magnetic aftereffect, $\Delta\mu/\mu$, is proportional to $\langle F_{me}^2 \rangle$, where $\langle \dots \rangle$ stands for an average over the whole material. After straightforward manipulation of (2), one gets $\Delta\mu/\mu \sim \lambda^2_{eff} \langle \tau^2 \rangle$, where $\langle \tau^2 \rangle$ is the second moment of the shear stress distribution [4] and the effective magnetostriction, λ_{eff}, is given by

$$\lambda^2_{eff} = \frac{8\pi}{25} (\lambda_s^2 + \langle m \rangle \lambda_a^2) \tag{4}$$

where $\lambda_a = - \frac{1}{4\pi} \sqrt{\frac{8\pi}{15}} \bar{C}_\gamma^{-1} \bar{z} [\frac{1}{2} a f'(a) - f(a)]$, $\lambda_s = (5/4\pi) \lambda_i + \lambda_a$.

The last quantity is coincident with the measured, isotropic magnetostriction constant of the alloy. The quantity $\langle m \rangle$ in (4) is the average value of a local order parameter, defined as

$$m_i = \frac{5}{z_i^2} \cdot \Sigma_{jk} [P_2(\cos\theta_{jk})]^2 - 1 \tag{5}$$

where $P_2(\cos\theta_{jk})$ is a Legendre polynomial, θ_{jk} is the angle between the j-th and the k-th bond direction around i. The double summation is to be performed over all neighbors of i. The terms with j=k are included.

Clearly the parameter m_i is a rotational invariant. Moreover, it is possible to show that it is a non-negative quantity. In fact, m_i can be written in terms of second- and fourth-order Legendre polynomials as

$$m_i = \frac{2}{7} \frac{1}{z_i^2} [5 \cdot \Sigma_{jk} P_2 (\cos\theta_{jk}) + 9 \cdot \Sigma_{jk} P_4 (\cos\theta_{jk})] . \tag{6}$$

By using the addition theorem for tesseral harmonics, [9], one finally gets:

$$m_i = \frac{2}{7} \frac{4\pi}{z_i^2} [\Sigma_\mu \eta^2_{2\mu} (i) + \Sigma_\mu \eta^2_{4\mu} (i)] \geqslant 0 \tag{7}$$

where $\eta_{l\mu}(i) = \Sigma_j Z_{2\mu}(\Omega_{ij})$, $l = 2,4$.

By using (6), it is possible to show that m_i is zero only in the case of perfect icosahedral symmetry ($z_i = 12$ atoms arranged at the vertexes of a regular icosahedron around i). Only in this case, in fact, both $\Sigma_{jk} P_2(\cos\theta_{jk})$ and $\Sigma_{jk} P_4(\cos\theta_{jk})$ are simultaneously zero. In all other cases, $m_i \neq 0$. This is not only true for arbitrary distortion of a set of twelve atoms away from the icosahedral configuration, but also in the case of different symmetrical environments possibly present in real metallic glasses, like f.c.c. configuration (for which $\Sigma_{jk} P_2(\cos\theta_{jk}) = 0$ but $\Sigma_{jk} P_4(\cos\theta_{jk}) = 21/4$), or Frank-Kasper's 14-atom cluster [13] (even if in this case m_i is very close to zero).

The order parameter m_i is then a measure of the departure of a given environment from pure icosahedral symmetry, either because of fluctuations in z_i, or because of distortions of the n.n. shell, or both. As a matter of fact, m_i is much more sensitive to the departure from icosahedral symmetry, than to fluctuations of z_i. In Figure 1, the average value of m_i has been plotted as a function of the distortion Δ of bond directions away from high-symmetry configurations ($z_i = 12$: icosahedron, f.c.c.; $z_i = 14$: Frank -Kasper's cluster). Each time a new set of distorted bond directions has been obtained by allowing each direction independently to fluctuate with uniform probability within a solid angle Δ around the direction corresponding to the starting, undistorted configuration. The reported value of $\langle m \rangle$ is an average over a set of 100 different distortions. As a general rule, $\langle m \rangle$ increases with increasing disorder.

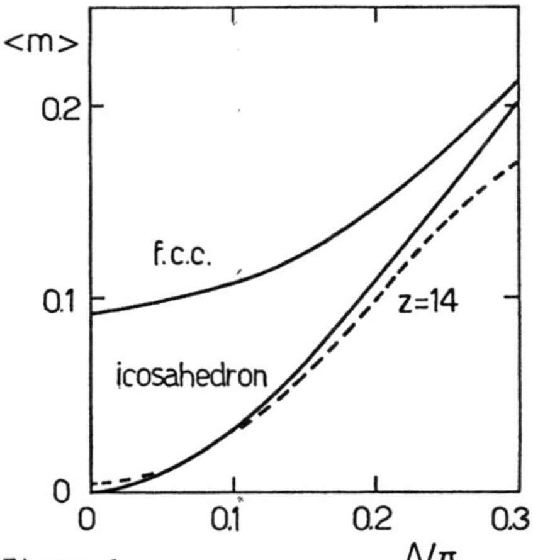

Figure 1.
Behavior of $\langle m \rangle$ as a function of distortion Δ/π
Full lines: icosahedral and fcc symmetry (z_i=12). Dashed line: Frank--Kasper 14-atom cluster.

4. Conclusion

The order parameter m_i is equally sensitive (by definition, see (7)) to the degree of dipolar, and of quadrupolar distortion of a set of z_i bond directions around atom i. This explain why $m_i = 0$ only in the case of pure icosahedral symmetry, giving neither dipolar, nor quadrupolar contribution.

Since m_i contains only information about directional order of bonds, it is more closely connected to the bond orientational order parameter Q_l introduced by STEINHARDT et al. [5], rather than to other local quantities, like local stresses and site-symmetry coefficients [3], involving also information on interatomic distances.

In real amorphous systems $\langle m \rangle$ is clearly nonzero because of the distortion existing in the structure. Changes of $\langle m \rangle$, reflecting different amorphicity conditions, can be predicted on the basis of different considerations [5], [6].

This leads to consider the permeability aftereffect, which is strictly related to $\langle m \rangle$ through (4), as a unique tool to get experimental information on processes of orientational bond ordering in metallic glasses.

1 J.D. Bernal, Nature 188, 510 (1960)
2 J.L. Finney, Proc. Roy. Soc. A319, 479 (1970)
3 D. Srolovitz, K. Maeda, S. Takeuchi, T. Egami, V. Vitek, J. Phys. F11, 2209 (1981)
4 T. Egami and D. Srolovitz, J. Phys. F12, 2141 (1982)
5 P.J. Steinhardt, D.R. Nelson, M. Ronchetti, Phys.Rev. B28, 784 (1983)
6 M. Kimura, and F. Yonezawa, J. Non-Cryst. Solids 61-62, 535 (1984)
7 P. Allia, D. Andreone, G. Riontino, R.Sato-Turtelli, F. Vinai, J. Appl. Phys. 53, 8798 (1982)
8 P. Allia and F. Vinai, Phys. Rev. B26, 6141 (1982)
9 M.T. Hutchings, in "Solid State Physics" 16 (Academic, New York 1964) p. 227
10 J. Herpin: "Théorie du Magnetisme", Presses Universitaires de France, Paris 1968, p.358
11 P. Allia and F. Vinai, to be published
12 P. Allia, P. Mazzetti, G.P. Soardo, F. Vinai, J. Magn. Magn. Mat. 15-18, 1361 (1980)
13 F.C. Frank, J.S. Kasper, Acta Cryst. 11, 184 (1958)

Frustration on Fractals

M. Cieplak

Institute of Theoretical Physics, Warsaw University
PO-00-681 Warsaw, Poland

J.R. Banavar

Schlumberger-Doll Research, P.O. Box 307, Ridgefield, CT 06877, USA

1. Fractals

MANDELBROT [1] has observed that shapes occurring in nature often look highly fragmented. Such shapes, known as fractals, can be attributed to coastlines, the mammalian brains, and so on. The length or volume of such objects depends on the length scale used. Consider, for instance, the Sierpinski gasket. To construct the 2-dimensional gasket we begin with a triangle. The midpoints of its edges are connected, creating 4 triangles. The central triangle is removed and the same procedure is continued for each of the new triangles. In the n-th step of the construction the length scale is $\varepsilon_n = 2^{-n}$ and the area within the gasket is proportional to ε_n^{2-D} where $D = \ln 3/\ln 2$. If no triangles were removed , the area would be scale invariant and $D = 2$. Note that the number of triangles in the gasket grows as $\varepsilon_n{}^{-D}$. On going from one length scale to the other the gasket looks the same, i.e. it is self-similar. Other fractals are usually self-similar in a statistical sense, but still they can be characterized by a fractal dimensionality D. In order to determine D of an object in a d-dimensional space, one may pick a length scale, a, and find the corresponding measure of the object within the volume $(aL)^d$. For large L's the measure will go as L^D.

Fractal geometries have recently received attention in condensed matter physics. KIRKPATRICK [2] and GEFEN et al. [3] have observed that a percolating cluster and its backbone are fractals on length scales not exceeding the correlation length. Thus at the percolation threshold the objects are all fractals. The fractal dimensionalities of 2d clusters and their backbones have been found to be close to 1.90 [4] and 1.6 [3] respectively. Since the backbone is finitely ramified and its D is close to ln3/ln2, GERFEN et al. [3] model the backbone by the Sierpinski gasket. This model, however, still has flaws. For instance, the average coordination number of the backbone is significantly smaller than the gasket's coordination number of 4 [5]. Among other fractal related research we mention studies of the diffusion-limited aggregation [6-8] , relevant for the process of coagulation, and the random walk decays [9] which may describe the corrosion process.

The fractal lattices have been used to deepen our comprehension of otherwise well-understood phenomena, such as random walks [10] . Here, we would like to point out advantages of using fractals in studies of the spin glass phenomenon which, on regular lattices, is not well understood.

2. Spin glasses

Spin glasses [11-13] are systems with random and competing exchange interactions. Such interactions do not allow for a spatial long range spin ordering and give rise to the phenomenon of frustration, i.e. to the presence of many degenerate or almost degenerate ground states. Spin glasses show no anomaly in the specific heat but have a cusp in the plot of the susceptibility versus temperature. The cusp is centered at a finite temperature which shows a weak dependence on the frequency of the applied field, see,e.g.,[14] . The existence of the cusp has been interpreted as related to a transition to either an equilibrium or a metastable state in which spins freeze in random directions.

The essential physics of spin glasses is contained in the Edwards-Anderson model [15] described by the Hamiltonian

$$H = - \sum_{<ij>} J_{ij} S_i S_j \, . \tag{1}$$

The exchange constants J_{ij} couple the neighboring Heisenberg or Ising spins and are drawn with probabilities corresponding to the Gaussian probability distribution. This model, however, is not amenable to an exact analysis and most of its properties are inferred from simulations of finite sized systems. In particular, it has not been clear what is the lower critical dimensionality, d_c,of the system. The quantity d_c is a dimensionality above which an ordered phase can exist, at $T \neq 0$, and below which it cannot because of too strong thermal fluctuations.

The present authors [16, 17] , see also [18] , have been among the advocates of the picture in which the 3d spin glass exhibits only a $T = 0$ equilibrium phase transition. We propose to make a virtue out of considering finite sized systems and to study scaling properties of a sensitivity of the system to changes in the boundary conditions (b.c.). This approach originates in EDWARDS and THOULESS's work [19] on the Anderson localization and generalizes the concept of the helicity modulus introduced by FISHER et al. [20] .

Briefly, we consider dividing a macroscopic sample of the material into blocks. In a scaling picture, one deduces properties of the sample by analyzing bigger and bigger subsystems constructed from finite blocks, including the interactions between the blocks. A measure of such interactions can be obtained by determining the scaling stiffness energy, ΔE_s [16] . In order to evaluate ΔE_s we envision placing the block between two walls which are meant to simulate the ends of neighboring blocks. The spins on the walls have some fixed configuration (for instance, all of these spins are either up or down) which imposes certain b.c. on the block. If we take two different wall "potentials", or equally well if we take the periodic and antiperiodic b.c., we determine the corresponding free energies. In an equilibrium calculation at $T = 0$, one calculates the true ground state energies.

In spin glasses, there is nothing to choose between the two b.c. and the difference between the two free energies, ΔF, averages out to zero, when many blocks are considered. This is not so in unfrustrated systems in which ΔE_s may be defined as the configurational average $\langle \Delta F \rangle_c$ and is identical to the usual exchange stiffness. In spin glasses, we take ΔE_S to be measured by the root mean square average of ΔF, or, equally well, by the average modulus of ΔF.

At high temperatures, ΔE_S is an exponential function of the size of the block. At T = 0, the spin glass is ordered in its lowest ground state and ΔE should scale with the size if the block algebraically. If it then scales to infinity, the ordering is strong and persists up to a critical temperature: the system is above its d_c. If it scales to zero, the ordering is weak and gets destroyed by thermal fluctuations: the system is below its d_c. Finally, if it is scale invariant, the system is at its d_c and it may (like the 2d XY model) or may not have (like the 2d Heisenberg system) a nontrivial transition.

Our numerical results [16, 17] on Heisenberg and Ising spin glasses are consistent with algebric decays of ΔE_S at T = 0. They also seem to indicate that the 2 and 3d systems are below the d_c whereas the 4d (at least Heisenberg) system is at d_c. Even with an extensive use of the computer time, these calculations had to be restricted to rather small systems, not exceeding 12x12x8 spins in the 3d case. It has thus been helpful to confirm the validity of the method by applying it to a toy model of a spin glass. This model consists of Ising spins located on the 2d Sierpinski gasket and interacting via nearest-neighbor forces [21].

3. Ising Spin Glass on the Sierpinski Gasket

The gasket is like a triangular lattice in which certain triangles are missing in a self-similar manner. Thus, unlike the Edwards-Anderson chain considered by GRINSTEIN et al.[22], it allows for frustration.

The uniform system of spins on the gasket with $J_{ij} = J_0 > 0$ has been analyzed by GEFEN et al. [23], by means of a decimation procedure. Due to the finite order of ramification, the system undergoes merely a T = 0 transition even though there is a nontrivial transition for Ising spins on the full triangular lattice [24]. Thus for random couplings we expect to obtain a T = 0 transition or no transition at all, depending on the type of randomness. In the former case this would correspond to the expected behavior of the 2 and 3d Edwards-Anderson spin glasses on regular lattices.

In the uniform Ising chain problem, the decimation of every other spin yields a renormalized exchange coupling $J' = 1/2\, kT \times \ln\cosh(2J/kT)$ which acts between spins 2 lattice constants apart . At T = 0, J'= J and this is an unstable ferromagnetic fixed point. On the other hand, at $T \neq 0$, the consecutive decimations induce a flow of J' to the stable paramagnetic fixed point $J^* = 0$ corresponding to a decoupled system.

Random systems are specified by a probability distribution, $P_0(J)$, of the couplings on the microscopic length scale. In this case, it is interesting to ask what is the fixed probability distribution (fdp) which describes the system on macroscopic length scales.

Consider a triangular fragment of the gasket which spans 2 lattice spacings. The fragment, henceforth referred to as a unit, consists of 3 corner and 3 midpoint spins. We carry out an exact decimation procedure in which the midpoint spins are eliminated and effective interactions J'_{ij} between the corner spins are found. In each such decimation step 9 exchange interactions are replaced by 3 effective exchange interactions and a constant term. The self-similarity of the gasket allows to iterate this procedure recursively and to obtain renormalized values of the interactions at each length scale. The calculations may be performed either at finite temperatures or at $T = 0$. In the former case a partial trace over the 8 states of the midpoint spins is performed, whereas in the latter case only the ground state contributes.

Note that the renormalized couplings depend on all of the couplings in a unit and are thus correlated within the unit. However, if one considers many units, it is meaningful to ascribe a new probability distribution for the renormalized couplings at a given length scale since there are no correlations between couplings in different units. Such correlation effects may, in principle, be studied by considering multiple-J probability distributions. The renormalized single-J distribution P' can be characterized by a mean value and its dispersion, and so can be the fpd that P' flows to.

Another quantity of interest is the scaling stiffness energy. Note that for any particular length scale the Hamiltonian for the gasket is equivalent to that for a triangle with renormalized couplings. The boundary conditions affect the corner spins of the triangle and have to "work" within the 8 states of the triangle. Therefore ΔE_s may be defined as the energy difference between the highest and the lowest of these states.

We have carried out a numerical analysis of the recursion relations obtained by starting from 3^9 triangles with 3^{10} exchange couplings (i.e. 29526 spins). The triangles were combined into units and the decimation was performed. The renormalized couplings were then combined into new units and so on.

In ref.[21] we report results obtained for several distributions $P_0(J)$. For uniform ferromagnetic couplings the system is at its d_c and it undergoes transition at $T = 0$. The $T = 0$ fpd is $\delta(J-J_0)$ with $J_0 > 0$. On the other hand, for $T \neq 0$ the fpd is the paramagnetic $\delta(J)$: the average J, the dispersion, σ, and ΔE_s tend to zero exponentially. A random ferromagnet is slightly below its d_c. Uniform antiferromagnetic couplings yield no transition at all since the system is fully frustrated and has only the paramagnetic fpd: the flow is completed in one decimation step. In the $\pm J$ system, with equal probabilities for either sign, on an average half of the triangles are fully frustrated. Thus σ and ΔE_s decrease exponentially at all T. At $T = 0$ they become zero within 5 iterations. The $\pm J$ model on the square lattice, considered by MORGENSTERN and BINDER [25], may behave in a similar way since it is paramagnetic even at $T = 0$.

A random antiferromagnet, with the exchange couplings chosen to be negative of the absolute values of Gsussian numbers, and the Gaussian spin glass always flow to the paramagnetic fpd. In the former system $\langle J \rangle$ iterates to zero in one step. For both these systems, σ and ΔE_s are proportional to each other and obey the law

$$\sigma,\ \Delta E_s \sim \exp(-N/N_0)\ /\ \sqrt{N} \quad , \quad \text{for } T > 0 \tag{2a}$$

$$\sim 1/\ \sqrt{N} \quad , \quad \text{for } T = 0\,. \tag{2b}$$

N denotes the number of triangles in a block at a given length scale L, i.e.

$N \sim L^D$, and is related to the number of spins $N_s = 1.5\ (N+1)$. N_0 plays the role of a correlation length. This length diverges when T tends to zero. Both systems are frustrated. They belong to the same universality class, undergo a T = 0 transition, and are below their d_c. The transition manifests itself in a switch from the algebraic to the exponential b.c. sensitivity, as proposed in [16] .

4. Dynamics of Spins on the Gasket

It is interesting to ask now about the dynamic properties of the spin glass on the gasket. This seems to be a hard problem. Therefore we have considered [26] 6 spins in a single unit and investigated its single-flip dynamics [27] exactly. Unlike rings of 4 spins discussed by KINZEL [28] , the units can be frustrated. Equations of motion for the single, triple and quintuple spin expectation values can be written in terms of a 32x32 matrix. The relaxation times of the 32 eigenmodes are studied as a function of T.

For ferromagnetic units one relaxation time diverges at T = 0. In this mode all spins are up. Other relaxation times are finite. For $J_{ij} = -1$ all modes decay even at T = 0. For the Gaussian and random antiferromagnetic couplings there are several non-decaying modes at T = 0 corresponding to each of the local energy minima. For T>0 the mode degeneracy is split. Even though the systems are then paramagnetic there is at least one mode with an almost infinite relaxation time up to finite temperatures.

Time delayed spin correlations show the Arrhenius behavior and the apparent freezing temperature depends on the time of observation. This agrees with the Monte Carlo results of MCMILLAN [29] for the $\pm J$ model on the square lattice and confirms the dynamical origin of the spin glass behavior.

1 B.B.Mandelbrot: Fractals: Form, Chance and Dimension (Freeman, San Francisco 1977)
2 S.Kirkpatrick: in Ill-condensed matter, ed.R.Balian, R.Maynard, and G. Toulouse (North Holland, Amsterdam 1979)
3 Y.Gefen, A.Aharony, B.B.Mandelbrot, and S.Kirkpatrik: Phys.Rev.Lett. 47 1771 (1981)
4 A.Kapitulnik, A.Aharony, G.Deutscher, and D.Stauffer: J.Phys. A 16, L269 (1983)
5 M.Z.Cieplak and M.Cieplak: Phys.Lett. (to be published)
6 T.A.Witten Jr. and L.M.Sander: Phys.Rev.Lett. 47, 1400 (1981)
7 P.Meakin: Phys.Rev. A 27, 604 (1983)
8 M.Muthukumar: Phys.Rev.Lett. 50, 839 (1983)
9 J.R.Banavar and M.Muthukumar (submitted for publication)
10 S.Alexander and R.Orbach: J.de Physique-Lett. 43, 625 (1982)
11 P.J.Ford: Contemp.Phys. 23, 141 (1982)
12 K.Binder: Z.Phys. B 48, 319 (1982)
13 K.H.Fischer: Phys.Stat.Sol. (b) 116, 357 (1983)
14 L.Lundgren, P.Svedlindh, and O.Beckman: Phys.Rev. B 26, 3990 (1982)
15 S.F.Edwards and P.W.Anderson, J.Phys. F 5, 965 (1975)
16 J.R.Banavar and M.Cieplak: Phys.Rev.Lett. 48, 832 (1982); J.Phys.C 16, L755 (1983); Phys.Rev.B 26, 2662 (1982)
17 M.Cieplak and J.R.Banavar: Phys.Rev.B 27, 293 (1983); Phys.Rev.B 29, 469 (1984)

18 J.R.Banavar, M.Cieplak, and M.Z.Cieplak: Phys.Rev.B 26, 2482 (1982)
M.Z.Cieplak and M.Cieplak: J.Phys. (to be published);
M.Cieplak and M.Z.Cieplak (submitted for publication)
19 J.T.Edwards and D.J.Thouless: J.Phys.C 5, 807 (1972)
20 M.E.Fisher, M.N.Barber, and D.Jasnow: Phys.Rev.A 8, 1111 (1973)
21 J.R.Banavar and M.Cieplak: Phys.Rev.B 28, 3813 (1983)
22 G.Grinstein, A.N.Berker, J.Chalupa, and M.Wortis: Phys.Rev.Lett. 36, 1508 (1976)
23 Y. Gefen, B.B.Mandelbrot, and A.Aharony: Phys.Rev.Lett. 45, 855 (1980)
24 Th.Niemeijer and J.M.J.Van Leeuwen: Physica 71, 17 (1974)
25 I.Morgenstern and K.Binder: Phys.Rev.Lett. 43, 1615 (1979)
26 J.R.Banavar, M.Cieplak, and M.Muthukumar: (unpublished)
27 R.J.Glauber: J.Math.Phys.: 4, 294 (1963)
28 W.Kinzel: Phys.Rev.B 26, 6303 (1982)
29 W.L.Mc Millan: Phys.Rev.B 28, 5216 (1983)

A Weakly Anisotropic Ising Model in a Random Field

R.A. Cowley and M. Hagen*
Department of Physics, University of Edinburgh, Mayfield Road
Edinburgh EH9 3JZ, Scotland

H. Yoshizawa and G. Shirane
Brookhaven National Laboratory, Upton, NY 11973, USA

R.J. Birgeneau
Department of Physics, Massachusetts Institute of Technology
Cambridge, MA 02129, USA

The application of a uniform field to a random antiferromagnet produces a random staggered field [1], and so enables the effect of applying a random field to be studied experimentally in a convenient and controllable manner. Detailed experiments have been performed on the good Ising models $Rb_2Co_xMg_{1-x}F_2$ [2], $Co_xZn_{1-x}F_2$ [3], and $Fe_xZn_{1-x}F_2$ and as reviewed in detail elsewhere [4], the results reveal a number of unexpected features which are still only partially understood. In particular when the samples are cooled from the paramagnetic phase in an applied field, the system shows unusual behaviour, characterised by a Lorentzian squared profile in the quasi-elastic scattering cross-section, and a lack of long range order. In contrast if the samples are cooled in zero applied field, and the field applied at low temperatures, the long range antiferromagnetic state persists. On heating, the long range order persists, in the case of the three-dimensional systems, up to the phase transition temperature.

Theoretically there has been considerable controversy about the effect of random fields. Imry and Ma [5] suggested from a simple domain energy balance argument that the lower critical dimension $d_\ell = 2$ and after considerable controversy more recent theoretical work [6] has confirmed this result. Consequently in three dimensions a transition might be expected to a state with long range antiferromagnetic order. Birefringence and heat capacity measurements [7] have shown a logarithmic divergence of the specific heat at low applied fields, which scales in agreement with these theoretical predictions. The field cooled neutron scattering results described above did not, however, yield a state of long range order and if interpreted naively would suggest that $d_\ell > 3$. Very recently Villain [8] has shown that there are very large energy barriers in the random field Ising model and has suggested that the field cooled results arise because these energy barriers prevent the motion of the domain walls between different microdomains, and hence the establishment of thermodynamic equilibrium and long range order. These energy barriers are predicted to be important for dimensionalities $d < 4$.

The experiments described below were performed on $Mn_xZn_{1-x}F_2$. The spins in MnF_2 interact with largely Heisenberg-like exchange interactions and smaller dipolar interactions. These latter lead to a small gap in the spin wave spectrum and Ising-like critical properties. This system was chosen for study because the small spin wave gap might be expected to permit the system to remain in thermodynamic equilibrium down to lower temperatures, and to give more mobile domain walls, while finally a preliminary experiment [9] with $x = 0.78$ gave results which were surprisingly different from those obtained on the good Ising models with much smaller values of x. The present experiments were performed with an excellent

---

*Present address: Institut Laue-Langevin, F-34082 Grenoble Cedex, France

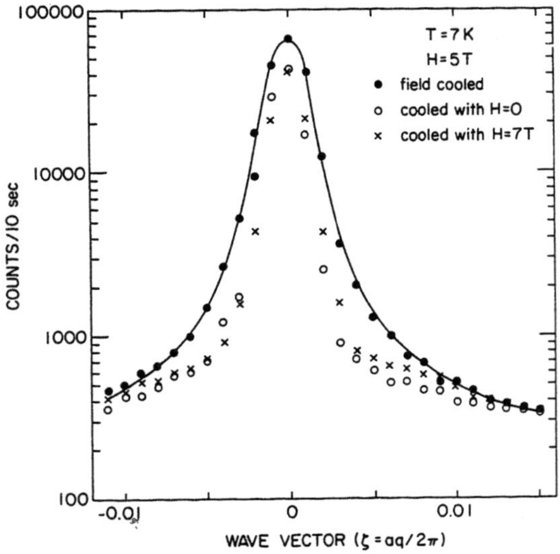

Fig. 1. The scattering observed for wavevector transfers (1,ζ,0) at 7 K and 5 K when $Mn_{0.65}Zn_{0.35}F_2$ was cooled in fields of 0, 5 and 7 T.

sample with x = 0.65 kindly supplied by D.P. Belanger, V. Jaccarino and A.R. King. The experiments were performed at the Brookhaven High Flux Beam Reactor using similar techniques to those employed in the other samples and will be discussed in detail elsewhere [10].

A novel feature of the low anisotropy is that it is possible to apply a sufficiently large field (> 5.2 T), such that the spins flop from being aligned along the c axis into the basal plane. The applied field then generates a random field largely perpendicular to the spin direction which is not expected to affect the long range order. In fig 1 we show the scattering obtained by scanning the wavevector transfer transversely through the (100) magnetic Bragg reflection, at T = 7 K in a field of 5 T. The sample was prepared in three ways: by field cooling, by cooling in zero field and raising the field at 7 K, and finally by cooling to the long range ordered spin-flop phase in a field of 7 T and then reducing the field to 5 T. The results show that the final state is dependent on the past history. The long range order obtained by cooling with H = 0 or H = 7 T persists while the field cooled measurement gave a definitely broader profile and higher intensity. Both aspects indicate a lack of long range order, due to relief of extinction, when the sample is field cooled. Surprisingly this hysteretic behaviour was found to persist up to T_N ~ 40 K at 5 T, when the spin wave gap is certainly less than 2.0 K. We do not understand why the spin waves are apparently unable to relax the system under these conditions.

At higher temperatures field cooled measurements were made of the scattering. Typical profiles are shown in fig. 2 for cooling in a field of 4 T. The results show a steadily decreasing line width with decreasing temperature which could be well described by a profile consisting of the sum of a Lorentzian and a Lorentzian squared, as found close to T_N for x = 0.78. The solid line in fig 1 shows that the same functional form also describes the data at low temperatures. These results show that the microdomain state

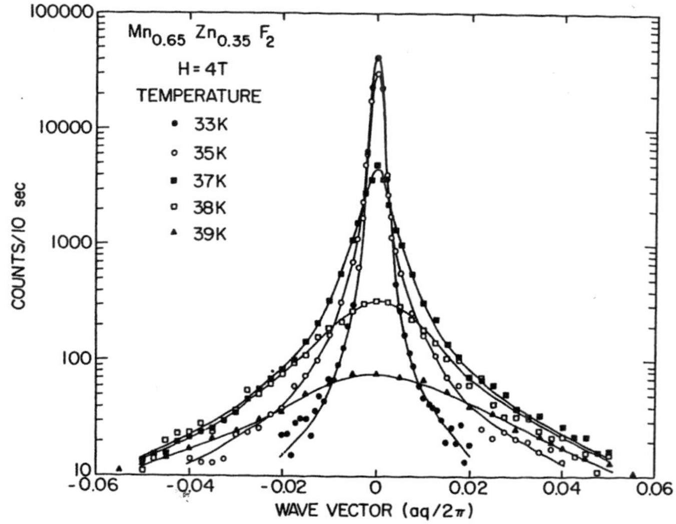

Fig. 2. The scattering observed for wavevector transfers $[1,\zeta,0]$ for field, 4 T, cooled measurements. The solid lines are fits to Lorentzian + Lorentzian squared profiles.

The temperature dependence of the inverse correlation function is shown in fig. 3 for field cooling in various different fields. The results are similar to those for the good Ising models but differ in one aspect. The arrows in fig. 3 show the temperatures at which specific heat measurements on the same sample [7] gave a critical peak. In $Fe_xZn_{1-x}F_2$, the inverse

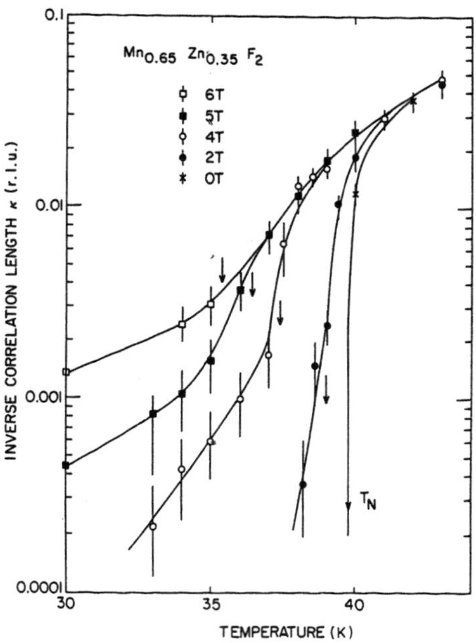

Fig. 3. The temperature dependence of the inverse correlation length for cooling in various fields.

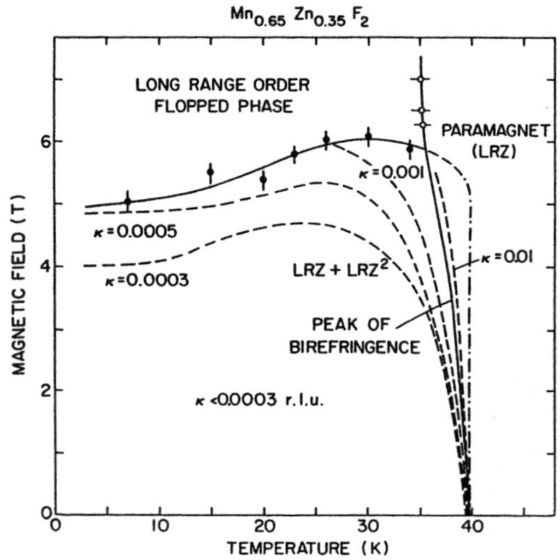

Fig. 4. The phase diagram. Solid lines show the phase boundaries and dotted lines show constant K. The dot-dash lines show the rough boundary beyond which the Lorentzian squared profile is not detectable.

correlation length was found to be almost independent of temperatures below this temperature, but in fig. 3 it is clearly not so. We do not understand this different behaviour.

Finally in fig. 4 we summarise the phase diagram obtained from these measurements. At high temperatures there is a normal paramagnetic phase. On cooling in high fields there is an apparently fairly normal transition to a spin-flopped phase. At lower fields, the results at low temperatures are dependent on the past history of the sample. In field cooled experiments the scattering is well described by a Lorentzian + Lorentzian squared form but the inverse correlation length becomes very small at low temperatures. It is nevertheless significantly larger than zero for applied fields larger than 3 T.

The phase diagram and the above conclusions are consistent with recent detailed measurements [11] of the $Mn_xZn_{1-x}F_2$ system by dilatometric and specific heat techniques in which the neighbourhood of the multi-critical point was probed more carefully than we did. Finally the most surprising aspect of these results is that there are history dependent effects in $Mn_xZn_{1-x}F_2$ and the other associated random field behaviour even though there are many spin waves present to relax the system.

The work at Brookhaven was supported by the Division of Basic Energy Sciences, U.S. Department of Energy, contract DE-AC02 76CH0-0016, at MIT by NSF contract No DMR-79-23206, at Edinburgh by SERC.

References

1. S. Fishman and A. Aharony: J. Phys. C 12, L729 (1978)
2. R.J. Birgeneau, H. Yoshizawa, R.A. Cowley, G. Shirane and H. Ikeda: Phys. Rev. B28, 1438 (1983)

3. M. Hagen, R.A. Cowley, S.K. Satija, H. Yoshizawa, G. Shirane and R.J. Birgeneau: Phys. Rev. B28, 2602 (1984)
4. R.J. Birgeneau, R.A. Cowley, G. Shirane and H. Yoshizawa: J. Stat. Phys. 34 (1984)
R.A. Cowley, R.J. Birgeneau, G. Shirane and H. Yoshizawa: p333 'Multicritical Phenomena' Plenum Press
5. Y. Imry & S.K. Ma: Phys. Rev. Lett. 35, 1599 (1975)
6. G. Grinstein and S.K. Ma: Phys. Rev. Lett. 49 685 (1982)
7. D.P. Belanger, A.R. King and V. Jaccarino: Phys. Rev. Lett. 48, 1050 (1982); D.P. Belanger, A.R. King, V. Jaccarino and J.L. Cardy: Phys. Rev. B29, 2522 (1983)
8. J. Villain: Phys. Rev. Lett. 52, 1543(1984)
9. R.A. Cowley and W.J.L. Buyers: J. Phys. C 15, L1209 (1982)
10. R.A. Cowley, H. Yoshizawa, G. Shirane, M. Hagen and R.J. Birgeneau: to be published
11. Y. Shapiro, N.F. Oliviera Jr. and S. Foner: Phys. Rev. B (in press)

Spin Dynamics Associated with Local Magnons in FeF_2:Mn

D.W. Hone

Physics Department, UCSB, Santa Barbara, CA 93106, USA

S.M. Rezende

UCSB and Universidade Federal de Pernambuco, Recife, Brasil

I. Introduction

It has long been recognized [1] that substitutional impurities in solids can be invaluable as local probes on an atomic scale. This is certainly the case in insulating magnets - ferro-, antiferro- and paramagnetic - where there are readily available resonant experimental techniques highly specific to individual chemical species or even to the local environment of a given atom. On the theoretical side initial interest was focused on isolated impurities and the powerful techniques available to handle localized perturbations in systems with crystalline translational symmetry, but there has also been increasing study of macroscopically disordered systems, with finite concentrations of impurities. Perhaps the remarkable thing is that after two decades or more of study there are still interesting things to be learned, even about the simplest prototype insulating magnets. There are, of course, the active and exciting investigations associated with the random field problem, an example of macroscopic disorder, which has been discussed earlier in this workshop. But what we will present here is a detailed picture of a remarkable variety of phenomena in a rather more conventional - or, at least familiar - system, FeF_2:Mn, which has recently been the subject of intensive experimental and theoretical investigation. This picture has evolved over a period of years to give what we believe to be a single consistent and reliable description of a very interesting system. Though perhaps not absolutely complete, it is certainly nearly so, at least in the independent impurity (low concentration) limit. Though there may remain some surprises in such systems, we would expect that exciting future physics in this general area is likely to be found in the exploration of alloys and other macroscopically disordered magnets.

One of the dramatic features frequently associated with point impurities is the appearance of localized excitations. The early models and systems studied [1,2] typically involved local modes above and well separated from the continuous energy bands, leading to strong localization. Of special interest in FeF_2:Mn, on the other hand, is the appearance of an s_0-symmetry local magnon just below the spin wave continuum. The absolute and relative position of these energies can be shifted significantly by the application of readily available laboratory magnetic fields. The energetic proximity of localized and band minimum modes implies relatively weak localization, or substantial participation of the Fe^{2+} spins in the neighborhood of the Mn impurity in the localized excitation. Further, the degree of this coupling to the neighborhood, or the spatial extent of the local mode, is sensitive to the applied magnetic field because of the different g factors of the host Fe and impurity Mn spins. With the advent of high resolution far infrared laser spectroscopy it has become possible to study in detail these low energy, strongly coupled excitations. The experimentally observed consequences of the weak localization, or strong involvement of host spins in the local magnon, include: a) large enhancement of the impurity mode intensity; b) "frequency pulling" of the host and impurity modes; c) substantial broadening of the impurity mode and d) asymmetric lineshape of the k≃0 host magnon. All these effects have been measured in detail in FIR transmission experiments. In these experiments the spectrum is obtained by sweeping the external magnetic field H_0 parallel to the c axis of the crystal, so that the impurity and host modes are brought into resonance with the laser frequency, as illustrated in Fig.1. Full details of the experiments are given by SANDERS et al.[3]. Because of the coupling between electromagnetic and long wavelength spin waves, the FIR transmission spectrum of the host mode exhibits a broad lineshape with a flat section characteristic of the forbidden gap of the polariton dispersion curve. As a consequence the data must be obtained by fitting the spectrum to the theoretical prediction of a polariton model, which can be done quite accurately, as shown in [3].

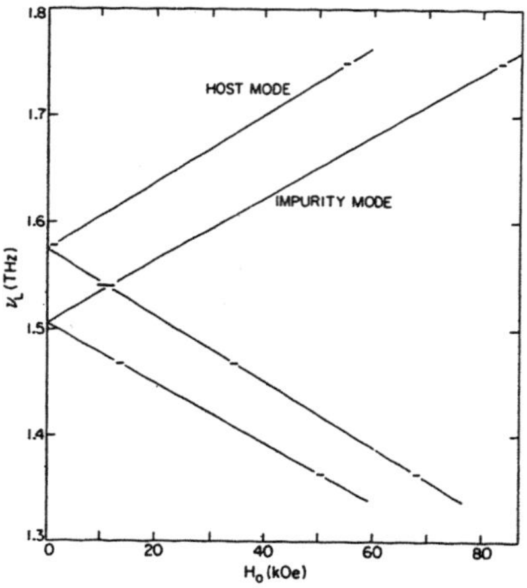

Fig. 1. Field dependence of the up- and down-going host and impurity modes in FeF_2:Mn with x ≃ 0.01 at.%. The horizontal dashes indicate typical FIR laser frequencies.

The effects described above have also been observed [4] in inelastic light scattering experiments. In these measurements one probes magnons with wavevector $k \sim 10^5 cm^{-1}$, which is far from the polariton region ($k \sim 10^3 cm^{-1}$), so that the scattering lineshapes reflect the true magnon behavior. In contrast to the earlier results obtained with a standard Raman scattering apparatus [5], the spectra obtained [4] by analysis of the scattered light with a Fabry-Perot interferometer exhibit clearly distinct lines for host and impurity modes. The light scattering experiments have been used to measure the relative intensities of the local and host modes and the linewidth for impurity concentrations x>0.3 at. %. However, due to its higher resolution, the FIR transmission technique has provided more detailed measurements of the relaxation of both modes. We now present a brief account of the main experimental results obtained with the two techniques and give their theoretical interpretation.

II. Enhancement of the Impurity Mode Response

The most striking feature of the FIR and optical spectra is the surprisingly large integrated intensity of the impurity mode response, typically comparable to that of the host mode even at impurity concentrations below 1 at.%. In the light scattering spectrum the enhancement of the impurity mode is even more surprising than in the FIR data in view of the weak photon coupling to the Mn^{2+} S-symmetry ground state. The high intensity here arises indirectly through the oscillating transverse component of the exchange fields of the Fe^{2+} neighbor spins which couple strongly to the radiation fields. This problem has been treated within a simple sublattice model which attributes an equal magnetization to every host spin (regardless of its position relative to an impurity), and similarly for the impurity spins, and solving the coupled equations of motion of these magnetizations. This approach explains the enhancement of the infrared and Raman impurity intensities only qualitatively. The correct calculation must take into account the spatially extended character of the local mode. This has been done using zero-temperature Green's functions in the limit of a single impurity in the crystal [6]. The theory is based on a simple model Hamiltonian for FeF_2:Mn which includes isotropic exchange, single-ion uniaxial anisotropy along the c axis of the lattice (chosen here as the z direction) and the Zeeman interaction with a magnetic field H_0 applied along the c axis,

$$H_{mag} = \mu_B H_0 \sum g_\ell S_\ell^z + 2J\sum \vec{S}_\ell \cdot \vec{S}_{\ell+\delta} - D\sum (S_\ell^z)^2 + 2J_2'\vec{S}_0 \cdot \sum \vec{S}_{\delta_2} - 2J_1'\vec{S}_0 \cdot \sum \vec{S}_{\delta_1} - D'(S_0^z)^2 \quad (1)$$

where sites are labeled by subscripts on the spin operators and a single impurity is located at the origin (ℓ=0). We have retained only the dominant exchange terms, between next-nearest neighbors on opposite sublattices of the body-centered tetragonal arrangement of magnetic atoms, for the pure FeF_2 host, or between Fe spins. The impurity spin is taken to couple with its eight next-nearest neighbors at positions δ_2, and its two nearest neighbors at positions δ_1. The magnetic response of the system to a uniform electromagnetic field (frequency ω) is given by the relevant susceptibility $\chi(q=0,\omega)$. In order to find χ one must solve the equations of motion for the Green's functions of the impure spin system. This is easily done within the harmonic (independent spin wave) approximation, where the spin operators are expressed in terms of the boson operators c_α and $c_\alpha^\dagger$ which diagonalize the quadratic part of the Hamiltonian (1). These are introduced by

$$c_\alpha = \sum \Gamma_\ell^\alpha S_\ell^\pm / (2S_\ell)^{\frac{1}{2}} \quad (2)$$

with S^+ appropriate to modes whose energy increases with the applied field ("up-going magnon branch") and S^- to the downgoing branch. For a single impurity the calculated susceptibility within this independent magnon approximation, at frequencies ω near the local mode frequency ω_s, is [6]

$$\chi(\omega \simeq \omega_s) = \frac{m^2 S}{\hbar V}\left(\frac{1}{\omega-\omega_s}\right)\left\{\sum (-1)^{\ell\ell'} \Gamma_\ell^{s*} \Gamma_{\ell'}^s + \left[\left(\frac{m'}{m}\sqrt{\frac{S'}{S}} - 1\right)\Gamma_0\left(\sum \Gamma_i^{s*} - \sum \Gamma_j^{s*}\right) + \text{c.c.}\right]\right\} \quad (3)$$

where $(-1)^{\ell\ell'} = -1$ for ℓ and ℓ' on opposite sublattices and +1 otherwise; i and j are summed over opposite sublattices, and m, m' are gyromagnetic factors for host and impurity spins. The strength of the pole at $\omega = \omega_S$ is enhanced by the factor in curly brackets, involving sums over the wave function site amplitudes Γ_ℓ^S appropriate to the local mode. Since the wave functions are normalized, one expects those sums to increase with increasing extension of the wave function; indeed, the sums can be done explicitly and this increase demonstrated [6], in quantitative agreement with experiment, with impurity-host exchange parameters $J_1' = 0.2\ cm^{-1}$ and $J_2' = 1.79\ cm^{-1}$, both antiferromagnetic.

III. Frequency Pulling

Another evidence of the strong coupling between the local mode and the q=0 magnon in FeF_2:Mn is the pulling of their frequencies relative to the values in the single impurity limit. Relative frequency shifts are of order a percent for 0.1 at.% impurity. Though the coupled magnetization model gives a good description of the frequency pulling as a function of x, it requires parameters which correspond to interaction constants $J_1' = 2.0\ cm^{-1}$ and $J_2' = 2.04\ cm^{-1}$. These are very different from the values which explain a wide range of other phenomena (including the local mode response just discussed). In particular, J_1' is far too big. Actually it is not surprising that the coupled magnetization model does not yield the correct parameters, because in the limit $x \to 0$ it gives the Ising model frequencies, a result which is known to fail if the impurity mode is not sufficiently localized. The correct theory for the energies at finite impurity concentrations has yet to be developed.

IV. Local Mode Linewidth

Fig. 2 shows the impurity mode linewidth data obtained from the low temperature FIR transmission spectra of several samples with different impurity concentrations and different thicknesses. Three main mechanisms have been identified [3] for the broadening of the local mode in FeF_2:Mn at low temperatures, namely: i) Radiation damping, ii) Single impurity processes and iii) Impurity-impurity interaction. The radiation damping mechanism [7] is greatly enhanced by the participation of the Fe^{2+} spins in the local mode and is the source of the linear dependence of the linewidth on the sample thickness, as shown in detail in Ref. [3]. Extrapolation of the data of Fig. 2 to zero thickness gives the intrinsic linewidth of the local mode. Impurity banding effects [8] are responsible for the concentration dependence of the linewidth but they cannot account for the large residual value of about 1kOe as $x \to 0$.

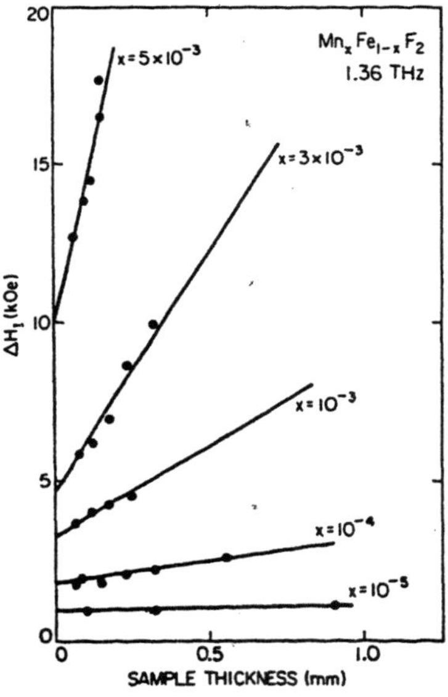

Fig. 2 Impurity mode linewidth measured by FIR laser spectroscopy for several Mn concentrations in FeF_2 vs. sample thickness. The straight lines are best fits to the data; their slopes determine the radiation damping contribution to ΔH_I[3].

The latter is due to single impurity damping processes, which have recently been studied theoretically [9,10]. For the up-going local mode, anisotropy and dipolar interactions which break the spin-rotational symmetry about the field axis provide [9] a mechanism for the decay of the local modes into band magnons in the frequency range 1.54-1.93 THz. However, since the down-going mode is energetically the lowest-lying magnetic excitation in the system, other excitations must be involved in the relaxation process. At least two arguments suggest that phonons play a dominant role in the damping: Acoustic phonons are the only collective excitations in FeF_2 with energies below the Mn impurity mode, and the magnetic moments ("spins") of the Fe^{2+} ions that participate in the local-mode excitation couple strongly to lattice vibrations. In pure FeF_2 phonons do not contribute to the relaxation of magnons because their dispersion relations are very different; thus energy and momentum cannot be conserved simultaneously. But in Mn-doped FeF_2 there is no translational symmetry around the impurities, and the requirement of momentum conservation is relaxed.

The coupling between spins and lattice vibrations arises from several sources, but in FeF_2 it has been shown [11] that the dominant mechanism is phonon modulation of the crystal field, which can be calculated from first principles. Using the spin-lattice interaction in FeF_2 and the results of the Green's function calculation of the impurity mode properties in FeF_2:Mn, it has been shown [10] that two phonon assisted processes can effectively relax the local mode: one is the direct decay into a degenerate acoustic phonon and the other is the decay into a magnon-phonon pair.

The interaction Hamiltonian for the decay of the local mode into an acoustic phonon can be written as [10]

$$H_{int} = -i(\hbar S^3/MN)^{\frac{1}{2}} b \sum \left[q/(\omega_q)^{\frac{1}{2}} \right] F(q) a_q^{\dagger} c_s + H.c. \tag{4}$$

where $a_q^{\dagger}$ is the creation operator of the phonon with frequency ω_q, the form factor $F(q)$ is essentially the Fourier transform of the magnon wave function at the phonon wave vector (it is given explicitly for the s_0 mode in Ref.[10]), and b is a magnetoelastic coupling coefficient.

The terms in $S^x S^y$, $(S^x)^2$, and $(S^z)^2$ in the spin-lattice Hamiltonian give rise to processes in which the local magnon decays into one phonon and one propagating magnon. The corresponding interaction Hamiltonian is given explicitly in Ref. [10].

The simplest process for local-magnon relaxation is direct decay into a single phonon. In the absence of crystalline translational symmetry (because of the impurity), the requirements of wave-vector conservation are relaxed. In general, the local magnon can decay into any phonon energetically degenerate with it (unless other symmetries prevent magnon-phonon coupling). In FeF_2 the transverse-acoustic phonon bands extend as high as 3THz, and above 2.04 THz the magnon also becomes degenerate with phonons in the B_{1g} optical band, but the contribution from the optical phonon to the linewidth has been shown to be negligible. At temperatures low compared with the local-magnon energy, where the number of thermally excited phonons degenerate with that magnon is negligible, the standard golden-rule calculation gives for the decay rate of the local mode into acoustic phonons a contribution

$$\Delta H_{TA} = \frac{S^3 b^2}{\pi \gamma \hbar \rho} \quad \frac{q_0^4 |F(q_0)|^2}{\omega_s (c_0 - 3c_1 q_0^2)} \tag{5}$$

where ρ is the mass density, $F(q_0)$ is the local mode form factor, q_0 is the wave number of the phonons degenerate with the local magnon, $\omega(q_0) = \omega_S$, $(c_0 - 3c_1q_0^2) = d\omega/dq$ characterize the acoustic phonon density of states in our approximation of the dispersion relation, and γ is the gyromagnetic ratio ($\gamma\hbar = g\mu B$).

When the up-going mode has frequency above 1.54 THz (H_0>11 kOe), it can decay into an acoustic phonon-down-going propagating magnon pair. The contribution of this process to the linewidth is

$$\Delta H(\text{m+ph}) \simeq \frac{S^2b^2\Omega}{4\pi^4\gamma\hbar\rho}\int d^3k\; q_0^3(k)\,\frac{|C(k,q_0)|^2}{c_0^2} \tag{6}$$

where $q_0(k)$ is the phonon wave vector demanded by energy conservation, C is another suitably defined form factor [10], and Ω is the unit cell volume. The linewidths given by (5) and (6) were evaluated numerically for FeF_2:Mn. Their dependences on the local mode energy are shown in Fig. 3 by curves A and B. The obvious ω^3 dependence (from q_0^4/ω_S) in (5) is dominated by the rapid falloff of the form factor $F(q_0)$ above q_0 of the order of the inverse of the local magnon size, so curve A decreases with ω. Note that the ranges below and above 1.507 THz correspond to the down-going and up-going local modes respectively. As pointed out in Ref. [10], near 1.7 THz there is a strong admixture of magnons and phonons and the actual normal modes are magnetoelastic excitations. But far from this region the bare phonon is a good approximation to the excitation into which the local mode decays. The data point at 1.36 THz in Fig. 3 is the residual impurity mode linewidth obtained from Fig. 2 and it is clearly accounted for by the calculation described above. The bars at 0.85 and 2.5 THz represent data taken at the High Magnetic Field laboratory of Osaka University [12] and both are also in good quantitative agreement with the theory.

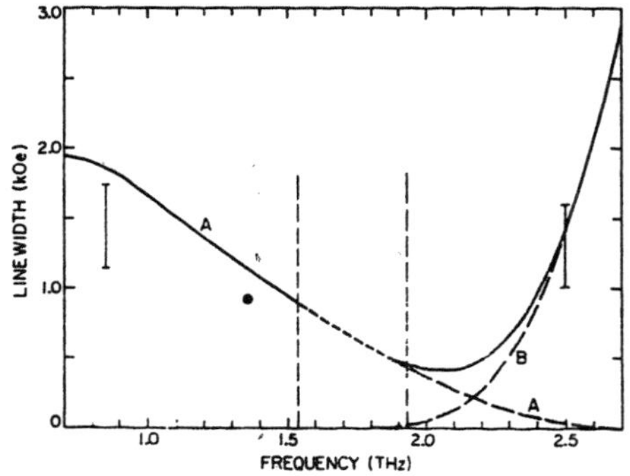

Fig. 3. Residual local-mode linewidth from phonon-associated relaxation processes. Theoretical curves are labeled A (decay to a single phonon) and B (decay into a phonon-magnon pair)[10].In the region between 1.54 and 1.93 THz S_Z-nonconserving magnetic decay also contributes to the linewidth [9]. Data are represented by the dot and the vertical bars [3].

Within the frequency range 1.54-1.93 THz marked by vertical dashed lines in Fig. 3, the up-going local mode is degenerate with the down-going magnon band. Thus in this region the local mode can decay into the magnetoelastic excitations not only via the coupling to phonons (Eq. (4)), as already discussed (and represented by the dashed extensions of curve A in the figure), but also in principle via explicitly magnetic interactions to the magnon component of the final states. This does not occur through the dominant magnetic Hamiltonian (1), because that operator is invariant to spin rotations about the z-axis, whereas the decay process here requires a single spin flip (the initial and final excitations being on the up-going and down-going branches, respectively). But there are small magnetic interaction terms, otherwise negligible, which do break this spin rotational symmetry. The most obvious of these is the magnetic dipole-dipole Hamiltonian

$$H_d = (\hbar^2/2)\gamma^2 \sum \left[\vec{S}_i \cdot \vec{S}_j - 3(\vec{S}_i \cdot \hat{r}_{ij})(\vec{S}_j \cdot \hat{r}_{ij}) \right] r_{ij}^{-3} . \tag{7}$$

In addition, there is an orthorhombic magnetocrystalline anisotropy term associated with the electrostatic fields of the ions surrounding a Fe^{2+} site. Although the dominant departure of the environment from cubic symmetry is the tetragonal distortion measured by the uniaxial term $D(S^z)^2$ in Eq.(1), there is also a smaller asymmetry in the x-y plane, leading to the term

$$H_{oa} = -(E/2S) \sum \eta_i \left[(S_i^x)^2 - (S_i^y)^2 \right] \tag{8}$$

where η_i is +1 for i on the up sublattice and -1 for i on the down sublattice, because the F^- cage is rotated by $\pi/2$ about the z axis in going from a magnetic site on one sublattice to one on the other. Thus, there is a local, but not global, orthorhombic anisotropy.

Independent calculations [9,10] of this S_Z-nonconserving magnetic decay mechanism and the phonon-associated mechanism discussed above demonstrate that the two contributions to the linewidths are expected to be comparable. Since the final states involved are the same, we should properly take into account the quantum interference between the amplitudes for the two processes in calculating the total probability for decay; the corrections from the interference terms may well be large. However, to carry out the amplitude calculation for the symmetry-breaking anisotropy interaction to find both magnitude and phase, we need the final-state wave function. The impurity offers primarily a magnetic contrast to the host atoms, and as opposed to the phonon final states we have considered above, which are well represented as plane waves, the continuum magnons are distorted from plane waves precisely in the neighborhood of the impurity, where their overlap with the initial local mode state is substantial. This difficulty is avoided in the approach of Ref.[9] where the local impurity susceptibility is calculated directly; its imaginary part measures the local mode linewidth contribution from the dipolar and orthorhombic anisotropy interaction included. As is common to Green's function calculations this one has the advantage of not demanding explicit knowledge of the form of the wave functions, but the price paid is that it also does not yield the essential information about the phase of the transition amplitudes. We do not know whether constructive interference, a slightly larger orthorhombic anisotropy constant E than estimated in Ref. 9, or a combination of these may explain the relatively large observed linewidth [3] of 3 kOe at 1.75 THz, as compared with a combined theoretical value of about 1 kOe from the two mechanisms treated independently. But we do note that a factor of 2 in the anisotropy constant or completely constructive interference would each essentially remove the discrepancy, and neither seems unreasonable.

At sufficiently low impurity concentrations it is reasonable and useful [8] to view the concentration dependence of the local mode linewidth as arising from the spatial overlap of the wave functions of local magnons centered on different impurities. To second order in the concentration one effectively sums the results of the dynamical interactions of impurity pairs over a random spatial distribution of impurities. The four Green's functions labeled by the sites of the single pair of the two impurity problem are coupled by the equations of motion. We are interested in the behavior of the system near the local mode frequency $\omega = \omega_S$; the parts of the Green's functions singular at $\omega = \omega_S$ can be reproduced by an effective interaction:

$$H_{eff} = \omega_s \sum c_i^\dagger c_i + \sum V_{ij}\, c_i^\dagger c_j \tag{9}$$

where c_i destroys a local mode associated with the impurity site i, and we have written the Hamiltonian for the full system as a sum of pairwise interactions. Clearly at higher concentrations one would have to modify the effective medium which moderates the interaction between local modes. We have essentially calculated [8] the uniform magnetic susceptibility appropriate to this Hamiltonian averaged over impurity distributions. More accurately, we have calculated the frequency moments of the imaginary part of the susceptibility, which is directly related to the absorption lineshape, again to second order in concentration.

The predictions of linewidth are compared with experiment [3] at a laser frequency of 1.36 THz in Fig. 4. With the inclusion of an independent single impurity phonon-related width of 1 kOe, as calculated earlier, we find excellent quantitative agreement. It is true that the concentration dependence is relatively weak, but we emphasize that there are no adjustable parameters. For completeness in Fig. 4 we compare the results with the predictions for a system with perturbation strength twice as large, which could not be reconciled with this experiment. More directly and unambiguously affected by impurity parameters are the positions of satellite resonance lines associated with Mn impurity pairs in various spatial positions. We have successfully identified [13] six such experimental lines with the predictions of theory. The agreement is quantitative with impurity parameters $J' \lesssim 0.5\ cm^{-1}$, and $J' \simeq 1.8\ cm^{-1}$, in agreement with the values found necessary to explain the enhancement and linewidth data above.

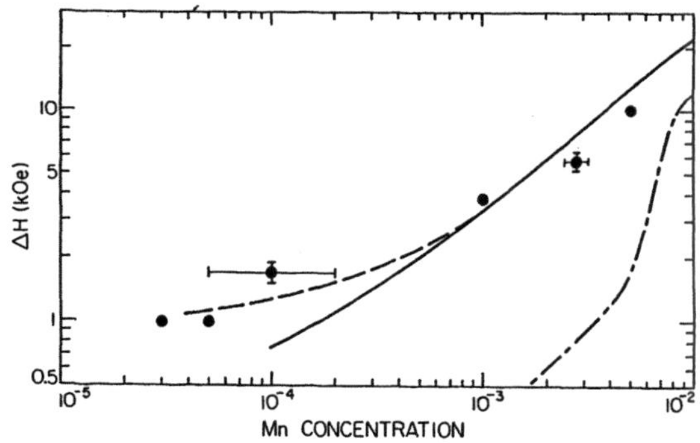

Fig. 4. Concentration dependence of the impurity banding contribution to the local mode linewidth [8]. Data are obtained at 1.36 THz [3].

V. Host Mode Lineshape

As we remarked earlier, the lineshape of the FIR transmission spectrum of the host mode in FeF_2:Mn is mainly dominated by the polariton nature of the propagating magnetic mode in the sample. Nevertheless, by letting $\omega \to \omega + \gamma\ \Delta H/2$ in the coupled magnetization equations of motion, the linewidth of the host mode can be obtained from the lineshape by carefully fitting the experimental spectrum to the calculated transmission. The most significant features of the data are: i) the relaxation rate of the host mode is lower than that for the impurity mode; ii) the linewidths increase rapidly with the impurity concentration and iii) the relaxation rate increases with increasing frequency. All of these features are confirmed by the light scattering measurements [4].

The first attempt to explain the host mode linewidth based on impurity exchange and anisotropy scattering led to values above the polariton gap which were more than two orders of magnitude smaller than the data. The reason for this failure is that the perturbation potential associated with the Mn impurity is so strong that the spatially extended local mode appears; hence, the usual perturbation theory (Fermi Golden Rule) is not directly applicable. It has been shown that inclusion of multiple scattering effects (from a single impurity) corrects this problem and accounts for the observed linewidths [14].

The broadening mechanism for the host mode is two-magnon scattering, in which a magnon of wavevector k decays via impurity scattering into degenerate magnons with $\vec{k}' \neq \vec{k}$. In the usual situation where the impurity mode is far from the spin wave band, the linewidth is calculated [15] by the standard Golden Rule expression

$$\Delta H(\omega) = \frac{\Omega}{\gamma\hbar^2(2\pi)^2}\int |F_{k_0k}|^2 \quad \delta(\omega-\omega_k)\ d^3k \tag{10}$$

where ω is the frequency of the mode k_0 and F_{k_0k} is the Fourier transform of the scattering potential V. In the presence of a strong perturbation V must be replaced here by

$$T = (1 - VG^o)^{-1}\ V, \tag{11}$$

which describes multiple scattering from the potential to all orders. Of course, far from a scattering resonance or a bound state $T \simeq V$ and the modification to ordinary first order perturbation theory is small. However, here the effect is enormous, as the existence of a local mode implies, and as a consequence, the linewidth is enhanced by three orders of magnitude [14]. The frequency dependence of the observed linewidth is accounted for by the increasing density of degenerate magnon states with increasing frequency. Fig. 5 shows the calculated linewidth [14] for x=0.5% as a function of the mode frequency relative to the bottom of the magnon band, in units of $\gamma\Delta_{FB}/2$

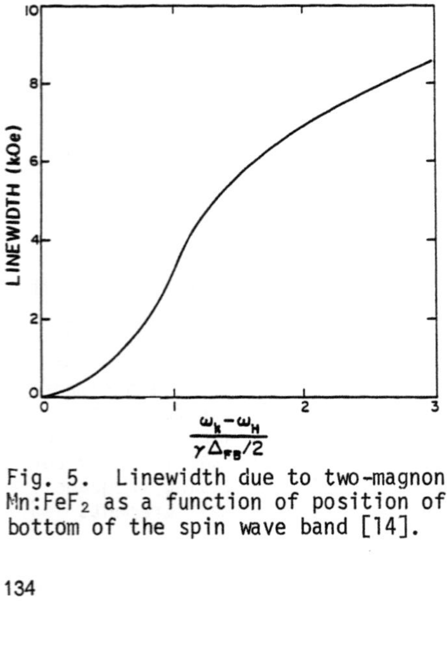

Fig. 5. Linewidth due to two-magnon impurity multiple scattering in 0.5% $Mn:FeF_2$ as a function of position of the magnetic mode relative to the bottom of the spin wave band [14].

(the width of the spinwave manifold). We note that the contributions from exchange, anisotropy and g-factor scattering are equally important for the linewidth, which has the magnitude of a few kOe and is strongly frequency dependent. In Fig. 6 the experimentally measured transmission host mode lineshape obtained for a sample with x = 0.5 at.% is compared to theoretical prediction of the polariton model with a frequency dependent relaxation parameter given by the above theory. As can be seen, the agreement between theory and experiment is remarkably good.

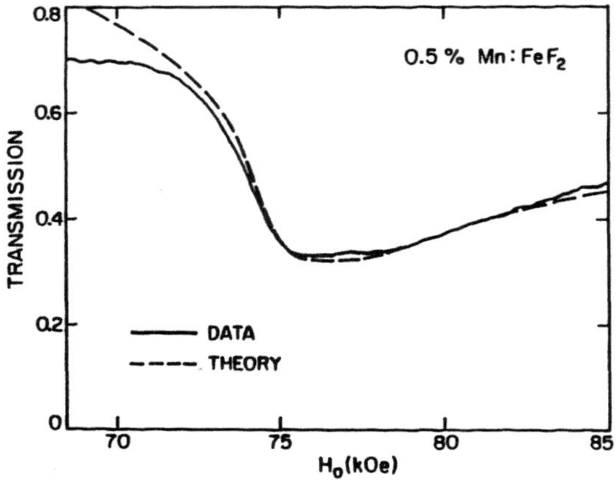

Fig. 6. Comparison of transmission data from a 80 µm thick sample of 0.5 at.% Mn:FeF_2 (solid curve) with the theoretical prediction of polariton model with a frequency dependent relaxation parameter obtained from the two-magnon impurity multiple scattering calculation [14].

VI. Concluding Remarks

We have seen in this review how valuable as a local probe a substitutional impurity can be in a simple magnetic system, by way of the particularly interesting and well-studied example, FeF_2:Mn. The variety of phenomena studied and information available is remarkable. We have indicated that a single, relatively simple, Hamiltonian (1) is quantitatively compatible with all available observations. This and closely related systems can continue to be used as quantitatively understood models of magnetic insulators.

We would like to thank all our collaborators explicitly named in the references below. Special thanks are due to Professor V. Jaccarino who inspired much of this work and contributed many helpful suggestions. This work was supported in part by the National Science Foundation through Grant Nos. DMR-80-08004 and DMR-80-17582. One of us (S.M.R.) acknowledges the support of Conselho Nacional de Desenvolvimento Científico e Tecnólogico (CNPq), Brazil.

VII. References

1. An early review of this area is in the book Localized Excitations in Solids, edited by R.F. Wallis, (Plenum, New York, 1968)
2. See, e.g., R.A. Cowley and W.J.L. Buyers: Rev. Mod. Phys. 44, 406 (1972) and M. Butler, V. Jaccarino, N. Kaplan, and H.J. Guggenheim: Phys. Rev. B 1, 3058 (1970)

3. R.W. Sanders, R.M. Belanger, M. Motokawa, V. Jaccarino, and S.M. Rezende: Phys. Rev. B 23, 1190 (1981)
4. S.M. Rezende and E.F. Silva, Jr.: submitted to Phys. Rev. B
5. S.M. Rezende, C.B. de Araujo, and E. Montarroyos: Solid State Commun. 35, 627 (1980)
6. S.M. Rezende, Phys. Rev. B 27, 3032 (1983)
7. R.W. Sanders, D. Paquette, V. Jaccarino and S.M. Rezende: Phys. Rev. B 10, 132 (1974)
8. C. Wiecko and D. Hone: J. Phys. C 13, 3883 (1980)
9. P. Thayamballi and D. Hone: Phys. Rev. B 27, 2924 (1983)
10. S.M. Rezende, D.W. Hone, and R.M. Toussaint: Phys. Rev. B 29, 1638 (1984)
11. S.W. Lovesey: J. Phys. C 5, 2769 (1972)
12. M. Motokawa and R.M. Toussaint (unpublished)
13. R.M. Belanger, D. Hone, and M. Motokawa: Phys. Rev. B 25, 3186 (1982)
14. R.M. Toussaint, D. Hone, V. Jaccarino, and S.M. Rezende, submitted to Phys. Rev. B
15. R. Loudon and P. Pincus: Phys. Rev. 132, 673 (1963)

2D Pure and Dilute Quantum Transverse Ising Model

G. Kamieniarz

Institute of Physics, A. Mickiewicz University, PO-Poznan, Poland

1. Abstract

The quantum critical behaviour of the bond-diluted transverse Ising model on the honeycomb and square lattices at zero temperature is investigated by approximate block renormalisation group transformations. The critical exponents and the critical fields are calculated for both pure and diluted systems.

2. Introduction

In this paper the quantum critical behaviour of the transverse Ising model at zero temperature is investigated through the approximate renormalisation group block method.

The model in question is of interest as a model for tunnelling ferroelectrics, cooperative Jahn-Teller systems and induced magnets [1,2]. In two or more dimensions it undergoes a phase transition at finite temperature which can be reduced to zero by increasing the transverse field Γ to a critical value Γ_c. The system therefore shows quantum critical behaviour as the field goes through Γ_c. The quantum critical behaviour as a function of $\Gamma-\Gamma_c$ of the zero temperature d-dimensional transverse Ising model is the same as the thermal critical behaviour as a function of $T-T_c$ of the zero-field Ising model in d+1 dimensions [3-5]. Recent real space renormalisation group studies have reproduced these properties [6-13].

Over the last few years the critical behaviour of the random transverse Ising model at zero temperature has been extensively studied [14-16]. For the two-dimensional bond-diluted quantum Ising model the percolative behaviour is expected [17] to yield a discontinuous jump in the critical field at the percolation concentration p_c. Below p_c there is no long-range order, so that the critical field changes discontinuously to take the finite value necessary to destroy the order in the chain-like percolating clusters. The behaviour predicted by HARRIS [17] has been confirmed in [9,10,18]. STINCHCOMBE [9] derived a decimation transformation using the bond-moving procedure. As this scheme neglects non-commutation aspects of the model hamiltonian, dos SANTOS [10] has attempted to incorporate these aspects coming to the results quite comparable to those of [9].

In this paper we give an alternative treatment of the pure and diluted planar quantum transverse Ising model described by the hamiltonian

$$\mathcal{H} = - \sum_j \Gamma_j S_j^x - \frac{1}{2} \sum_{i,\delta} J_{i,i+\delta} S_i^z S_{i+\delta}^z \qquad (1)$$

where Γ_i stands for the field, $J_{i,i+\delta}$ is the nearest-neighbour exchange parameter, $S_i^{x,z}$ are the spin $\frac{1}{2}$ operators and the sums run over sites on the planar hexagonal and square lattices. Contrary to the previous treatments [9,10], we apply a real space renormalisation group method elaborated for quantum systems [6,16,19] at zero temperature. Our aim is to obtain the critical behaviour of the diluted system, to check the preservation of the scaling relations in the scheme proposed and to extend the analysis beyond the critical region.

3. Method

The renormalisation group method applied in this paper is a block one. We divide the lattice into independent blocks of n_s sites. Each block consists of a central site and z adjacent sites, where z is the coordination number. The blocks form a new lattice identical with the original one [20]. Then the hamiltonian (1) is divided into two parts. The first part includes all the interactions occurring inside the isolated blocks and is exactly diagonalised. However, as in [16], we remove from this part the field-dependent terms that refer to the central sites of the blocks. The second 'interacting' part treated approximately includes all the remaining terms of the original hamiltonian. The eigen states of the block hamiltonian belonging to the lowest energy level are found to be doubly degenerate. For the block k we define new spin operators ('block operators') $S_k'^x$, $S_k'^z$ in the basis of these doublet states. The first non-interacting part projected into the subspace of the direct product of the lowest doublet states leads to the truncated hamiltonian which can be considered as a constant [16]. In terms of block spins S_k' the corresponding projection of the remaining 'interaction' part is of the same form as the initial hamiltonian (1) but with the renormalised parameters Γ_k', $J_{k,k+\delta}'$.

4. Results for the Bond-Diluted Case

The exchange parameter $J_{i,i+\delta}$ is a random variable for bond-diluted systems and causes both renormalised parameters Γ_k' and $J_{k,k+\delta}'$ to be random. We introduce the new random variable

$$x_i = J_{i,i+\delta} / \Gamma. \qquad (2)$$

The reduced variable is distributed according to the binary distribution

$$P(x_i) = (1-p)\,\delta(x_i) + p\,\delta(x_i - x_o) \qquad (3)$$

if the bonds of the hamiltonian (1) are present ($J_{i,i+\delta} = J$) or absent ($J_{i,i+\delta} = 0$) with probability p and 1-p respectively and $x_o = J/\Gamma$. The distribution of the scaled variable x_k' loses its binary form (3) and evolves to a more general form consisting of a series of delta-functions. This renormalised distribution is approximately replaced [21] by a distribution of the initial binary form with renormalised parameters p' and x_o' chosen so that the approximate form preserves the zeroth and first moments of the renormalised distribution. For the honeycomb lattice the scaled parameters p' and x_o' become

$$p' = 2p^3 - p^6, \tag{4}$$

$$p'x_o' = x_o^3 \left[2p^3 (1-p)^2 / (1+x_o^2)^{\frac{1}{2}} + 4p^4 (1-p) + 2p^5 (1+x_o^2)^{\frac{1}{2}} \right]. \tag{5}$$

In both z=3 and z=4 cases there are two solutions for the fixed point p^* : $p^*=1$, referring to the pure nonrandom system, as well as $p^*=p_c<1$, interpreted as the percolation concentration. The eigenvalues λ_p of the linearised transformations at $p^*=p_c$ are greater then 1 and lead to the percolation correlation length exponent

$$\nu_p = \tfrac{1}{2} \ln n_s / \ln \lambda_p . \tag{6}$$

The numerical values p_c and ν_p are presented in Table 1.

Table 1. Percolation concentrations and correlation length exponents

	p_c		ν_p	
	z = 3	z = 4	z = 3	z = 4
series [22,23]	0.653	0.500	1.34	1.34
Stinchcombe [9]	-	0.618	-	1.62
dos Santos [10]	-	0.618	-	0.82
Saxena [18]	0.500	0.333	-	-
present	0.848	0.755	1.33	1.51

The non-trivial fixed points $x_o^* = 1/y_c$ corresponding to $p^* = 1$ refer to the pure transverse Ising system. In Table 2 the critical fields $y_c = (\Gamma/J)_c$ and the corresponding critical correlation-length exponents ν are presented for the squarelattice as well as references to other results.

For $p^* = p_c$ we obtain new non-trivial fixed points $x_o^* = 1/y_o^*$ and new critical exponents ν_d reported in Table 2. In our calculations the critical concentrations p_c are overestimated. Regarding y_o^* we can also expect overestimation, the more so as the critical field Γ_c for z = 4 pure system exceeds the series result [3] .

The renormalization group transformations [4,5] can be numerically iterated yielding the flow line diagram [20] . Qualitatively the flow line

Table 2. Critical fields and exponents ν for the pure and random system on the square lattice

	series	6	8	9	10	13	18	present
y_c	1.52	1.32	1.64	0.78	0.50	1.66	1.67	1.95
ν	0.63	1.1	1.2	0.50	1.0	0.97	-	0.66
y_o^*	-	-	-	0.44	0.31	-	0.71	1.48
ν_d	-	-	-	0.59	1.0	-	-	0.65

picture is the same as found in the previous renormalization schemes [9,10]. The critical curve joins the three non-trivial fixed points and exhibits the discontinuous jump at the percolation concentration in agreement with the Harris conjecture. The phase boundary was determined as the separatrix between the domains of attraction of the trivial fixed points. In the region below the critical curve the system displays long-range order. This can be readily seen as the magnetization σ is given by the formula

$$\sigma = \lim_{n\to\infty} \prod_{m=0}^{n} \left[(1 + zp^{(m)}/\lambda_o^{(m)})/n_s \right] \tag{7}$$

where

$$\lambda_o^{(m)} = (1 + (y_o^{(m)})^2)^{\frac{1}{2}} \quad , \quad y_o^{(m)} = 1/x_o^{(m)} \tag{8}$$

whereas $p^{(m)}$ and $x_o^{(m)}$ are determined in the iteration step m from the corresponding renormalization group transformations. In the domain of attraction of the trivial fixed point $p^* = 1$, $y_o^* = 0$ the magnetization does not vanish as after a sufficiently great number of iterations $p^{(m)} \to 1$ and $\lambda_o^{(m)} \to 1$. Above the critical curve $\sigma = 0$.

The fixed point $p^* = 1$, $y^* = y_c$ describes the critical behaviour of the pure non-random system while the fixed point $p^* = p_c$, $y^* = y_o^*$ can be interpreted [9,10,11] as describing the chain-like critical behaviour of the percolating cluster. The flow from the latter to the former fixed point indicates the crossover between these two types of critical behaviour. For the bond diluted system with $p > p_c$ the critical behaviour is expected to be determined by the pure fixed point. By using (7) we can calculate σ numerically for an arbitrary p and extract the critical index β. It turns out that σ is reduced for $p < 1$ but β remains constant until the vicinity of the percolation threshold, and equal to the pure value. In addition to that, this value of the exponent β coincides with that calculated from the scaling relations for $p = 1$. For the sake of the limited space we omit here the problem of other critical exponents and more detailed analysis of the global picture of the bond-diluted systems.

5. Conclusions

In this paper we have presented the critical behaviour of the diluted transverse Ising model at T = 0 by having recourse to the real-space renormalization group elaborated for quantum systems. Our scheme is not limited to the critical region but can be also applied to the entire ferromagnetic and paramagnetic regions. As concerns the critical properties our results agree qualitatively with those found previously [9,10,12]. Although the numerical values of the exponents ν_p and ν are promising, the remaining exponents are not so accurate. For the pure system, as a whole, our results are comparable to those given in [6]. Moreover the flow line picture is oversimplified although displays the discontinuity of the critical curve. As $\alpha > 0$ for our model at $T = 0$ [4], we could expect the randomness to be relevant [24,25]; however this shortcoming is also presented in the previous calculations [9,10,12]. The scheme adapted here enables us to recover in the unified way the real-space renormalisation group results for the pure and diluted quantum Ising system.

Acknowledgements

I thank the organizers of the Workshop for the invitation and Prof.R.B. Stinchcombe for clarifying discussions.

1 R.Blinch and B.Zeks: Adv.Phys. 21, 693 (1972)
2 R.B.Stinchcombe: J.Phys. C6, 2459 (1973)
3 P.Pfeuty and R.J.Elliott: J.Phys. C4, 2370 (1971)
4 M.Suzuki: Progr.Theor.Phys. 56, 1454 (1976)
5 J.Oitimaa and G.J.Coombs: J.Phys. C14, 143 (1981)
6 K.A.Penson, R.Jullien and P.Pfeuty: Phys.Rev. B19, 4653 (1979)
7 L.Sneddon and R.B.Stinchcombe: J.Phys. C12, 3761 (1979)
8 D.C.Mattis and J.Gallardo: J.Phys. C13, 2519 (1980)
9 R.B.Stinchcombe: J.Phys. C14, L263 (1981)
10 R.R.dos Santos: J.Phys. A14, LI 79 (1981)
11 R.R.dos Santos, L.Sneddon and R.B.Stinchcombe: J.Phys. A14, 3329 (1981)
12 R.R.dos Santos: J.Phys. C15, 3141 (1982)
13 A.L.Stella, C.Vanderzande and R.Dekeyser: Phys.Rev. B27, 1812 (1983)
14 P.Pfeuty: Phys.Lett. 72A, 245 (1979)
15 R.B.Stinchcombe: J.Phys. C14, 2193 (1981)
16 K.Uzelac, R.Jullien and P.Pfeuty: J.Phys. A13, 3735 (1980)
17 A.B.Harris: J.Phys. C7, 3082 (1974)
18 V.K.Saxena: Phys.Rev. B27, 6884 (1983)
19 P.Pfeuty, R.Jullien and K.A.Penson: Real-Space Renormalization, ed.T.W. Burkhardt and J.M.J. van Leeuwen (Berlin, Heidelberg, New York: Springer Verlag, 1982), p.119
20 G.Kamieniarz: J.Phys. C16, L1021 (1983)
21 J.M.Yeomans and R.B.Stinchcombe: J.Phys. C12, 347 (1979)
22 M.F.Sykes and J.W. Essam: Phys.Rev. 133A, 310 (1964)
23 A.G.Dunn, J.W.Essam and D.S.Ritchie: J.Phys. C8, 4219 (1975)
24 A.B.Harris: J.Phys. C7, 1671 (1974)
25 R.B.Stinchcombe: Phase Transition and Critical Phenomena, ed.C.Domb and J.L.Lebowitz (Academic Press, London, New York, Paris 1983) p.151

The Random Field Ising Model. Equilibrium and Non-Equilibrium Properties

J. Villain
DRF/DN, Centre d'Etudes Nuclêaires, 85X, F-38041 Grenoble Cedex, France

1 A Controversial Problem

The Random Field Ising Model (RFIM) is defined by the Hamiltonian

$$\mathcal{H} = -J\sum_{\langle ij\rangle} S_i S_j - \sum_i H_i S_i \quad (1)$$

where i, j denote lattice sites, $S_i = \pm 1$ and $\overline{H_i H_j} = H^2 \delta_{ij}$ (2)
and J and H are constant parameters.

In the limit $H \to 0$, at temperature $T < J$, there is long range ferromagnetic (if $J > 0$) or antiferromagnetic (if $J < 0$) order if the space dimension D is larger than some lower critical value D_{cl}. The behaviour near the transition temperature T_c follows mean field theory if D is larger than the upper value $D_{cu} = 6$ [1] . The lower dimension D_{cl} is controversial [2] .If one writes that domains of radium R are forbidden if the wall energy ($\sim R^{D-1}$) is larger than the random field (RF) energy ($\sim R^{D/2}$), one obtains $D_{cl} = 2$ [3] . This evaluation, although probably correct [4, 5, 6, 7] is questionable because walls may be rough. More complicated arguments based on the replica trick [8] or perturbation expansion [9] yield $D_c = 3$. Birefringence data [10] are consistent with $D_{cl} < 3$, but neutron experiments in 2 or 3 dimensions [2, 11] yield a correlation length

$$R_c \simeq \text{Const} \times H^{-\nu_H} \quad (3)$$

and this result is consistent with a lower critical dimension $D_c > 3$. Indeed, at the critical dimension, one expects $R_c \sim \exp{(H_0/H)^2}$ [2] .

2 Who is Right ?

The treatments which yield $D_c = 3$ are probably wrong. i) Methods based on the "most divergent diagrams" of the Wilson-Fisher expansion [9] fail when the Landau free energy functional has many minima. Indeed the various minima, extrema and saddle-points are not weighted by the Boltzmann factor but by completely unrealistic weights which are not even positive ! The reader can easily check this statement if he considerers formula (4) of Ref. [12] in a zero-dimensional space, i.e. for a Hamiltonian $V(\phi) - h\phi$ with a single variable ϕ . ii) Methods based on the replica trick lead to wrong results if corrections to scaling are neglected -and this is unfortunately an unavoidable or hardly avoidable simplification.

Though the result $D_c = 2$ has not been exactly proved either, it is supported by intuitive domain arguments [3, 4, 5] which will be briefly recalled below and can be made "nearly" rigorous [7] . They are also in agreement with computer simulations [13] . Now, why is this result in disagreement with neutron scattering experiments ?

The most plausible explanation will be presented in this paper. The theoretical result $D_{cl} = 2$ is correct at equilibrium, but systems prepared by cooling in a constant random field from the paramagnetic region are in a metastable state. This is in agreement with experiment by Wong et al. [14] .

3 Nature of Metastable States and Potential Barriers

Metastability is necessarily related to potential barriers. According to very old ideas [15] potential barriers in the RFIM can be attributed to domain wall pinning by heterogeneities [16,17,18,19]. Fig. 1 shows a typical case. Our picture of a metastable state of a RFIM is as follows. The system is essentially constituted by metastable domains, the radius of which has a given order of magnitude R_c. The number of domains with spin up and spin down is the same. In 3 dimensions, domains of both type percolate, so that there is an infinite domain of spin up, made of entangled tubes of radius about R_c, and one similar domain of spin down. Both infinite domains contain a (probably small) number of finite domains of opposite sign.

Two cases should be distinguished. i) when the wall thickness is much larger than the interatomic distance [16]. ii) when both lengths are comparable [19,20]. Surprisingly, the results will be found similar in both cases. In this Section, broad walls will be considered, as is correct not too much below T_c.

Figure 1 shows a wall which is obliged to make a bump of radius ρ and height ξ in order to avoid an unfavourable region. This unfavourable region constitutes a potential barrier. The bump shape results from a competition between the random field (RF) energy, and the wall tension energy. For large values of ρ and ξ these two contributions are of order

$$W_{RF} \approx m_o H \sqrt{\rho^{D-1}\xi} \qquad (4) \qquad\qquad W_{WT} \approx g\rho^{D-1}(\xi/\rho)^2 = g\rho^{D-3}\xi^2 \qquad (5)$$

where the wall tension constant g is of order J at low temperature while the spontaneous magnetisation m_o is of order 1 in appropriate units. Note that formula (5) is not applicable to the RFIM at too low temperature, where the alternative treatment of Grinstein and Fernandez [19] should be used.

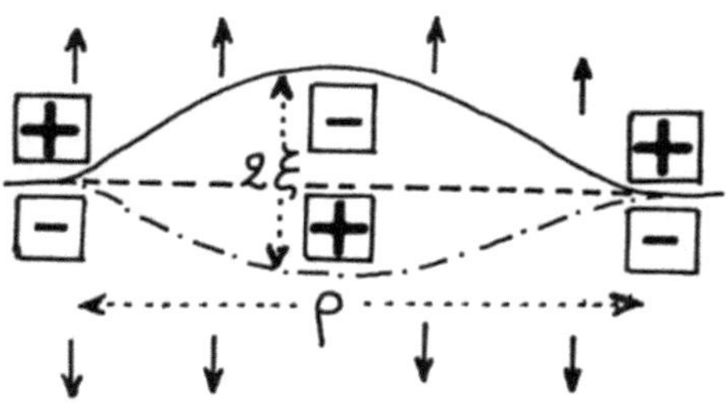

Figure 1. Two metastable positions (full line and dash-dotted line) of a wall separating a spin up domain from a spin down domain. The dashed line shows an unstable position corresponding to a potential barrier. ± signs within squares represent random fields.

We now make the assumption that the RF energy (4) cannot be smaller than the surface tension energy (5), otherwise the bump would be unstable. The typical value of ξ has to maximise (4) within this constraint. Insertion of this value into (7) yields the typical barrier height ξ for a given bump radius ρ. One obtains

$$\xi \approx (Hm_o/g)^{2/3}\,\rho^{(5-D)/3} \qquad (6)\,, \qquad\qquad \omega \approx Hm_o(Hm_o/g)^{1/3}\,\rho^{(D+1)/3}\,. \qquad (7)$$

If a barrier cannot be jumped during a time t, its height ω must satisfy the Arrhenius law $\exp(\omega/T) > t/\tau$, where the order of magnitude of the microscopic time τ is expected to be $\tau = \hbar/J$. Insertion of (7) yields the minimum value of ρ after a waiting time t, namely (for $g = J$ and $m_o = 1$)

$$(H/T)(H/J)^{1/3}\,\rho^{(D+1)/3} \approx \ln(t/\tau)\,. \qquad (8)$$

Can this value be identified with the domain size R_c ? We do not believe so. Indeed formula (6) implies a curvature radius

$$R \approx \rho^2/\xi \approx (J/H)^{2/3}\ \rho^{(D+1)/3}. \tag{9}$$

Insertion of (8) yields the minimum value R_c or R after a time t :

$$R_c \approx (JT/H^2)\ln(t/\tau)\ . \tag{10}$$

This is in agreement with the experimental formula (3), with $\nu_H = 2$, and a time-dependent multiplicative constant. These results are not in contradiction with existing experimental data [21] . In some experiments [2] ν_H was found to be about 3.6, but this large value may be attributed to excessively slow cooling. The time dependence in (12) is weak and cannot easily be checked experimentally. Finally, assuming $\tau \approx \hbar/J$, relation (10) yields the correct order of magnitude within a factor 3.

Remarks. i) What happens when the bump size ρ is larger than $\sqrt{R\xi}$, in contradiction with (9) ? Probably, such big bumps do not contribute to domain metastability, since the barrier height is overwhelmed by surface tension energy. Metastability is completely due to smaller bumps. ii) The assumtion $W_{RF} < W_{WT}$ only applies to the highest barriers. As pointed out by Bruinsma and Aeppli [17] , there are barriers which correspond to bump heights much larger than (6), but to much lower bump height variations and to barrier energies much lower than (7). In a realistic treatment where jumps over barriers are taken into account, it is reasonable to assume that the domain size at time t is determined by the highest barriers since lower barriers have been jumped over.

4 The case of narrow walls [19]

If the walls are narrow, it is easily seen [20] that the square $(\xi/\rho)^2$ should be replaced by ξ/ρ in (5). The same argument as before yields

$$\xi \approx (H/J)^2\, R^{3-D} \tag{11}$$

$$\omega \approx H^2 R/J \tag{12}$$

where g and ρ have been replaced by J and R, respectively. Surprisingly, the Arrhenius law, together with (12), yields relation (10) again ! This result is true for all dimensions, although relation (11), and therefore the above derivation, is obviously incorrect for D > 3. What happens is that the elementary precesses are two-dimensional for any value of D, as seen from Fig. 2. The decay of a metastable domain (assumed to be cubic in Fig.2) results from peeling its successive layers. Peeling is a two-dimensional process for which relations (11) and (12) should hold, at least approximately. There may be weak corrections because two is the lower critical dimension, in both cases of broad and narrow walls. The waiting time is proportional to the number of layers, but this factor is negligible in comparison with the exponential in the Arrhenius law.

Of course, formula (10) is only applicable for long time. The precise condition [19] is $\ln(t/\tau) > 8J/T$ for a simple cubic lattice, and may be quite restrictive at very low temperature. Grinstein and Fernandez [19] have evaluated R_c for shorter times too.

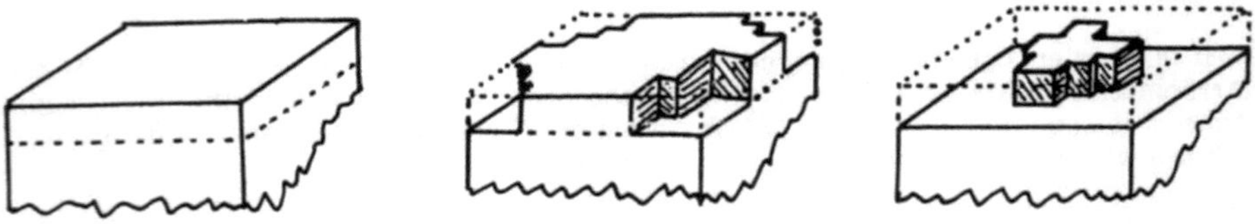

Figure 2. Decay of a domain by peeling. The cubic shape has been assumed so as to make drawing easier.

Formula (12) was checked by computer simulations [22]. The logarithmic behaviour is a well-known feature of after-effect phenomena [15,23].

5 Correlation length at the zero field transition temperature T_{co}

The correlation length R_o at T_{co} is among the available experimental data [2,11]. It can be roughly determined from the Ansatz that the system, near T_{co}, is characterized by a typical length R_o (the typical radius of regions with positive or negative magnetisation m (the average magnetisation inside these regions). Then, the Landau free energy functional

$$\mathcal{F} = \int d^D r \left[\frac{1}{4} A m^2(r) + \frac{1}{4} B m^4(r) + \frac{1}{2} J(\vec{\nabla} m)^2 - H(r)m(r)\right] \tag{13}$$

yields a free energy density of order

$$F \approx \frac{A}{2} m^2 + \frac{B}{4} M^4 + \frac{1}{2} J_R m^2 R_o^{-2} - HmR_o^{-D/2} . \tag{14}$$

In the mean field approximation (MFA) $B \approx J_R \approx J$, and $A \approx J(T-T_{co})/T_{co}$. m is determined for given R_o by minimisation of (14) with respect to m, then R_o is determined by maximisation of m with respect to R_o (*). At T_{co}, $A = 0$ and minimisation of (14) yields

$$m \approx \operatorname{Inf}\left[(H/B)^{1/3} R_o^{-D/6} , \quad (H/J_R) R_o^{(4-D)/2}\right] \tag{15}$$

and maximisation of this expression yields, for $D < 4$,

$$Bm^3 \approx J_R m R_o^{-2} \approx HR_o^{-D/2} . \tag{16}$$

In the MFA, relations $J_R \approx B \approx J$ yield $R_o \approx m^{-1} \approx (J/H)^{2/(6-D)}$. If scaling theory is used instead of MFA, J_R and B are given by

$$J_R R_o^{-2} \approx J R_o^{\eta-2} \quad (17\text{ a}) \qquad\qquad Bm^3 \approx Jm^{\delta} . \quad (17\text{ b})$$

Relation (17a) results from the expression of the spin pair correlation function at T_{co} in zero field, namely $1/J_R k^2 = 1/Jk^{2-\eta}$. Expression (17 b) results from the equation of state at T_{co} in a uniform field H_o and zero random field, which can alternatively be written $Bm^3=H$, or $m^{\delta} \sim H$. The scaling relation $\delta = 1+2(2-\eta)/(D-2+\eta)$ and formulae (16) and (17) yield

$$R_o \approx (J/H)^{1/(1-\eta/2)} \quad (18) \qquad\qquad m \approx (H/J)^{(D-2+\eta)/(2-\eta)} . \quad (19)$$

The exponent in (18) is larger than the published experimental value, which is about 0.8 [2, 11]. At T_{co} there are no domains and no walls, so that metastable states are not expected to exist. This statement is in agreement with experiment. Somewhere below T_c potential barriers should appear. This is the subject of the next Section.

6 Genesis of metastable states

The problem addressed so far was the following one : if a domain structure is existing at time 0, how do domains grow with time ? In this Section, the question to be answered is : "is a metastable domain structure formed when a RFIM is cooled in a

---

(*) It would be a mistake to minimise (14) with respect to both R_o and m because the energy is dominated by short wavelength components. The suggested procedure is consistent with the Ansatz that the local magnetisation is dominated by Fourier components of wavelength about R_o. This procedure is in agreement with linear response theory in the cases when it can be applied.

constant random field from the paramagnetic region ?" The first thing required is a definition applicable at any temperature. A metastable state will be defined as a relative minimum of the functional (13). The absolute minimum is the stable state. Stable and metastable states satisfy the Euler-Lagrange equation

$$\nabla^2 m(r) = Am(r) + Bm^3(r) - H(r) . \tag{20}$$

Before worrying about metastable states, the stable state will be considered. Below T_c, the ordered phase is stable with respect to the formation of domains of radius R [3] if the RF energy $W_{RF} \approx Hm_o R^{D/2}$ is smaller than the domain wall energy $W_{DW} \approx gR^{D-1}$, where m_o is the spontaneous magnetisation and R may not be smaller than the wall thickness λ. Scaling theory [24] yields

$$\lambda \sim m_o^{2/(D-2+\eta)} \tag{21}$$

$$g \sim J\, m_o^{(D-1)\,2/(D-2+\eta)} . \tag{22}$$

From the stability condition $W_{FR} < W_{DW}$ for $R > \lambda$, one deduces that, in equilibrium, the system breaks into domains when m_o is smaller than some threshold of order (19). For this value, the order of magnitude of λ is given by (18), and the domain size R has the same order.

The possibility of metastability can be investigated by the argument of section 3. Conditions $\lambda < \xi < \rho$ and $\xi > \rho^2/R$, together with relations (11), (21) and (22), show that metastability is possible if λ is smaller than R_o (given by 18), or alternatively if m_o is larger than (19). Thus, metastability appears at the temperature where, at equilibrium, long range order appears. The system may therefore remain out of equilibrium. The size of spin up regions (and spin down as well, of course) increases so that barrier heights become higher and higher with increasing time, and metastability is established.

This scheme is plausible and in agreement with experiment. However, it is not the only possible one. Another possibility is that a spontaneous magnetisation, superimposed to disorder-induced fluctuations, appears at some transition temperature. This probably happens above 4 dimensions, because the random field can then be treated by linear response theory at all temperatures without any inconsistency. Thus to lowest order in H, one goes continuously from the paramagnetic state to a long-range ordered state.

In other words, two types of evolutions toward equilibrium are a priori possible. A. Domain growth. B. Appearance of a uniform magnetization, followed by progressive disappearance of ill-oriented domains. The two schemes are illustrated by the following non-random problem. Consider a ferromagnet in a field of uniform modulus with periodically alternating sign, $H(r)=H\sin Qx \sin Qy \sin Qz/|\sin Qx \sin Qy \sin Qz|$. The period $L = 2\pi/Q$ should be large enough, so that the domain wall concept makes sense at low temperature. For small H, a continuous transition occurs. For large H there is no order at any temperature. For intermediate field a first order transition occurs when the magnetization m_o is large enough, so that the domain wall energy $gL^{(D-1)/2}$ compensates the Zeeman energy Hm_oL^D. Some randomness can now be introduced by allowing fluctuations in the field modulus. Then, all domains do not flip at the same temperature and a type A process is expected. In weak field a type B process is expected from perturbation theory. Thus, a type A process (where no long range order appears at finite time) is possible at least for random, but not independent fields, having a sufficiently large correlation length L. In the case of independent random fields, neutron scattering experiments [2, 11] are consistent with type A, but theory is unable to rule out type B. This is because, at T_c, the domain size L and the wall thickness λ have the same order of magnitude (18).

References

1 A. Aharony, Y. Imry and S.K. Ma, Phys. Rev. Lett. 37, 1367 (1976)
2 M. Hagen, R.A. Cowley, S.K. Satija, H. Yoshizawa, G. Shirane, R.J. Birgeneau, H.J. Guggenheim, Phys. Rev. B28, 2602 (1983)
3 Y. Imry, S.K. Ma, Phys. Rev. Lett. 35, 1399 (1975)
4 J. Villain, J. Physique Lettres, 43, L551 (1982)
G. Grinstein, S.K. Ma, Phys. Rev. Lett. 49, 685 (1982)
5 G. Grinstein, S.K. Ma, Phys. Rev. B28, 2588 (1983)
J. Villain, B. Séméria, F. Lançon, L. Billard, J. Phys. C16, 6153 (1983)
6 T. Nattermann, J. Phys. C.16, 4113 (1983)
7 D.S. Fisher, J. Fröhlich and T. Spencer, to appear in J. Stat. Phys. (1984)
8 D. Mukamel, E. Pytte, Phys. Rev. B25, 4779 (1982)
9 H.S. Kogon, D.J. Wallace, J. Phys. A14, L527 (1981)
10 D.P. Belanger, A.R. King, V. Jaccarino, J.L. Cardy, Phys. Rev. B28, 2522 (1983)
11 R.A. Cowley, R.J. Birgeneau, G. Shirane, H. Yoshizawa : Multicritical Phenomena (Ed. R. Pynn, Plenum, New-York 1983)
12 G. Parisi, N. Sourlas, Phys. Rev. Lett. 43, 744 (1979)
13 J. Fernandez, G. Grinstein, Y. Imry, S. Kirkpatrick, Phys. Rev. Lett. 51, 2O3 (1983)
D. Andelman, H. Orland, L.C.R. Wijewardhana, Phys. Rev. Lett. 52, 145 (1984)
14 P.Z. Wong, J.W. Cable, Phys. Rev. B28, 5361 (1983)
15 L. Néel, J. Phys. Rad. 12, 339 (1951) and Oeuvres Scientifiques, pp. 353-434 (Editions du CNRS, Paris 1978)
16 J. Villain, Phys. Rev. Lett. 52, 1543 (1984)
17 R. Bruinsma, G. Aeppli, Phys. Rev. Lett. 52, 1543 (1984)
18 H. Yoshizawa, D.P. Belanger, Brookhaven preprint (1984)
19 G. Grinstein, J.F. Fernandez, IBM Yorktown Heights preprint (1984)
20 K. Binder, Z. Phys. B50, 343 (1983)
21 R. Birgeneau, R. Cowley, M. Hagen, private communication (1984)
22 E.T. Gawlinski, K. Kaski, M. Grant, J.D. Gunton, Temple University
M. Grant, J.D. Gunton, Phys. Rev. B29, 1521 (1984)
23 J.J. Prejean, J. Souletie, J. Physique 41, 1335 (1980)
24 B. Widom, J. Chem. Phys. 43, 3892 (1965)

Part V

Excitations in Ordered Magnets

Theory of the Raman and Brillouin Scattering on the Excited Electronic States and Magnons in Europium Chalcogenides

W. Cieslewicz

Instytut Fizyki Politechniki Poznanskiej, PO-Poznan, Piotrowo 3, Poland

A. Stasch

Istitudo di Elettronica Ouantistica CNR, I-50127 Firenze, Italy

I. Introduction

Recently considerable progress has been made in the investigations of the excited electronic states by using the Raman spectroscopy [1]. In the present paper we examin the possibility of applying of this method in the measurement of the electronic properties of magnetic semiconductors. Among those the europium chalcogenides seem to be of particular interest for the following reasons: a) the free Eu^{2+} ion is in the orbital singlet state and the structure of the 4f band is very simple, b) the optical absorption data are well explained on the basis of the d-f exchange model proposed by GUNTHERODT, WACHTER and IMBODEN [2,3], which will be called GWI model throughout the paper. On the other hand, the absorption and reflection experiments don't give information about the internal structure of the electronic bands, values of the d-f exchange integral and spin orbit coupling in the 5d multiplet. In the present paper we show that electronic Raman processes and one-magnon Brillouin scattering may be useful sources of such information .

2. Electronic Wave Functions and Energies in the GWI Model

In the framework of the GWI model the wave functions ψ_i and eigenenergies E_i of the conduction electrons in the 5d band are given by [2,4] :

$$\psi_1 = |-1,+\tfrac{1}{2}\rangle ,$$

$$\psi_2 = (1-g)^{\frac{1}{2}}|-1,-\tfrac{1}{2}\rangle + (g/2)^{\frac{1}{2}} (|+2,+\tfrac{1}{2}\rangle - |-2,+\tfrac{1}{2}\rangle) ,$$

$$\psi_3 = f^{\frac{1}{2}}|+1,+\tfrac{1}{2}\rangle + ,(1-f)/2^{\frac{1}{2}} (|+2,-\tfrac{1}{2}\rangle - |-2,-\tfrac{1}{2}\rangle ,$$

$$\psi_4 = |+1,-\tfrac{1}{2}\rangle ,$$

$$\psi_5 = -(1-f)^{\frac{1}{2}}|+1,+\tfrac{1}{2}\rangle + (f/2)^{\frac{1}{2}} (|+2,-\tfrac{1}{2}\rangle - |-2,+\tfrac{1}{2}\rangle) ,$$

$$\psi_6 = g^{\frac{1}{2}}|-1,-\tfrac{1}{2}\rangle - (1-g)/2^{\frac{1}{2}} (|+2,+\tfrac{1}{2}\rangle - |-2,+\tfrac{1}{2}\rangle) , \quad (1)$$

$$E_1 = -\lambda(B+1) , \quad E_2 = \tfrac{1}{2}\lambda(1-D_+) , \quad E_3 = \tfrac{1}{2}\lambda(1-D_-) ,$$

$$E_4 = \lambda(B-1) , \quad E_5 = \tfrac{1}{2}\lambda(1+D_-) , \quad E_6 = \tfrac{1}{2}\lambda(1+D_+) , \quad (2)$$

where

$B = J_{d-f}/\lambda$, $D_\pm = (9 \pm 4B^2 + 4B)^{\frac{1}{2}}$, $f = (D_- + 2B - 1)/D_-$, $g = (D_+ + 2B + 1)/D_+$

and λ is the spin orbit constant in the 5d band. J_{d-f} denotes the d-f exchange integral. It should be noted now, that in the description of the electronic state in the valence band we take, instead of the atomic functions of the references [2,4,5], the following linear combinations of the a_{2u} , t_{1u} and t_{2u} symmetry:

$$f_1 = (3/8)^{\frac{1}{2}} |+1,+\tfrac{1}{2}\rangle + (5/8)^{\frac{1}{2}} |-3,+\tfrac{1}{2}\rangle , \quad f_2 = |0,+\tfrac{1}{2}\rangle ,$$

$$f_3 = (3/8)^{\frac{1}{2}} |+1,+\tfrac{1}{2}\rangle + (5/8)^{\frac{1}{2}} |+3,+\tfrac{1}{2}\rangle , \quad f_4 = (5/8)^{\frac{1}{2}} |+1,+\tfrac{1}{2}\rangle - (3/8)^{\frac{1}{2}} |-3,+\tfrac{1}{2}\rangle ,$$

$$f_5 = (1/2)^{\frac{1}{2}} |+2,+\tfrac{1}{2}\rangle + (1/2)^{\frac{1}{2}} |-2,+\tfrac{1}{2}\rangle , \quad f_6 = (5/8)^{\frac{1}{2}} |-1,+\tfrac{1}{2}\rangle - (3/8)^{\frac{1}{2}} |+3,+\tfrac{1}{2}\rangle ,$$

$$f_7 = (1/2)^{\frac{1}{2}} |+2,+\tfrac{1}{2}\rangle - (1/2)^{\frac{1}{2}} |-2,+\tfrac{1}{2}\rangle . \tag{3}$$

3. The Differential Cross Sections and Polarization Selection Rules for the Brillouin and Raman Processes

Having obtained the wavefunctions ψ_i and f_i we are able to calculate the differential cross sections using the following formula:

$$\frac{\delta^2\sigma}{\delta\omega\delta\Omega} = \frac{e^4\omega_1}{mc^4} \left| \frac{\langle f_i | p_1 \cdot e_1 | \psi_j \rangle \ \langle \psi_j | p_2 \cdot e_2 | f_i \rangle}{E_i - E_f - \hbar\omega} \right|^2 \delta(\hbar\omega_1 - \hbar\omega_2 - \hbar\omega_m) , \tag{4}$$

where the indices 1 and 2 refer to the incident and scattered photons. E_i and E_f denote the energy of the electrons in the d and f bands, respectively. ω_m is the magnon frequency. The one-magnon scattering is treated here as a transition in which an electron from the state f_i is virtually excited into the state ψ_j affected by the strong spin orbit interaction and then scattered into f_i, generating the magnon. Now, taking into account all possible virtual processes we obtain for the value of the differential cross section $8.8 \cdot 10^{-8} cm^2 sr^{-1} s$ in the case of the linear polarization of the incident photons along z axis and scattered light polarized right circularly in the x,y plane (i.e in the e_1^z , e_2^+ configuration). In the e_1^+ , e_2^z geometry its value is equal to $4.2 \cdot 10^{-8}$ in the same units and all diagonal elements of the Brillouin tensor vanish. In our numerical estimations the typical values $J_{d-f} = \lambda = E_g - \hbar\omega_1$ have been used [2,3,5] . Note that the predicted selection rules don't depend on the values of λ and J_{d-f} as long as both differ from zero. Using Eq. (4) it is possible to calculate the differential cross section for the excited state electronic scattering in the very similar way [4] . The results for all allowed $\psi_1 \rightarrow \psi_i$ processes are presented in the Table 1. In the calculations the concentration of 10^{16} cm^{-3} for the conduction electrons was assumed.

Table 1. The Selection Rules for the Excited State Electronic Raman Scattering in Europium Chalcogenides

Transition	Photon polarization incident	scattered	$\hbar\omega_1 - \hbar\omega_2$ [eV]	Cross section $[cm^2 sr^{-1} s]$
$\psi_1 \rightarrow \psi_2$	e_1^z	e_2^+	0.04	$1.4 \cdot 10^{-12}$
$\psi_1 \rightarrow \psi_3$	e_1^+	e_2^+	0.11	$9.8 \cdot 10^{-14}$
$\psi_1 \rightarrow \psi_5$	e_1^+	e_2^+	0.30	$1.2 \cdot 10^{-14}$
$\psi_1 \rightarrow \psi_6$	e_1^z	e_2^+	0.47	$1.3 \cdot 10^{-14}$

4. Conclusions

We performed an analysis of the possibility of using Raman and Brillouin spectroscopy in the investigations of the electronic structure of the europium chalcogenides. By estimating excited state electronic Raman scattering and one-magnon Brillouin processes we provide a new searching test of the validity of the GWI model. With this aim the Raman and Brillouin tensors have been calculated. Our results show that it is possible to measure the values of the d-f exchange integral and spin orbit constant in the six independent ways by investigating the energies and relative intensities of the allowed Raman and Brillouin processes.

References

1. L.V. Haley, J.A. Koningstein: Chem. Phys. 73 , 263 (1982)
2. G. Guntherodt, P. Wachter, I. Imboden: Phys.cond. Matter 18 , 292 (1971)
3. G. Guntherodt: Phys.cond. Matter 18 , 37 (1974)
4. A. Stasch, P. Wachter: Sol. St. Comm. 40 , 451 (1981)
5. A. Stasch: J. Phys. C , 16 , 4057 (1983)

Effect of a Surface on Magnon-Magnon Interactions in Ferromagnets

D.E. Kontos and M.G. Cottam

Department of Physics, University of Essex, Colchester CO4 3SQ, United Kingdom

1 Introduction

In recent years there has been considerable interest in the surface properties of Heisenberg ferromagnets, and it has been shown that in addition to the bulk spin wave excitations at $T \ll T_C$ there may be surface spin waves (for reviews see [1-3]). These latter excitations are localised near the surface, having amplitudes which decay (usually exponentially) with distance from the surface. At non-zero temperatures interactions between the spin waves will occur, leading to an energy renormalisation and damping of the modes. In infinite Heisenberg ferromagnets the interactions between bulk spin waves have been treated rigorously by DYSON [4]. However, the presence of a surface is expected to give rise to richer and more complicated schemes of spin wave interactions, since scattering processes may occur involving surface spin waves as well as bulk spin waves. In this paper we use a Green's function method to study spin wave interactions in semi-infinite Heisenberg ferromagnets at low temperatures $T \ll T_C$. Specifically we deduce expressions for the energy renormalisation and damping of surface spin waves due to their interaction with either bulk spin waves or surface spin waves. The results are illustrated by numerical examples, and comparisons are made with the renormalisation for bulk spin waves.

The ferromagnet is assumed to occupy the half-space $z < 0$ and to be represented by the Hamiltonian

$$\mathcal{H} = -\frac{1}{2}\sum_{i,j} J(i,j)\vec{S}_i.\vec{S}_j - g\mu_B \sum_i [H_0+D(i)]S_i^z \qquad (1)$$

where $\vec{S}_i$ is a spin operator at site i, J(i,j) is an exchange interaction, and H_0 is an applied magnetic field. Surface anisotropy is taken into account by means of the effective field D(i), which we assume to be zero except in the surface layer ($z = 0$) where it has the value D. Here we restrict attention to the case of nearest neighbour exchange interactions, assumed to have the value J_S if both spins are in the surface layer and the bulk value J otherwise. Such a model has been used in previous calculations for surface excitations (see [1]). More generally we have also considered the effect of next-nearest neighbours. The theory will be applied to ferromagnets with simple cubic (s.c.), b.c.c. and f.c.c. lattices, all with (001) crystallographic orientation of the surface; more generally we have also derived results for (001) surfaces.

With the aid of the Holstein-Primakoff representation the Hamiltonian (1) can be transformed to boson operators, keeping terms up to fourth order. As is well known this type of approach can lead to inconsistencies for low spin values, but these difficulties are avoided if 1/S is formally treated as a small parameter and calculations are performed consistently to each power in

1/S (see [5,6]). The linear spin wave part of $\mathcal{H}$ can then be diagonalised by a further transformation to normal mode operators. If $\vec{q} = (q_x, q_y)$ denotes a two-dimensional wavevector parallel to the surface, we then have

$$\mathcal{H} = \sum_{\vec{q},\mu} E(\vec{q},\mu) a^{+}_{\vec{q},\mu} a_{\vec{q},\mu} + \mathcal{H}_{int} . \tag{2}$$

Here μ labels the normal modes of the semi-infinite system which consist of a quasi-continuum of bulk spin waves with wavevector $\vec{Q} = (\vec{q}, q_z)$ and a discrete surface spin wave branch characterised by wavevector $\vec{q}$. The eigenvalues $E(\vec{q},\mu)$ have the usual form for the energies $E_B(\vec{Q})$ and $E_S(\vec{q})$ of bulk and surface spin waves respectively in a linear approximation [1]. The dispersion relations are

$$E_B(\vec{Q}) = g\mu_B H_o + SJ[\gamma_{\parallel}(0) - \gamma_{\parallel}(\vec{q}) + 2\gamma_{\perp}(0) - 2\gamma_{\perp}(\vec{q})\cos(q_z c)] \tag{3}$$

$$E_S(\vec{q}) = g\mu_B H_o + SJ[\gamma_{\parallel}(0) - \gamma_{\parallel}(\vec{q}) + 2\gamma_{\perp}(0) + \gamma_{\perp}(\vec{q})\{\Delta(\vec{q}) + \Delta^{-1}(\vec{q})\}] \tag{4}$$

where $\Delta(\vec{q})$ is a parameter which depends on the surface characteristics:

$$\Delta(\vec{q}) = -\{(1 - J_S/J)[\gamma_{\parallel}(0) - \gamma_{\parallel}(\vec{q})] + \gamma_{\perp}(0) - [g\mu_B D/SJ\gamma_{\perp}(0)]\}/\gamma_{\perp}(\vec{q}) . \tag{5}$$

The geometrical factors $\gamma_{\parallel}(\vec{q})$ and $\gamma_{\perp}(\vec{q})$ are defined in Table 1, where a is the cubic lattice distance and c is the spacing between atomic layers parallel to the surface.

Table 1. The geometrical factors for some lattices with (001) surfaces

Lattice	c	$\gamma_{\parallel}(\vec{q})$	$\gamma_{\perp}(\vec{q})$
s.c.	a	$2[\cos(q_x a) + \cos(q_y a)]$	1
b.c.c.	a/2	0	$4\cos(q_x a/2)\cos(q_y a/2)$
f.c.c.	a/2	$4\cos(q_x a/2)\cos(q_y a/2)$	$2[\cos(q_x a/2) + \cos(q_y a/2)]$

The existence conditions for surface modes are $\Delta(\vec{q}) < -1$ (acoustic surface spin wave) or $\Delta(\vec{q}) > 1$ (optic surface spin wave), and these restrict the values of the parameters. In either case the penetration depth is $c/\ln|\Delta(q)|$. The dispersion relations (3) and (4) are illustrated in Fig. 1 taking $g\mu_B H_o/SJ = 0.8$. Curves S_1 ($J_S/J = 0.5$ and $g\mu_B D/SJ = 0$) and S_2 ($J_S/J = 0.5$

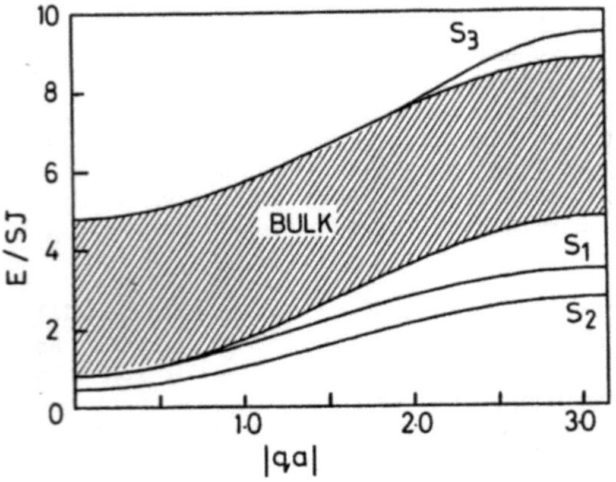

Figure 1. Example of linear spin wave dispersion relations for a s.c. (001) ferromagnet showing the bulk continuum and three surface branches (see text)

and $g\mu_B D/SJ = -0.8$) represent acoustic surface branches, whilst S_3 ($J_S/J = 1.8$ and $g\mu_B D/SJ = 0$) represents an optic surface branch. The term $\mathcal{H}_{int}$ in (2) describes the leading effect of magnon-magnon interactions and is quartic in the normal mode operators. Here we discuss the resulting energy renormalisation and damping of the surface spin waves. The calculations can be conveniently performed by evaluating Green's funtions of the form $\langle\langle a;a^+\rangle\rangle$ within a diagrammatic perturbation expansion in $\mathcal{H}_{int}$, and then analysing their complex poles.

2 Spin Wave Energy Renormalisation

In first order of perturbation there is a shift ΔE in the spin wave energies. The result for surface mode is found to be

$$\Delta E_S(\vec{q}) = \frac{1}{N_\perp}\sum_{\vec{k}} W_{SS}(\vec{q},\vec{k})\, n\left[E_S(\vec{k})\right] + \frac{1}{N_\perp N_\parallel}\sum_{\vec{K}} W_{SB}(\vec{q},\vec{K})\, n\left[E_B(\vec{K})\right] . \qquad (6)$$

The first and second terms are the contributions to ΔE_S due to interactions with surface modes and bulk modes respectively and will be denoted by ΔE_{SS} and ΔE_{SB}. In (6) $\vec{k}$ is a two-dimensional wavevector parallel to the surface, $\vec{K} = (\vec{k},k_z)$ is a three-dimensional wavevector, $n(E) = 1/\left[\exp(E/k_BT)-1\right]$, and $N_\parallel$ and $N_\perp$ denote respectively the number of atomic layers parallel to the surface and the number of sites in each layer (both numbers macroscopically large). The interaction vertices W_{SS} and W_{SB} are

$$\begin{aligned}W_{SS}(\vec{q},\vec{k}) = &- J\{\left[\Delta^2(\vec{q})-1\right]\left[\Delta^2(\vec{k})-1\right]/\left[\Delta^2(\vec{q})\Delta^2(\vec{k})-1\right]\}\{\alpha(\vec{q},\vec{k})\\ &\times \left[1-\sigma(1-\Delta^{-2}(\vec{q})\Delta^{-2}(\vec{k}))\right]+\left[\gamma_\perp(\vec{k})\Delta^{-1}(\vec{k})(1+\Delta^{-2}(\vec{q}))+\gamma_\perp(\vec{q})\Delta^{-2}(\vec{q})(1+\Delta^{-2}(\vec{k}))\right.\\ &\left.+\gamma_\perp(0)(\Delta^{-2}(\vec{q})+\Delta^{-2}(\vec{k}))+2\gamma_\perp(\vec{q}-\vec{k})\Delta^{-1}(\vec{q})\Delta^{-1}(\vec{k})\right]\} , \end{aligned}\qquad (7)$$

denoting $\sigma = 1-J_S/J$ and $\alpha(\vec{q},\vec{k}) = \left[\gamma_\parallel(0)+\gamma_\parallel(\vec{q}-\vec{k})-\gamma_\parallel(\vec{q})-\gamma_\parallel(\vec{k})\right]$, and

$$\begin{aligned}W_{SB}(\vec{q},\vec{K}) = &-J\{\left[1-\Delta^{-2}(\vec{q})\right]/\left[1-2\Delta^{-2}(\vec{q})\cos(2k_zc)+\Delta^{-4}(\vec{q})\right]\}\{\left[\alpha(\vec{q},\vec{k})+\gamma_\perp(\vec{q})\Delta^{-1}(\vec{q})\right.\\ &\left.+\gamma_\perp(0)\Delta^{-2}(\vec{q})\right]\left[\Delta^{-2}(\vec{q})\cos(2\theta-2k_zc)-\cos(2\theta)\right]+\left[\gamma_\perp(\vec{k})(1+\Delta^{-2}(\vec{q}))+2\gamma_\perp(\vec{q}-\vec{k})\Delta^{-1}(\vec{q})\right]\\ &\times\left[\cos(2\theta+k_zc)-\Delta^{-2}(\vec{q})\cos(2\theta-k_zc)\right]+\left[\gamma_\perp(0)+\gamma_\perp(\vec{q})\Delta^{-1}(\vec{q})\right]\left[\Delta^{-2}(\vec{q})\cos(2\theta)\right.\\ &\left.-\cos(2\theta+2k_zc)\right]\}-J\alpha(\vec{q},\vec{k})\{1-2\sigma\left[1-\Delta^{-2}(\vec{q})\right]\sin^2\theta\}-J\left[\gamma_\perp(0)-\gamma_\perp(\vec{k})\cos(k_zc)\right]\\ &\times\left[1+\Delta^{-2}(\vec{q})\right]-2J\gamma_\perp(\vec{q})\Delta^{-1}(\vec{q})+2J\gamma_\perp(\vec{q}-\vec{k})\Delta^{-1}(\vec{q})\cos(k_zc) .\end{aligned}\qquad (8)$$

The phase angle θ, which depends on the surface reflection properties of a bulk spin wave with wavevector $\vec{K}$, is defined by

$$\exp(2i\theta) = \left[\Delta(\vec{k})+\exp(ik_zc)\right]/\left[\Delta(\vec{k})+\exp(-ik_zc)\right] . \qquad (9)$$

The quantities W_{SS} and W_{SB} play an analogous role to the Dyson interaction vertex [4] in the renormalization of bulk spin waves in an infinite ferromagnet. They incorporate the statistical weighting of the various modes near the surface and the effects of the phase angle θ, and are consequently more complicated than in the infinite-medium case. Due to the Bose factors the summations in (6) are dominated by the behaviour at small $|\vec{k}|$ and $|\vec{K}|$, where W_{SS} and W_{SB} simplify. If these wavevector summations are replaced by integrals, taking care to use an appropriate density-of-states function for the semi-infinite medium [7], they can then be evaluated either analytically in various limiting cases or more generally using numerical methods.

We first discuss some analytic results for long wavelengths ($a^2q^2 << 1$) and for zero surface anisotropy (D = 0) in the case of s.c. (001) ferromagnets. If $0 < \sigma < 1$ there is a surface branch in the unrenormalised spin wave spectrum occurring just below the continuum of bulk modes, e.g. see curve S_1 in Fig. 1. The lower edge of the continuum has energy $E_B(\vec{q},0) \approx g\mu_B H_0 + SJa^2q^2 + O(q^4)$, and the surface mode with energy $E_S(\vec{q})$ is split off below this by a small amount proportional to $\sigma^2a^4q^4$. As a consequence the surface spin wave is only weakly bound to the surface in this case. On introducing reduced variables $\tau = k_BT/SJ$ and $h = g\mu_B H_0/SJ$, we obtain the following leading order contributions, assuming $a^2q^2 << \tau << 1$:

$$\Delta E_{SS}(\vec{q}) \approx -(J/8\pi)(1-8\sigma^2)\sigma a^2q^4\tau^2 F(2,h/\tau) \tag{10}$$

$$\Delta E_{SB}(\vec{q}) \approx -(J/32\pi^{3/2})a^2q^2\tau^{5/2}F(5/2,h/\tau) + O(\sigma^2q^4\tau^{3/2}) \tag{11}$$

where $F(n,h/\tau)$ is the Bose-Einstein integral function [6], which reduces to the Riemann zeta function $\zeta(n)$ if h = 0. The dominant contribution to ΔE_S in this case comes from the first term in (11) and is due to interactions with bulk spin waves. We note that this leading order contribution to ΔE_S has the same dependence on q and τ as the energy correction for a bulk spin wave with wavevector $(\vec{q},0)$ in an infinite ferromagnet [5,6]. This seems reasonable in view of the earlier comment concerning the proximity of the surface branch to the lower edge of the bulk continuum. The condition $a^2q^2 << \tau$ assumed in deriving (10) and (11) implies that the surface mode penetration depth ($\sim 1/\sigma aq^2$) is relatively large, e.g. larger than the thickness ($\sim a/\tau^{\frac{1}{2}}$) over which the magnetisation is appreciably perturbed by the presence of a surface [8]. In Fig. 2 we show some further results for s.c. (001) ferromagnets obtained by numerical integration from (6). The energy shifts are plotted as a function of k_BT/zSJ (z = number of nearest neighbours), which is of order T/T_c.

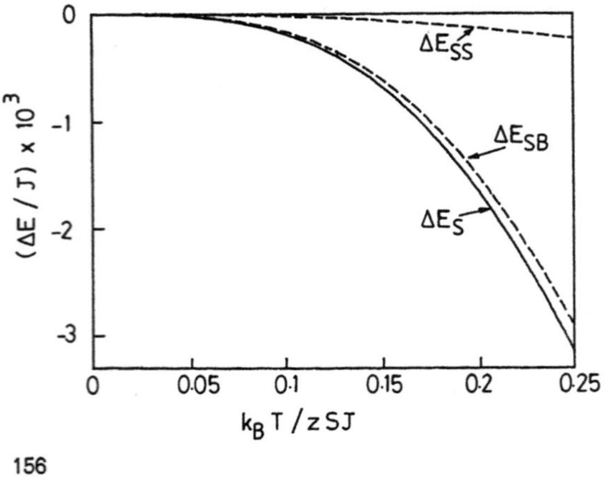

Figure 2. The calculated temperature dependences of ΔE_{SS}, ΔE_{SB} and ΔE_S for a s.c. (001) ferromagnet, taking $a\vec{q} = (0.3,0)$, h = 0.1, D = 0 and $\sigma = 0.2$

Similar calculations have been carried out analytically and numerically for f.c.c. and b.c.c. ferromagnets with (001) surfaces and D = 0. For example, in the long wavelength limit we again find that ΔE_{SB} provides the dominant contribution to the surface spin wave energy shift and that its dependence on q and τ takes the same form as in (11). However, the numerical factors and the σ-dependence are different. The results are consistent with recent calculations by MAZUR and MILLS [9] for the special case of a f.c.c. ferromagnet with a free surface ($\sigma = 0$). The behaviour for ΔE_S is modified if there is a surface anisotropy field. As an example we consider $D < 0$ and $0 < \sigma < 1$ for a s.c. (001) ferromagnet. Under these conditions there will be an acoustic surface spin wave split off below the bulk continuum, as for S_2 in Fig. 1. For anisotropy fields satisfying $\tau \ll d^2 \ll 1$ and $a^2q^2 \ll d^2$ (where $d = g\mu_B D/k_B T$) we find that $\Delta E_S(\vec{q})$ is approximately proportional to $d^3 \tau F(1, E_S(0)/k_B T)$. By contrast with the d = 0 case it comes mainly from interactions with surface waves. This difference is partly due to there being an approximate mode in the bulk spin wave amplitudes at the surface in the present case, and partly due to a relative enhancement in the number of thermally excited surface spin waves.

3 Spin Wave Damping

The damping is obtained by renormalising the spin waves to second order in $\mathcal{H}_{int}$. The mechanism is the low-temperature scattering process involving four spin waves, except that these may now be bulk modes or surface modes. Hence the damping $\Gamma_S(\vec{q})$ of a surface spin wave will be the sum of four contributions denoted by $\Gamma_S(\vec{q};4S)$, $\Gamma_S(\vec{q};3S,1B)$, $\Gamma_S(\vec{q};2S,2B)$ and $\Gamma_S(\vec{q};1S,3B)$, according to how many surface (S) and bulk (B) modes are involved. For simplicity we discuss here just the contribution $\Gamma_S(\vec{q};4S)$ which is given by

$$\Gamma_S(\vec{q};4S) = \frac{\pi}{8N^2} \sum_{\vec{k},\vec{p}} \Phi_{4S}(\vec{q},\vec{k},\vec{p})\,\delta\left[E_S(\vec{q})+E_S(\vec{k})-E_S(\vec{p})-E_S(\vec{q}+\vec{k}-\vec{p})\right]$$
$$\times\{n[E_S(\vec{k})]\left[1+n[E_S(\vec{p})]+n[E_S(\vec{q}+\vec{k}-\vec{p})]\right]-n[E_S(\vec{p})]\,n[E_S(\vec{q}+\vec{k}-\vec{p})]\} \qquad (12)$$

where Φ_{4S} is a vertex function [10]. The summations may be converted to integrals and performed analytically for certain cases. For example, in the absence of surface anisotropy (d = 0) and for $\tau \ll a^2q^2 \ll 1$ we eventually obtain for s.c. (001) ferromagnets

$$\Gamma_S(\vec{q};4S) \approx (J/32\pi S)\sigma^2 a^4 q^4 \tau^3 F(3/\tau)\,. \qquad (13)$$

This agrees with a previous calculation by TARASENKO and KHARITONOV [11] specifically for this structure and with d = 0. We find that the same dependences on q and τ are predicted for f.c.c. and b.c.c. ferromagnets. Equation (12) also applies if the surface anisotropy is non-zero; for example, if $d < 0$ such that $\tau \ll d^2 \ll 1$ and $a^2q^2 \ll d^2$ we estimate that for all three lattice structures $\Gamma_S(\vec{q};4S)$ is proportional to $d^6\tau F(1,E_S(0)/k_B T)$. The other contributions $\Gamma_S(\vec{q};3S,1B)$, $\Gamma_S(\vec{q};2S,2B)$ and $\Gamma_S(\vec{q};1S,3B)$ to the surface spin wave damping are given by analogous expressions to (12), but in general they require numerical evaluation. However, if the surface anisotropy field is sufficiently large ($d^2 \gg \tau$ and $d^2 \gg a^2q^2$) it may be shown that $\Gamma_S(\vec{q};3S,1B)$ and $\Gamma_S(\vec{q};1S,3B)$ are negligibly small.

The spin wave interaction vertices derived in this work for $T \ll T_c$ can also be obtained employing the perturbation formalism of VAKS, LARKIN and PIKIN [12], and this would provide a means of extending the results to higher temperatures. Details of the present calculations will be published elsewhere.

References

1 T. Wolfram and R.E. Dewames: Prog. Surface Science 2, 233 (1972)
2 H. Puszkarski: Prog. Surface Science 9, 191 (1979)
3 J.C.S. Lévy: Surface Science Reports 1, 39 (1981)
4 F.J. Dyson: Phys. Rev. 102, 1217 (1956)
5 T. Oguchi: Phys. Rev. 117, 117 (1960)
6 F. Keffer: Handbuch der Physik 18, 1 (1966)
7 D.L. Mills: Phys. Rev. B1, 264 (1970)
8 D.L. Mills and A.A. Maradudin: J. Phys. Chem. Solids 28, 1855 (1967)
9 P. Mazur and D.L. Mills: to be published (1984)
10 D.E. Kontos: Ph.D. Thesis, University of Essex (1983)
11 V.V. Tarasenko and V.D. Kharitonov: Sov. Phys. Solid State 16, 1031 (1974)
12 V.G. Vaks, A.I. Larkin and S.A. Pikin: Sov. Phys. JETP 26, 647 (1968)

Correlation Theory of Magnetic Excitations

P.-A. Lindgård

Risø National Laboratory, DK-400 Roskilde, Denmark

1. Abstract

The basic ideas of the correlation theory for magnetic excitations are presented. It includes self-consistently pair correlation effects and a self-consistent renormalization and damping of the excitations. Applications to the ideal Heisenberg ferromagnet EuO in the paramagnetic phase, to the singlet-doublet magnet Pr, and an antiferromagnetic quantum spin S=1 chain in a field are discussed. Central peak phenomena occur in the last two examples.

2. Introduction

In recent years magnetic excitations in many systems in which correlation effects play a significant role have been investigated experimentally. These are systems in which the long range order (LRO) is weak or absent, so only short range order (SRO) persists, for example, excitations in the paramagnetic phase of 3-d Heisenberg magnets, e.g., EuO [1] or in the low dimensional magnets, e.g., TMMC [2] or in systems with competing interactions, such as singlet ground state systems like Pr [3] and the low dimensional $CsNiF_3$ [4] and $CsFeCl_3$ [5]. Another class of systems which have so far not been fully covered experimentally is 2-d quantum antiferromagnets in which frustration effects might be expected. The competition between antiferromagnetic interactions and an external magnetic field for low dimensional chains or planes may give rise to a number of interesting phenomena. This class of systems is not yet fully experimentally investigated, either.

On the theoretical side, there exist several detailed investigations of particular models which more or less represent the experimental systems. Especially for the 1-d chain, the classical case ($S=\infty$) [6] and the extreme quantum case ($S=1/2$) [7] have received considerable attention; however, often only the continuum limit is considered. The main motivation for studing the 1-d systems is perhaps not to just understand these in their own right, but to use them as test systems on how to deal with strong correlation effects in order to be able to theoretically treat more common and realistic systems. Unfortunately, many of the theoretical model calculations do not seem possible to generalize to quantum spins $1/2 < S < \infty$ or to two and three dimensions. The random phase approximation theory [8] is such a general theory, but it fails in the present cases because it is designed to cover the LRO case with negligible correlation effects.

For this reason the correlation theory [9] was developed in which pair correlation effects are self-consistently included and the dynamics is treated in a realistic but simple and self-consistent way. Because SRO-effects are included, the theory gives results which depend on the lattice dimensionality d, the spin dimensionality n, and the spin length S. Furthermore, it can easily be applied to any lattice structure and realistic Hamiltonian. Quantum effects and the discreteness of the lattice are exactly included, as well as a good deal of the non-linear effects. However, it may not treat too well the extreme non-linearities as described by narrow wall solitons. Since this paper was intended to be a short review, no details will be given, but the main idea and the results of a few applications of the correlation theory will be discussed.

3. Correlation Theory: Idea and Examples

First, let us emphasize that the RPA theory is tremendously successful in explaining experimental data when correlation effects are small or when (LRO) is dominant. It would be good to preserve this fundament as a basis for more advanced theories. (LRO) is here taken in a generalized sense as when all spins on the average are in a particular state, i.e., not only a maximum spin state $|S\rangle$, but for example, also a non-magnetic singlet state $|0\rangle$. The physics of the random phase approximation can be described as follows. We assume a spin S_R and its neighbors at $R+\delta$ are participating in a certain spin wave mode with wave vector and frequency q and ω_q. If the neighbor spins participate in other spin waves, one assumes that these have an arbitrary phase, wave vector k, and frequency $\omega_{k\neq q}$ which average out in the ensemble average for S_R. Therefore, to a first approximation, S_R only sees the local average of its neighbor spins $\langle S_{R+\delta}\rangle = \langle S^z\rangle_{local}$. A diminished (LRO) means that there are many spin wave modes excited in the system and this means that the neighbor spins $S_{R+\delta}$ commonly participate in spin wave modes ω_k other than the ω_q we were considering for S_R (in each member of the ensemble). The RPA effective medium approximation is no longer valid because S_R now sees a time dependent molecular field and there is a coupling between ω_k and ω_q, which does not random out in the ensemble average.

The first step beyond the RPA is therefore to consider the effect of coupling two modes self-consistently. The self-consistent modes are not harmonic oscillation modes, but to a first approximation damped or overdamped oscillations and diffusion modes. This approximation we call the correlation theory [9]. It has the virtues of the RPA theory of being quite simple and applicable to many systems, now also disordered systems, because (SRO) effects are taken into account by including self-consistency of both static and dynamic properties. In the traditional application of the mode-mode coupling theory, as advocated in particular by KAWASAKI [10], LOVESEY et al. [11] and discussed by DE RAEDT et al. [12], the dynamical variables are usually taken to be the operators (and derivatives thereof) which in the hydrodynamic limit satisfy conservation laws. The rationale is that these modes should be slowly varying with time and therefore, dominating the dynamical behavior. For the antiferromagnet these variables are the magnetization density M_q and the energy density E_q. A coupling between these should occur for finite external fields. Clearly this basis is designed to treat the long wavelength limit $q \to 0$ and $\omega \to 0$. However, this limit is not easily observable by either neutron scattering or in numerical simulation studies. In the correlation theory [9], an alternative set of dynamical variables is chosen, namely dynamical variables (and derivatives thereof), which allow the local or short range properties to be calculated exactly. This provides a description of the normal modes at high q and ω. It is, therefore, a theory designed to obtain properties which can be tested by the above mentioned measurements. This basis turns out to be the same as is convenient for the description of the ordered phase. Consequently, all temperatures can be treated in the same framework. The correlation theory does not use the hydrodynamic concept energy modes, but gives a rather simple picture of the dynamics in terms of correlation functions. For the antiferromagnet the interplay between these on different sublattices is important. If no approximations were made, the choice of dynamical variables is only a matter of taste and convenience. However since approximations are necessary, the most physical basis (i.e., the one giving the best non-interacting normal modes) should be the most adequate.

The Heisenberg paramagnet is the simplest example to consider because we need only one dynamical variable S^x_q. The lineshape function is then [9] the damped harmonic oscillator lineshape

$$S^{xx}(q,\omega) = \frac{1}{\pi} \chi^{xx}_q \frac{1}{1-e^{-\omega/kT}} \frac{2\beta_q\langle\omega_q^2\rangle}{(\omega^2-\langle\omega_q^2\rangle)^2+4\beta_q^2\omega^2} . \tag{1}$$

This is also called the two-pole-approximation because it represents two complex poles at $\omega = \pm\,\alpha_q + i\beta_q$ such that $\alpha_q^2 + \beta_q^2 = \langle\omega_q^2\rangle$. When the damping β_q reduces to zero, $\alpha_q = \omega_q$, the first moment frequency; for finite β_q, one has $\alpha_q \neq \omega_q$. The peak position for (1) is not even at α_q for finite β_q; it goes to zero already for $\alpha_q = \beta_q$. This is due to overdamping of the oscillations. Knowledge about $\langle\omega_q\rangle$ and $\langle\omega_q^2\rangle$ is not sufficient to determine α_q and β_q separately unless their ratio is assumed fixed, which may be quite reasonable in some cases. In general, the damping β_q is determined by the mode-mode decoupling using the Mori theory [13] for the relaxation function. For a general dynamical vector $\underline{A}$, it reads

$$\underline{\underline{\chi}}(q,z) = (\underline{A}|\underline{A}^+)_{q,z} = \underline{\underline{\chi}}_q \frac{1}{\underline{\underline{z}}+i\langle\underline{\underline{\omega}}_q\rangle+(\underline{X}_1|\underline{X}_1^+)_{q,z}} \tag{2}$$

where $z = i\omega$ and the random force $X_1 = \dot{A}(t) - i\langle\omega_q\rangle A(t)$ is the force acting in a rotating coordinate system following the spin wave frequency. For the Heisenberg ferromagnet, $(X_1|X_1^+)$ involves terms of the kind

$$(S^x_k(t)S^y_{q-k}(t)|S^x_{-k}(0)S^y_{k-q}(0)) \sim (S^x_k(t)|S^x_{-k}(0))(S^y_{q-k}(t)|S^y_{k-q}(0)) \tag{3}$$

which is decoupled in the mode-mode coupling approximation. Here each relaxation function $(S^\alpha(t)|S^\alpha)$ is expressed in the two-pole approximations and the equations are closed by inserting back into (2). The damping parameter β_q is expanded in a Fourier series, the coefficients of which are determined self-consistently.

It is important to treat both the static and dynamic properties at the same level of accuracy. The t=0 limit of (3) is related [9] to the second moment $\langle\underline{\underline{\omega}}_q^2\rangle$, which can also be exactly written as $\langle\underline{\underline{\omega}}_q^2\rangle = \langle[\dot{\underline{A}},\underline{A}^+]\rangle/\underline{\underline{\chi}}_q$. From this the susceptibility $\underline{\underline{\chi}}_q$ can be found at the same level of approximation. The theory for one dynamical variable is closely related to Hubbard's theory [14]; however, in [14] dynamical self-consistency was not attempted and only nearest neighbor interactions were considered.

When calculating the static and dynamic properties by the correlation theory for the ideal Heisenberg magnet, EuO, good agreement was found [9] for $T > 1.02\ T_C$ with the considerable amount of available experimental data using no adjustable parameters. A few examples are shown in Fig. 1. No central peak is expected near T_C in this case. The sloppy spin waves persist even when the static correlation functions vanish at $T = \infty$. The criterion

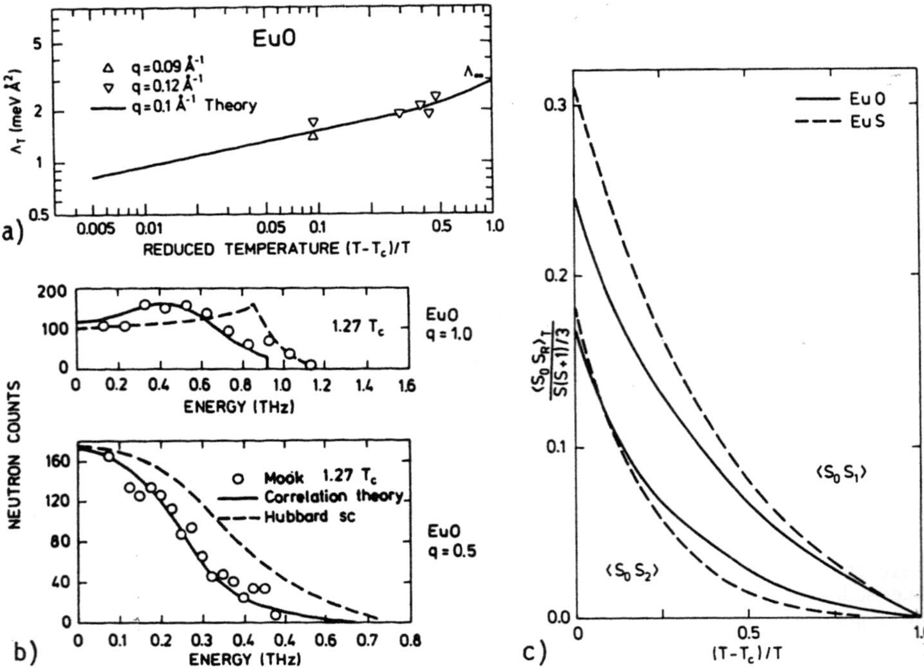

Fig. 1. a) Spin diffusion constant Λ_T. b) One half and zone boundary lineshapes showing sloppy spin waves for $T > T_C$. c) Calculated nn and nnn correlations (from [9])

for their existence is not the static one, $q > \kappa$, the inverse correlation length, but involves the relaxation time for a correlated region τ_κ such that the spin wave frequency $\omega_q > 1/\tau_\kappa$.

The next step is to consider the case with several dynamical variables. For the S=1 singlet doublet model, there are nine different local transition operators. However, one need only consider two coupled operators at a time, say S^x_q and $L^x_q = (S^y_q\, S^z_q + S^z_q\, S^y_q)$. Solving for these (and their derivatives) gives in the two-pole approximation a transverse susceptibility function with four complex poles, such that the two at $\pm\Omega_q + i\Gamma_q$ are renormalized to a frequency higher than the RPA frequency ω_q, $\Omega_q > \omega_q$ and the other two complex polés are at low frequencies giving a central peak response. See Fig. 2a. The central peak corresponds to spin diffusion in regions with locally induced magnetism in the surrounding singlet ground state matrix. The intensity of the central peak is approximately proportional to $\chi_q/\Omega_q^2 \sim \omega_q^{-4}$ and the width to $1/\chi_q \sim \omega_q^2$. This low frequency response is due to both quantum and non-linear effects, and is expected for any lattice dimension, with a strength proportional to the correlation effects. Qualitative agreement was obtained [15] with Pr (see Fig. 2b). A detailed comparison for $CsNiF_3$ and $CsFeCl_3$ is in progress [16].

A similar situation as discussed for the singlet ground state system occurs for a Heisenberg antiferromagnet in a magnetic field. Solving the

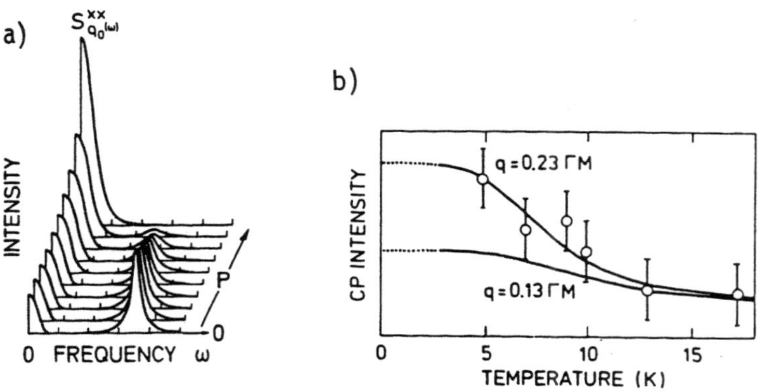

Fig. 2. a) Pressure dependence of $S^{xx}(q_0,\omega)$ for a singlet doublet model, showing a soft mode and a central peak. b) Temperature dependence of the central peak in Pr compared with measurements [3] (from [15])

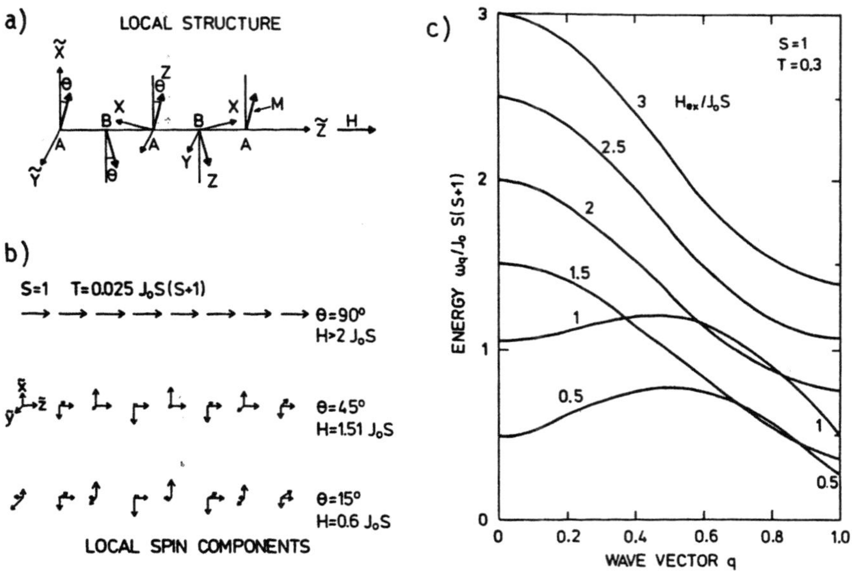

Fig. 3. a) Classical structure for AFM chain in a field H. b) Calculated complex cone structures for $H < H_c$ (classical) = $2J_0S$. c) First moment frequency for a AFM S = 1 quantum chain

local properties in a canted two sublattice model (A,B) gives six different dynamical operators (see Fig. 3a). However, again one need only consider two coupled operators at a time, say $S^x_{qA} \pm S^x_{qB}$ and $S^y_{qA} \pm S^y_{qB}$, corresponding to the local uniform and staggered x,y variables. As above, the local transverse susceptibility $\chi^{xx}(q,z)$ will show upwards renormalized spin wave poles and low frequency poles of non-linear origin. In the original coordinate system both the transverse and the longitudinal $S^{\alpha\alpha}(q,\omega)$ will, in addition, have a central peak contribution from the local longitudinal $\chi^{zz}(q,z)$. This contribution one may call a linear contribution, since it originates just from the canting effect. For the 1-d chain and S=∞ the correlation theory reproduces the classical, which is the same as the RPA result. For lower spin values, interesting effects occur [17]. The canting angle changes abruptly from $\theta = 90°$ at high fields to $\theta < 90°$ at a critical field H_C which is smaller than the classical critical field. In the canted regime several phase changes occur between complex cone structures with incommensurate spiral vectors. The local structure deduced from the calculated pair correlation functions is shown schematically on Fig.3 b. Contrary to the classical chain, which has a gapless spin wave spectrum in the canted regime, the quantum $S > 1$ chain shows a field dependent gap for $H < H_C$ (see Fig.3c). It may be of interest to study experimentally low dimensional quantum spin systems for various small spin values. According to the present study, the same type of phenomena happens for small spin values $S > 1/2$ as happens for the extreme quantum case $S = 1/2$. It is well known that there exist many more realizations of $S > 1/2$ systems.

4. References

1. H.A. Mook: Phys. Rev. Lett. 46, 508 (1981); L. Passell, O.W. Dietrich and J. Als-Nielsen: Phys. Rev. B 14, 4897-4923 (1976)
2. I.U. Heilmann, J.K. Kjems, Y. Endoh, G. Reiter, G. Shirane and R.J. Birgeneau: Phys. Rev. B 24, 3939 (1981)
3. S.K. Burke, W.G. Stirling and K.A. McEwen: J. Phys. C L976 (1981); K.A. McEwen and W.G. Stirling: J. Mag. Magn. Mat. 30, 99 (1982)
4. M. Steiner, K. Kakurai and J.K. Kjems: Z. Phys. B 53, 117 (1983)
5. M. Steiner, K. Kakurai, W. Knop, B. Dorner, R. Pynn, U. Kappek, P. Day, and G. McLeen: Solid State. Commun. 38, 1179 (1981)
6. H.J. Mikeska (this volume and references therein). See also Physics in 1-Dimension, ed. J. Bernasconi and T. Schneider (Springer Verlag) (1981)
7. J. des Cloizeaux and J.J. Pearson: Phys. Rev. 128, 213 (1962); N. Ishimura and H. Shiba: Progr. Theor. Phys. 57, 6 (1980), 64, 479; J.C. Bonner (this volume and references therein)
8. D.N. Zubarev: Usp. Fiz. Nauk 71, 71 (1960) [Sov. Phys. Usp. 3, 320 (1960)]
9. P.-A. Lindgård: J. Appl. Phys. 53, 1982 and Phys. Rev. B 27, 2980 (1983)
10. K. Kawasaki: Phase Transitions and Critical Phenomena, ed. C. Domb and M.S. Green (Academic Press), (1976) p. 166
11. S.W. Lovesey and J.M. Loveluck: J. Phys. C 12, 4015 (1979)
12. H. de Raedt and B. de Raedt: Phys. Rev. B 21, 304 (1980) and Phys. Rev. B 23, 4597 (1981)
13. M. Mori: Prog. Theor. Phys. 33, 432 and 39, 399 (1965)
14. J. Hubbard: J. Phys. C 4, 53 (1971)
15. P.-A. Lindgård: Phys. Rev. Lett. 50, 630 (1983) and J. Mag. Magn. Mat. 31-39, 1099 (1983)
16. P.-A. Lindgård: (to be published)
17. P.-A. Lindgård: (to be published)

A Rescaling Approach to Excitations in Magnetic Systems

B.W. Southern and P.D. Loly

Department of Physics, University of Manitoba
Winnipeg, Manitoba, R3T 2N2, Canada

1. Introduction

Real-space rescaling methods have been quite successful in not only describing the singularities which occur in thermodynamic quantities at a phase transition but also in determining detailed phase diagrams of magnetic systems. Recently, these scaling methods have also been employed to obtain information about the spectral properties of Green's functions (AOKI[1-3], Goncalves da Silva and Koiller[4]). Since many of the properties of crystalline, imperfect, and disordered magnets are conveniently calculated using Green's functions, these methods can also be useful in obtaining information about the excitation spectrum of these systems. We have developed an exact rescaling transformation for systems described by Hamiltonians in the tight binding form with arbitrary range of interaction in any dimension. The transformation contains all the information about the nature of the excitation spectrum and can be used to calculate the lattice Green's functions. The basic idea of the method is to take a system of equations which describe N degrees of freedom and perform a transformation on the system which reduces the number of degrees of freedom to a fraction of N but leaves the equations invariant in form with modified (renormalized) parameters. The relationship between the renormalized parameters and the original set is used to define a scaling transformation on the equations of motion. This procedure can then be repeated indefinitely by simply iterating the transformation many times until there is only one degree of freedom left. In this limit the problem can be trivially solved in terms of the limiting values of the renormalized parameters.

Consider a nearest-neighbour linear chain whose spectral properties are obtained by solving the following set of equations for the lattice Green's functions G(i,j):

$$(E-\varepsilon)G(i,j)= \delta_{i,j} + V\,G(i+1,j) + V\,G(i-1,j). \quad (1)$$

If we divide the sites into two groups which we simply label as "even" or "odd", then we can eliminate the "odd" sites from (1) and the equations for the "even" sites can be written in the same form as (1) but with the following renormalized parameters:

$$V'=V^2/(E-\varepsilon)\ ,\ (E-\varepsilon')=(E-\varepsilon)-2V'\ . \quad (2)$$

In terms of the reduced variable $Z=(E-\varepsilon)/V$, the transformation becomes

$$Z'=Z^2-2\ . \quad (3)$$

The properties of this map in the complex(energy) plane provide information about the excitation spectrum. If Z is real and $|Z|>2$, then Z′ goes to infinity under iteration which corresponds to V′ approaching zero. In this limit, the Green′s function G(0,0) approaches a limiting value given by $1/(E-\varepsilon^{\infty})$. However, for Z real and $|Z|<2$, the map is chaotic under iteration and G(0,0) does not converge. These two regions of energy are separated by a fixed point Z*=2 which corresponds to the band edge. The eigenvalue obtained by linearizing (3) about this point gives information about the singularities of the Green′s functions at the band edge. For complex values of Z, the imaginary part acts as a convergence factor with Z′ going to infinity and G(0,0) approaching a limiting complex value. Figure 1a shows the region of the complex plane which does not iterate to infinity and represents the distribution of zeroes of the Green′s functions while Fig.1b shows the calculated results for G(0,0).

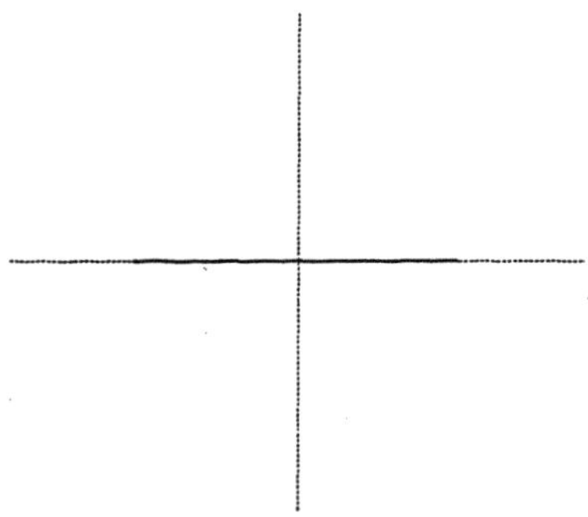

Fig.1a Complex Z-plane

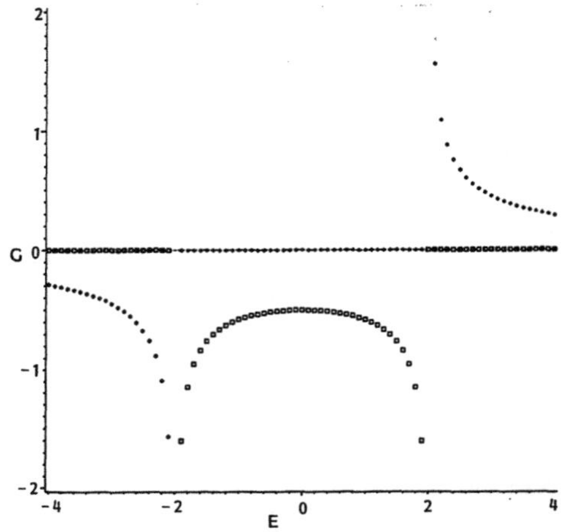

Fig.1b G(0,0) for 1-d chain

We have previously used this method to study the spectral properties of 1-d systems (with arbitrary range of interaction) which were either ordered or disordered as well as hierarchical and fractal lattices (SOUTHERN et al.[5,6] , LANGLOIS et al.[7] , TREMBLAY and SOUTHERN[8]). In all cases the initial range of coupling does not increase under the rescaling transformation. However, for conventional lattices in higher dimensions, even if the couplings are restricted to nearest neighbours initially, each successive elimination causes the range of interaction to increase. For this reason it is necessary to construct a transformation which is valid for arbitrary range of the couplings. We have derived a family of transformations which are exact for arbitrary range on any lattice.

In the study of critical phenomena, exact transformations are rare and one usually replaces the map defined in the whole space of Hamiltonians by an approximate map in a subspace of finite dimension (NIEMEIJER and van LEEUWEN [9]). Our exact transformations are infinite dimensional maps but in obtaining the results in the next section we have truncated the exact map to allow only interactions up to some maximum range and we set all further interactions equal to zero. However, as we shall demonstrate, the truncated map is adequate to determine the different regions of energy (i.e. phases) and obtain accurate values for the exponents which characterize the singularities of the Green′s functions at the boundaries between these regions.

2. Square Lattice

Consider the square lattice shown in figure 2. In order to calculate the Green's function G(r,s) we choose the site r as our origin and the remaining sites can be labelled with two integers l,m which specify the x and y coordinates in units of the lattice spacing a. The sites can be partitioned into two equal groups according to whether the sum (l+m) is even or odd. Both the 'even' and 'odd' groups form a 2-d square lattice which is rotated by 45^0 with respect to the original lattice and has a new lattice spacing $a'=(2)^{\frac{1}{2}}a$. The transformation can be written as follows:

$$\tilde{X}(p,q)= \{2X(p-q,p+q)+ \sum_{l,m} (-1)^{l+m+1}X(p-q-l,p+q-m)X(l,m)\}/D$$

$$\tilde{\alpha}(p,q)= \{ \alpha(p-q,p+q) + \sum_{l,m} (-1)^{l+m+1} \alpha(p-q-l,p+q-m)X(l,m)\}/D$$

$$D= 1 - \sum_{l,m} (-1)^{l+m+1}X(-l,-m)X(l,m) \tag{4}$$

where p and q label the sites on the new lattice in units of a'. The parameters $X(i,j)=V(i,j)/(E-\varepsilon)$ and $\alpha(i,j)$ is the inhomogeneous term in the equation for G(i,j).

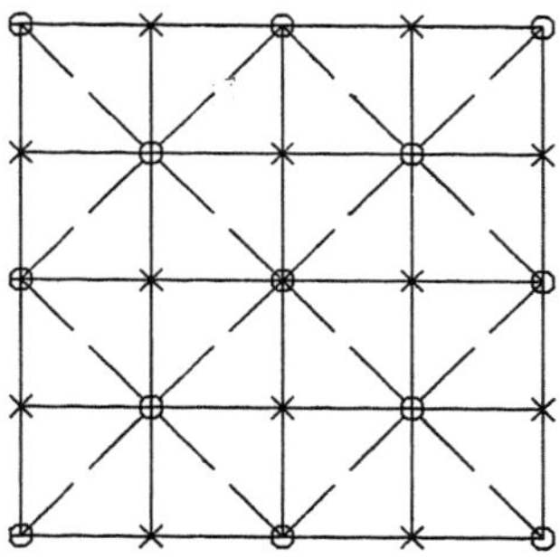

Fig.2 The Square lattice. The odd sites (X) are eliminated and the even sites(O) are relabelled

The transformation above is an exact transformation which can be used to calculate any Green's function G(r,s) for any initial range of the interactions. For real values of E outside the band of extended states, the iteration process converges rapidly. Although the interaction range initially increases, the X(p,q) soon decrease to zero in a monotonic fashion and $\alpha(r,s)$ reaches a limiting value. In consequence it is realistic to use a finite sized cluster. On the other hand, for real values of E inside the band, the behaviour of the transformation is chaotic under iteration. The range of the interactions increases indefinitely and $\alpha(r,s)$ does not reach a limiting value. Convergence is only obtained by using an imaginary part in the energy which acts as an attenuation factor. The significant range of the interactions now depends upon the size of the attentuation factor and thereby defines the size of cluster that must be used.

3. Finite Cluster Approximations and Complex Maps

Consider the transformation for the square lattice in (4). If we restrict the range to first neighbours only, then we have a one-dimensional map which can be written in the form

$$Z' = Z^2 - 1 \tag{5}$$

with the change of variable: $Z=\{2X(1,0)\}^{-1}=(E-\varepsilon)/(2V)$. This map is a member of a family of maps that have been studied by many authors both on the real line and in the complex plane. Julia[10] gave a complete treatment of one-dimensional complex iterative maps of this form. He discovered the existence of an exceptional set, the Julia set , which is the closure of the set of all unstable periodic cycles. This set is a separatrix which separates the attracting domains of the stable fixed points or cycles. More recently, Mandelbrot[11-13] has studied these maps in the complex plane and showed that the Julia sets have a fractal structure. The map (5) has a stable fixed point at infinite Z and an attracting 2-point cycle Z={0,-1} on the real axis. Figure 3 shows the set of points in the complex plane which do not iterate to the infinite energy fixed point. The Julia set is the boundary curve (plus its preimages) which separates the initial points which are either attracted to infinity or to the stable cycle on the real axis. In our case, the Julia set represents the region in the complex plane where the Green's functions are non-analytic.

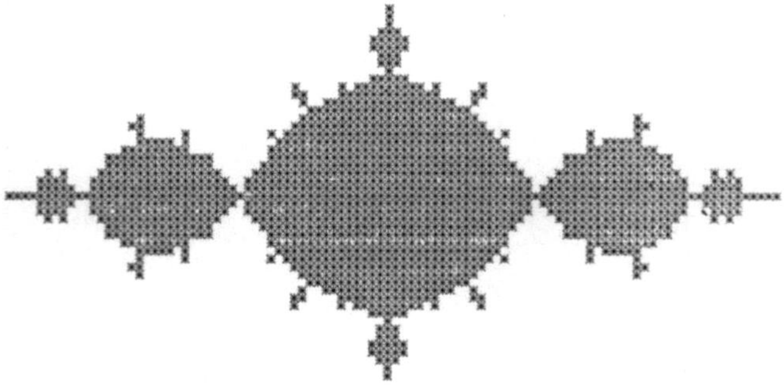

Fig. 3 Set of points not mapped to infinity by (5)

If we allow the range of interactions to extend to ±N along the x and y axes, then for N=1 we have only first and second neighbours. The map is a two parameter map:

$$\tilde{X}(1,0)= \{2X(1,1)+ 2X(1,0)^2\}/\{1-4X(1,0)^2+4X(1,1)^2\}$$

$$\tilde{X}(1,1)= \{ X(1,0)^2 - 2X(1,1)^2\}/\{1-4X(1,0)^2+4X(1,1)^2\}. \tag{6}$$

Note that even if X(1,1)=0 initially, it becomes non-zero under iteration. The truncation of the exact transformation prevents the range from increasing further. Figure 4 shows the region of the complex plane which is not mapped to infinity for increasing values of N. As the truncation procedure is improved, the region collapses to a line on the real axis which is the branch cut singularity of the exact Green's functions.

For our procedure to converge, the initial values of the complex energy must lie outside these regions and thus there is a minimum value of the imaginary part required. In calculations of the Green's functions

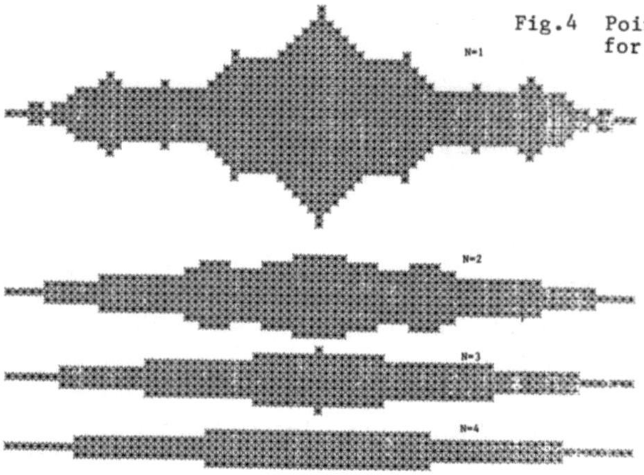

Fig.4 Points not mapped to infinity for N=1,2,3,4

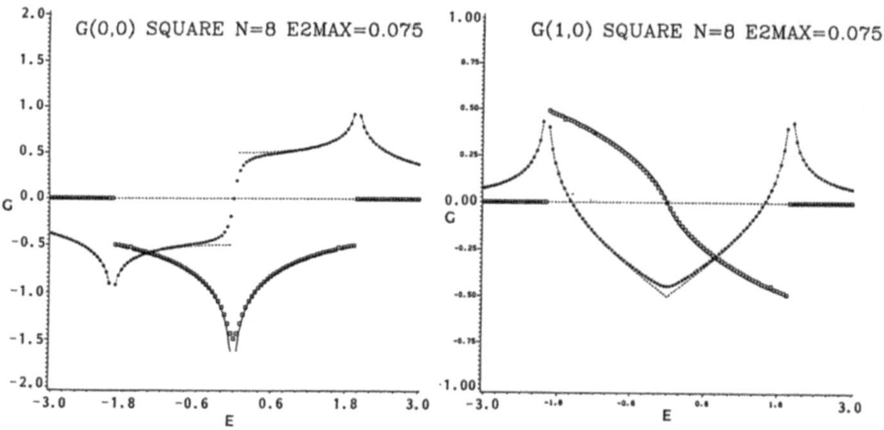

Fig.5 G(0,0) and G(1,0) for the nearest neighbour square lattice

we find that our best results are obtained when the complex part of E is close to the fractal curve. Figure 5 shows our results for G(0,0) and G(1,0) in the case of the nearest neighbour square lattice for N=8. Outside the band, 10 iterations are typically required to obtain an accuracy of 10^{-6}. Inside the band, 30 iterations are required when the imaginary part increases linearly from the band edges to a maximum value of 0.075 at the band center. The transformation (4) has fixed points which describe the singular behaviour of the Green's functions at the band edges. The exact exponents are obtained for N>3 since only interactions out to fifth neighbours are significant.

We have derived exact transformations for the triangular and honeycomb lattices and in both cases the results are in excellent agreement with known analytic expressions. The transformation can be generalized to any dimension and used to calculate the Green's functions of all three cubic lattices. In each case, the fixed points of the transformation give exact exponents for the singular behaviour of the Green's functions at the band edges.

References

1. Aoki H. : Solid St.Commun. 31 999-1002 (1979)
2. Aoki H. : J. Phys. C13 3369-3386 (1980)
3. Aoki H. : Physica 114A 538-542 (1982)
4. Goncalves da Silva C.E.T. and Koiller B. : Solid St. Comm. 40 215-219 (1981)
5. Southern B.W., Kumar A.A., Loly P.D. and Tremblay A.-M.S. : Phys. Rev. B27 1405-1408 (1983)
6. Southern B.W., Kumar A.A. and Ashraff J.A. : Phys. Rev. B28 1785-1791 (1983)
7. Langlois J.-M., Tremblay A.-M.S. and Southern B.W. : Phys. Rev. B28 218-231 (1983)
8. Tremblay A.-M.S. and Southern B.W. : J. de Physique Lett. 44 L843-L852 (1983)
9. Niemeijer Th. and van Leeuwen J.M.J. : in Phase Transitions and Critical Phenomena, ed. by C.Domb and M.S.Green (Academic,London) vol 6 (1976)
10. Julia G. : J.de Math .Pure et Appl. 4 47-245 (1918)
11. Mandelbrot B.B. : in Nonlinear Dynamics,R.H.G.Helleman, ed., Ann.N.Y.Acad.Sci. 357 ,249-259 (1980)
12. Mandelbrot B.B. : The Fractal Geometry of Nature (Freeman,San Francisco) (1982)
13. Mandelbrot B.B. : Physica 7D 224-239 (1983)

Part VI

Novel Magnetic Models and Systems

Monte Carlo Study of the 2-Dimensional Spin-1/2 *XY* model

H. De Raedt, B. De Raedt, and J. Fivez

Physics Department, University of Antwerp, Universiteitsplein 1
B-2610 Wilrijk, Belgium

A. Lagendijk

Natuurkundig Laboratorium, University of Amsterdam, Valckenierstraat 65
NL-1018 XE Amsterdam, The Netherlands

The generalized Trotter formula is used to map the two-dimensional spin-1/2 XY model onto a three-dimensional Ising model with complicated many-spin interactions. This relationship is used to construct an efficient Monte Carlo algorithm. Simulation data give strong evidence for the existence of a phase transition.

Although it is well-known that there cannot exist long-range order in the 2D(imensional) spin-1/2 XY model at non-zero temperature [1], high-temperature series expansions [2] suggest the existence of critical behavior. Little is known about the quantum model mainly because most techniques which have been succesfully applied to the classical model (also referred to as planar rotator model) are very difficult to extend to the quantum case. In particular, quantum renormalization group analysis of the 2D spin-1/2 XY model have given either contradictory or unreliable results.[3 - 8]

An alternative method to calculate the thermodynamic properties of this model is to exploit the formal analogy between d-dimensional quantum spin systems and (d+1)-dimensional Ising-spin models.[9] Using the generalized Trotter formula [9], Suzuki showed that $Z = \mathrm{Tr}\exp(-H) = \lim_{m\to\infty} Z_m$ where

$$Z_m \equiv \mathrm{Tr}\left[\prod_{<ij>} \exp\left(-\frac{\beta}{m} H_{i,j}\right)\right]^m , \qquad (1)$$

is the m-th approximation to the partition function Z, $H_{i,j}=-J(\sigma_i^x\sigma_j^x+\sigma_i^y\sigma_j^y)$, σ_i is the Pauli spin-operator at site i and the product in (1) runs over some ordered set of nearest neighbor bonds of the square lattice. The basic idea of our approach is to calculate the thermodynamic properties of the approximation (1) and to study the convergence of the results as a function of m. Two of us have already given a rigorous solution of the simplest approximation Z_1 for a large class of orderings and it was shown that in this approximation there is a phase transition of unconventional nature (i.e. no spontaneous magnetization but a logarithmic divergence of the specific heat). [10] A disadvantage of the special class of exactly solvable m=1 approximations is that the construction [10] that maps them onto staggered 8-vertex models [11] is difficult to generalize to m>1 and, more importantly, it is very difficult and inefficient to perform Monte Carlo simulations with this type of representation. An extensive discussion of the fundamental and technical problems that arise in simulations of 1D spin-1/2 models (XY and Heisenberg models) can be found in ref.[12].

For the d-dimensional XY model we have found representations of (1) that do not suffer from the problems encountered in simulations of 1D spin-1/2 systems. [12] First we note that (1) remains invariant if we rotate all spins over $\pi/2$ about the x-axis. Working with the XZ instead of XY model solves the problem of constructing algorithms that generate ergodic Markov chains. As this step is essential for the construction of an efficient algorithm, it is clear that our approach for the XY model cannot be used to study the isotropic Heisenberg system. We partition the Hamiltonian in two parts $H = A+B$ where A contains all horizontal and B all vertical bonds. Clearly A(B) is a sum of non-interacting XZ chains. We obtain an expression of the form (1) by decomposing A and B into blocks of two sites. Having specified the ordering in (1) the next step is to introduce complete sets of states between all the exponential factors in (1). As each spin interacts with four neighbors we need $4mN^2$ (N is the linear size of the system) Ising-spin variables. To proceed further we calculate the explicit form of the matrix elements of the two-site factors

$$<S_iS_j|\exp[K_m(\sigma_i^x\sigma_j^x + \sigma_i^z\sigma_j^z)]|\sigma_i\sigma_j> =$$

$$\delta_{S_iS_j,\sigma_i\sigma_j}\left(\tfrac{1}{2}\sinh 2K_m\right)^{1/2}\exp(\bar{K}_m S_i\sigma_j + K_m S_i S_j) \ , \qquad (2)$$

where $K_m = \beta J/m$, $\bar{K}_m = \frac{1}{2}\ln\coth K_m$, $\sigma_i^z|S_i> = S_i|S_i>$ and $\sigma_i^z|\sigma_i> = \sigma_i|\sigma_i>$. Finally we sum out $2mN^2$ of the Ising spins analytically. [13] In our simulations we have used two different decompositions for the XZ chains; Suzuki's [9] real-space and Barma and Shastry's [14] checkerboard break-up. For m=1 we can show that none of these two choices is in the class of orderings that lead to the rigorous m=1 solution. [10] In the case of the real-space decomposition the final expression for Z_m reads

$$Z_m = c(\sinh 2K_m)^{mN^2} \sum_{\{S_{ij}^k\}}{}' \sum_{\{\sigma_{ij}^k\}}{}' \prod_{k=1}^{m}\prod_{j=1}^{N}\prod_{i=1}^{N} h(j,k)v(i,k) \ , \qquad (3)$$

where for periodic boundary conditions

$$h(j,k) = \cosh[\bar{K}_m(S_{1,j}^{(k)}\sigma_{1,j}^{(k)} + \dots + S_{1,j}^{(k)}\dots S_{N,j}^{(k)}\sigma_{1,j}^{(k)}\dots\sigma_{N,j}^{(k)})$$

$$+ K_m(S_{1,j}^{(k)}S_{2,j}^{(k)}S_{3,j}^{(k)}\sigma_{1,j}^{(k)} + \dots + S_{1,j}^{(k)}\dots S_{N,j}^{(k)}\sigma_{1,j}^{(k)}\dots\sigma_{N-2,j}^{(k)}]$$

$$\times \exp[K_m(S_{1,j}^{(k)}S_{2,j}^{(k)} + S_{1,j}^{(k)}\dots S_{N,j}^{(k)}\sigma_{2,j}^{(k)}\dots\sigma_{N-1,j}^{(k)}] \ , \qquad (4)$$

c is an unimportant numerical constant and v(i,k) can be obtained from h(j,k) by substituting $S_{1,j}^{(k)} \to \sigma_{i,1}^{(k)}$ and $\sigma_{1,j}^{(k)} \to S_{i,1}^{(k+1)}$. The 3D lattice model (3-4) has complicated many-spin interactions and a coupling that depends on the lattice size in the new direction (labeled by the superscript k). The checkerboard representation for Z_m looks similar to (3-4). As for the 1D case [13] we find that the real-space decomposition converges faster than the checkerboard break-up and therefore we have used the real-space representation for most of our simulation work. The prime on the summation symbols in (3) indicate that the sums over the Ising spin variables are restricted by the 2N constraints $S_{1,j}^{(k)}\dots S_{N,j}^{(k)} = \sigma_{1,j}^{(k)}\dots\sigma_{N,j}^{(k)}$ and $\sigma_{i,1}^{(k)}\dots\sigma_{i,N}^{(k)} = S_{i,1}^{(k+1)}\dots S_{i,N}^{(k+1)}$. We have not been able to remove all constraints originating from conservation laws but we have reduced the number of constraints so drastically that they can be handled extremely efficiently. It is straightforward to write down the explicit form of any kind of correlation function in terms of functions of Ising spins. To check the algorithm we studied 3 x 3 systems as a function of m and found excellent agreement with results obtained by summing all terms in (3) and by diagonalizing the full Hamiltonian. At high temperatures ($T \geqslant 4$), our simulation results for the energy and specific heat are in very good agreement with those obtained from high-T series. [3] All data presented in this paper have been obtained from at least two independent runs of 10000 Monte Carlo steps per Ising spin each. Simulation of a N = 16, m = 8 (16 x 16 x 8, 4096 Ising spins) system takes 90 min. of CPU-time on a CDC 170/750.

Guided by the rigorous m = 1 result (for a different ordering in (1)!) that there is a phase transition at $\sinh 2K^c_f = 1$, we first perform simulations for m = 1 for systems of different size. In Fig. 1 we depict simulation data for the m = 1 approximant (C_1) to the specific heat. At $K \approx K^c$ the specific heat exhibits a maximum that grows slowly with the lattice size. This is in concert with the exact m = 1 solution which predicts a logarithmic divergence of the specific heat. [9] From our m = 1 simulation data we conclude that the phase transition in the m = 1 approximation is not the result of choosing a particular ordering. The first attempt to simulate the 2D XY model was reported by Suzuki et al. [16] who only studied the m = 1 case. Their simulation data for the specific heat disagree with ours because their algorithm cannot generate ergodic Markov chains of spin states.

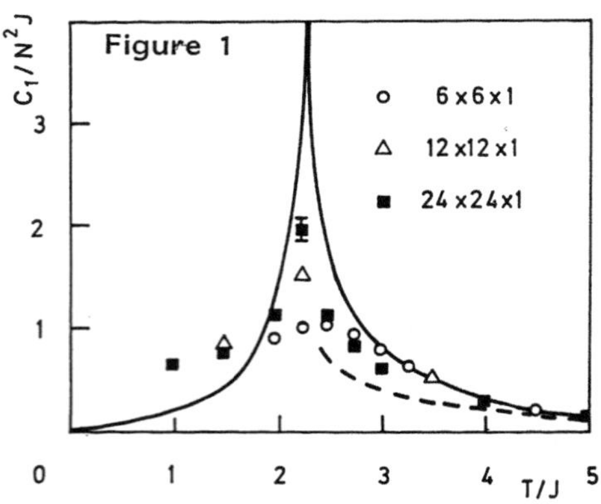

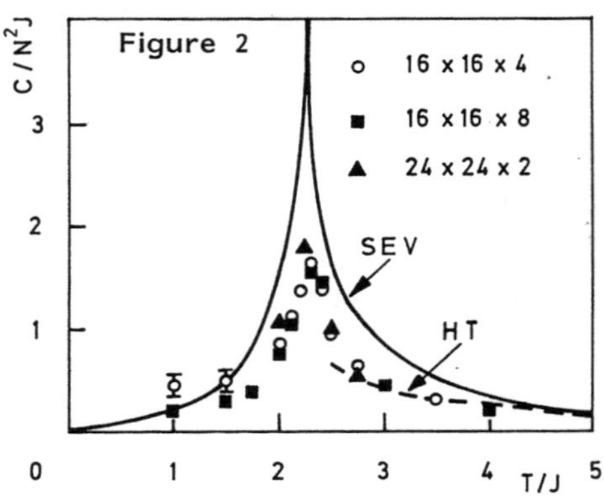

Fig. 1. Specific heat per site of the 2D spin-1/2 XY model in the m = 1 approximation for lattices of different size. Solid line: rigorous staggered 8-vertex solution; broken line: high-temperature expansion. For most temperatures full squares cover open triangles Statistical errors are of the size of the symbols.

Fig. 2. Specific heat per site of the 2D spin-1/2 XY. Solid line: rigorous staggered 8-vertex solution; broken line: high-temperature expansion. Comparison with the data of Fig. 1 shows that in the critical region the m-dependence is very weak.

The power of our approach is that by increasing m we can improve approximation (3) systematically. From the m > 1 data shown in Fig.2 we conclude that for T/J > 2 the specific heat, which from point of view of convergence is the most difficult quantity to calculate, [13] depends weakly on the particular value of m. Except for $2 \leqslant T \leqslant 2.5$ the size dependence of the energy per site and specific heat per site is small. More convincing evidence that the convergence of the energy and specific heat is very good is given in Fig. 3. In the critical region ($T/J \approx 2.27$) the m-dependence of physical quantities is very weak and therefore we have to conclude that at least quantitatively the critical properties of the 2D spin-1/2 XY model are

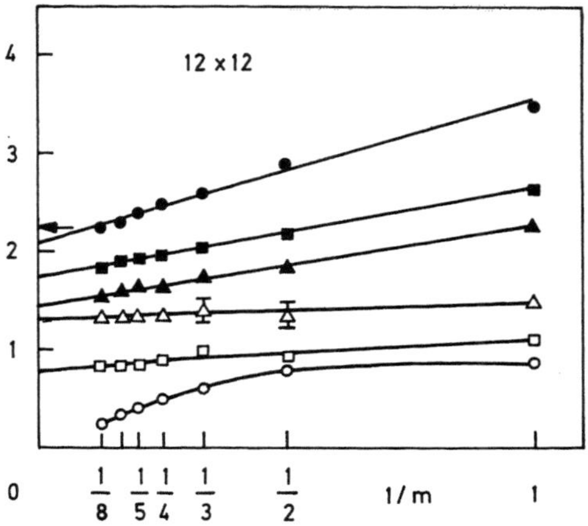

Fig. 3. Minus the energy per site (full symbols) and specific heat per site (open symbols) as a function of 1/m. Circles: T/J = 1; squares: T/J = 2 and triangles T/J = 2.25. The arrow gives the Pearsons estimate of the ground-state energy.

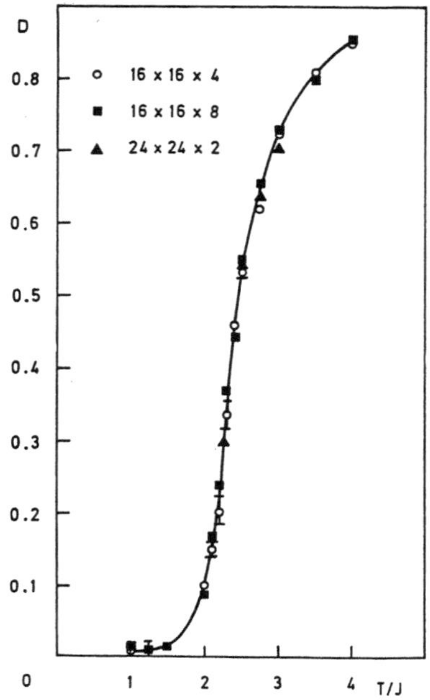

Fig. 4. The vortex correlation function, which measures the degree of disorder in the system, for lattices of different size and different m. In the critical region the m-dependence of this correlation function is very weak.

that of the m = 1 representation. To elucidate the nature of the transition we calculate spin-correlation functions which are sensitive to the occurrence of vortex-like excitations. In particular we study the quantum version of Swendsen's vortex detector [17] $D = N^{-2}\Sigma_{\text{plaquettes}} < (1 - \sigma_i^x\sigma_k^x - \sigma_i^z\sigma_k^z)(1 - \sigma_j^x\sigma_l^x - \sigma_j^z\sigma_l^z)>$ where i(j) and k(l) label the opposite corners of an elementary square. Simulation data for the approximant D_m are shown in Fig. 4. As for the energy and specific heat the m-dependence of D_m is very weak. The rapid change of the disorder-parameter D as K approaches K_1^c strongly suggests that there is phase transition.

To demonstrate the versatility of our method we have also carried out simulations for systems with N = 8, 16 and $1 \leqslant m \leqslant 32$ for temperatures down to T/J = 0.5. Within the statistical accuracy of the simulation data (which for the energy is less than one percent), we find that for $0.5 \leqslant T/J \leqslant 1.5$, the energy is constant. The extrapolated ground state energy per site $\approx$-2.1 which is in agreement with estimates given by Pearson . [18]

Three of us (H.D.R., B.D.R. and J.F.) thank the National Fund for Scientific Research, Belgium for financial support. This work is supported by the National Fund of Scientific Research, Belgium in the framework of the project "Supercomputers" and the Dutch Stichting voor Fundamenteel Onderzoek der Materie. We thank the Stichting Algemeen Rekencentrum Amsterdam for granting us a lot of computer time on their CDC machines.

References

1. N.D. Mermin and H. Wagner, Phys. Rev. Lett. 17, 1133 (1966).
2. D.D. Betts, Phase Transitions and critical Phenomena, Vol. 3, eds. C. Domb and M.S. Green (Academic Press, New York 1974).
3. D.D. Betts and M. Plischke, Can. J. Phys. 54, 1553 (1976).
4. J. Rogiers and R. Dekeyser, Phys. Rev. B 13, 4886 (1976)
5. R. Dekeyser, M. Reynaert, A.L. Stella and F. Toigo, Phys. Rev. B 18, 2343 (1978).
6. A.L. Stella and F. Toigo, Phys. Rev. B 17, 2343 (1978).
7. H. Takano and M. Suzuki, J. Stat. Phys. 26, 635 (1981).
8. T. Tatsumi, Prog. Theor. Phys. 65, 451 (1981).
9. M. Suzuki, Prog. Theor. Phys. 56, 1454 (1976).
10. A. Lagendijk and H. De Raedt, Phys. Rev. Lett. 49, 602 (1982).
11. C.S. Hsue, K.Y. Lin and F.Y. Wu, Phys. Rev. B 12, 429 (1975).
12. J.J. Cullen and D.P. Landau, Phys. Rev. B 27, 297 (1983).
13. H. De Raedt, A. Lagendijk and J. Fivez, Z. Phys. B 46, 261 (1982).
14. M. Barma and B.S. Shastry, Phys. Rev. B 18, 3351 (1978).
15. K. Binder, Monte Carlo Methods in Statistical Physics, ed. K. Binder (Springer, Berlin 1979).
16. M. Suzuki, S. Miyashita and A. Kuroda, Prog. Theor. Phys. 58, 1377 (1977).
17. R. Swerdsen, Phys. Rev. Lett. 49, 1302 (1982).
18. R.B. Pearson, Phys. Rev. B 16, 1109 (1977).

Quenching of Spin Fluctuations in Highly Enhanced Magnetic Materials

K.A. Gschneidner, Jr. and S.K. Dhar

Ames Laboratory* and Department of Materials Science and Engineering
Iowa State University, Ames, IA 50011, USA

Abstract. Low temperature heat capacity measurements from 1 to 20 K as a function of magnetic fields from 0 to 10 T (100 kOe) and magnetic susceptibility measurements at ~1.5 T from 1 to 300 K have been made on materials which exhibit unusual properties due to spin fluctuations. These include: the itinerant ferromagnet Sc_3In; the mixed valent compounds $CeSn_x$ (2.95 < x < 3.05), $CeSi_{1.85}$ and $CeSi_{1.9}$; RCo_2 (R = Sc, Y, Lu), Sc, and $Pd_{1-x}Ni_x$ (0 < x < 0.0275) alloys. Usually the application of high magnetic fields leads to a depression of the electronic specific heat constant. Results reported by other investigators on Pd, UAl_2, $TiBe_2$ and $U_{0.85}Al_{0.15}Al_2$ are also discussed. To date six different types of behaviors have been identified.

1. Introduction

In late 1980 IKEDA and GSCHNEIDNER [1] were the first to unequivocally observe the quenching of spin fluctuations in highly exchange enhanced Pauli paramagnets at high magnetic fields. Although an earlier claim had been made in 1975 in the case of UAl_2 [2] the scatter of the experimental data at their highest field of 4.3 T made such a conclusion quite tenuous; indeed subsequent studies on UAl_2 showed that fields of at least 8 T are required to see a measurable change in the heat capacity [3].

To date most of the experimental evidence for the quenching of spin fluctuations is a result of heat capacity measurements made at high magnetic fields, generally up to 10 T, but occasionally up to 17 T. Usually magnetic susceptibility and/or magnetization measurements at high fields do not give any evidence for this quenching [4].

Since 1980 the quenching of spin fluctuations have been observed in eleven different materials and a variety of behaviors have been observed. Below we briefly review the state-of-the-art of the quenching of spin fluctuations in highly exchange enhanced Pauli paramagnetic materials.

2. High Field – Low Temperature Heat Capacity

$$C = \gamma T + \beta T^3 + DT^3 \ell n(T/T_s) \tag{1}$$

where γ is the electronic specific heat constant; β is the coefficient of the T^3 (lattice) term and is related to the Debye temperature (θ_D); D is the coefficient to the $T^3 \ell nT$ term; and T_s is the spin fluctuation temperature. The experimental electronic specific heat constant is enhanced by many-body effects and is related to the electronic specific heat constant calculated from the bare density of states (γ_o) by

*Operated for the U.S. Department of Energy by Iowa State University under contract no. W-7405-ENG-82. This work was supported by the Office of Basic Energy Sciences.

$$\gamma = \gamma_o[1 + \lambda_{ep} + \lambda_{spin}(H,T)] \quad (2)$$

where λ_{ep} is the electron-phonon enhancement factor and λ_{spin} is the paramagnon (or spin fluctuation) enhancement factor. A sufficiently high magnetic field will reduce λ_{spin} (i.e. quench spin fluctuations) and thus γ. At sufficiently high temperatures, usually between 15 and 20 K, $\lambda_{spin} = 0$ because of renormalization of the spin fluctuations [5].

The coefficient of the T^3 term in Eq. (1), β, may be written as

$$\beta = \beta_L + \beta_M(H) \quad (3)$$

where β_L is the normal lattice contribution and $\beta_M(H)$ is the contribution due to an induced moment on the magnetically active atom(s) in the solid. At $H = 0$, $\beta_M(0) = 0$, but at values of $H > 2$ T, $\beta_M(H) > 0$ and saturates at $H \approx 7$ T. Initially this magnetic contribution goes as T^3, reaches a maximum at ~6 K and then falls off rapidly. From the magnetic entropy associated with this induced moment we estimate that from ~0.001 to ~0.1 electron/atom are involved in $CeSn_3$ and Pd-1.0 at.% Ni, respectively.

The last term in Eq. (1) is actually a negative contribution to the low temperature heat capacity since $T < T_s$ and causes an upturn in the heat capacity as $T \to 0$. According to theory this contribution is valid for $T \ll T_s$, but when one compares the theoretical shape of this contribution with the experimental data it appears to hold up to $\sim 0.8T_s$. Furthermore, the application of field suppresses the upswing; and this is probably due to the field dependence of the coefficient D of the $T^3 \ell nT$ term in Eq. (1), which is proportional to $\gamma_o(S-1)^2/ST_s{}^2$, where S is the Stoner enhancement factor and is equal to $\chi(0)/\chi_o$, where $\chi(0)$ is the measured magnetic susceptibility at 0 K and χ_o is the susceptibility calculated from the bare density of states. An analysis indicates that if the Stoner factor decreases with increasing field it would lead to a reduction in the upswing of the low temperature tail in the heat capacity. A reduction in the Stoner enhancement will only occur if $\chi(0)$ decreases with increasing field. The available high field magnetic susceptibility data for UAl_2[6] shows that $\chi(0)$ decreases by 16% as the field is increased from ~4 to ~28 T which is in good agreement with the observed heat capacity results. But for $TiBe_2$ the susceptibility increases with increasing field [7], suggesting that the above argument does not hold. However, for $TiBe_2$, the development of an induced moment with increasing field may explain why the expected correlation does not hold.

3. Systematization and Classification of Behaviors

The observed quenching of spin fluctuations in various materials can be categorized into six types of behaviors based on the variation of low temperature heat capacity as a function of temperature as the applied magnetic field is increased from 0 to at least 10 T, see Fig. 1. The important physical properties of these materials and their type behavior are listed in Table 1.

Also listed in Table 1 are a few type "0" materials ($CeSn_{3.05}$ [12], pure Pd [9,13] and Pd-0.05 and -0.1 at.% Ni[9]) whose heat capacities were found to be independent of magnetic field up to 10 T.

There also appears to be a rough correlation between the field dependence of the heat capacity and the temperature variation of the magnetic susceptibility. The χ vs. T dependences can be divided into four different behaviors, which are shown in Fig. 2. The relationship between the heat capacity and susceptibility behaviors are noted below.

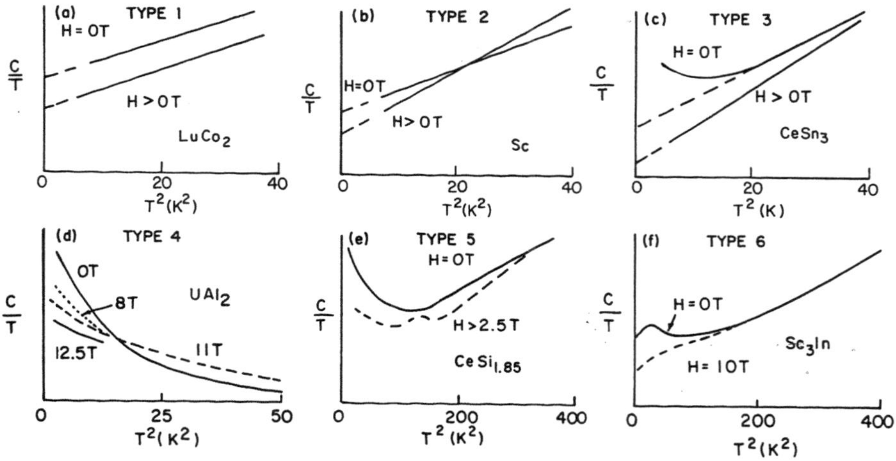

Fig. 1. The five types of behaviors of the low temperature high magnetic field heat capacities of highly exchange enhanced Pauli paramagnets, and the itinerant "ferromagnet" Sc_3In

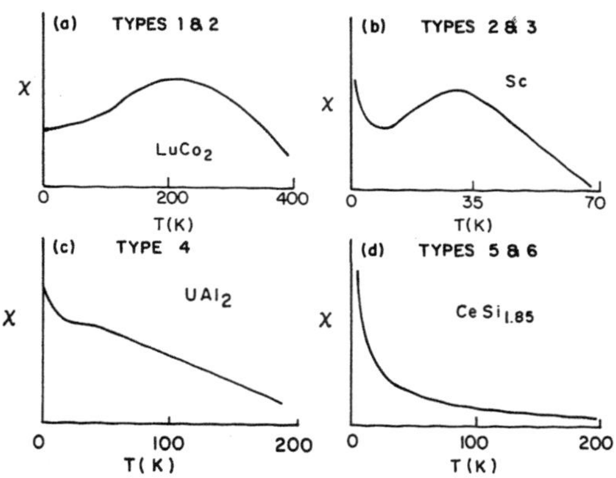

Fig. 2. The temperature variation of the magnetic susceptibility of the various highly exchange enhanced Pauli paramagnets

3.1 Type 1 Behavior

The temperature dependence of the heat capacity of a type 1 material at zero field is the same as that of a normal metallic material. The only effect of high magnetic fields is to suppress γ (and λ_{spin}) — as is shown in Fig. 1a. The known type 1 materials are $LuCo_2$ [1,19], $ScCo_2$ [19], and a "pure" Pd sample [15] (which probably contained Co impurities [16]). The magnetic susceptibility of a type 1 material is shown in Fig. 2a.

3.2 Type 2 Behavior

For type 2 materials γ is lowered due to the suppression of λ_{spin} and the slope β is increased due to a contribution from $\beta_M(H)$ (Fig. 1b). The known type 2 materials include pure Sc [20,21], YCo_2 [19], and Pd-alloys doped with nickel (0.47 to 2.25 at.%) [9,17,18]. The χ vs. T behavior can vary as shown in Fig. 2a (YCo_2) or as in Fig. 2b(Sc).

Table 1. Summary of the properties of exchange enhanced paramagnets

Material	Type	γ (at 0 K) $\left[\frac{mJ}{g\text{-}at.K^2}\right]$	$\frac{-\Delta\gamma}{\gamma(0)}$ at 10 T	$\frac{\Delta\beta}{\beta(0)}$ at 10 T	λ_{spin}	$\chi \times 10^4$ (at 0 K) [emu/g-at.]	S	T_s [K]	Ref.
$CeSi_{1.90}$	5	65	4.6	-	11.5[a]	24	43	33	8,9
$CeSi_{1.85}$	5	105	17.4	-	19.8[a]	140	262	33	8,9
$CeSn_{3.0}$	3	18.15	26.9	113	5.07	10	28	5.8	10
$CeSn_{3.0}$ [H_{100}]	3	18.75	15.4	65	5.67	7.2	20	5.6	11
$CeSn_{3.0}$ [H_{110}]	3	18.75	21.1	120	5.67	8.0	22	5.6	11
$CeSn_{3.05}$	0	15	0	0	3.88	5.4	15	-	12
Pd	0	9.45	0	0	0.37	7.36	10	250	9,13,14
Pd(Co?)	1	9.36	6.8	0	-	-	-	-	15,16
Pd(0.05%Ni)	0	9.48	0	0	-	-	-	-	9
Pd(0.10%Ni)	0	9.58	0	0	-	-	-	-	9
Pd(0.47%Ni)	2	9.90	13.4	27	0.56	11.76	15.3	20	17
Pd(0.97%Ni)	2	10.85	11.9	31	0.65	18.36	23.9	20	17
Pd(1.65%Ni)	2	12.53	11.1	20	-	-	-	-	18
Pd(1.95%Ni)	2	13.10	11.8	28	-	-	-	-	18
Pd(2.25%Ni)	2	13.41	11.9	35	-	-	-	-	18
Pd(2.75%Ni)	2	13.98	16.0	47	-	-	-	-	18
RCo_2 { $ScCo_2$	1	6.13	7.2	0	2.36	3.67	15.5	20	19
RCo_2 { YCo_2(36%Y)	2	10.19	3.8	18	4.64	5.33	22.5	35	19
RCo_2 { YCo_2(35%Y)	2	11.02	-	-	5.13	6.00	25.3	35	19
RCo_2 { $LuCo_2$	1	8.85	10.3	0	3.85	3.96	16.7	16	1,19
Sc	2	10.33	11.5	134	0.96	3.93	6.2	~10	20,21
$TiBe_2$	5	18.77	9.7	-	0.51[a]	30.7	65	50±25	22-24
UAl_2	4	47.3	5.7	-	0.62[a]	14.7	4.35	29	3,25
$U_{0.85}La_{0.15}Al_2$	4	79.4	5.0	-	0.71[a]	30.3	5.4	39	25
Sc_3In(24.4%In)	6	10.7[b]	-	-	1.12	c	∞	6.3[d]	26-28

[a] Calculated by reviewers from data given by authors and assuming λ_{ep}=0.3 if not given
[b] At 10 T [c] "Ferromagnetic" [d] Authors assumed T_c (Curie temperature) = T_s

3.3 Type 3 Behavior

All the remaining spin fluctuation materials (types 3, 4 and 5) exhibit a $T^3 \ell nT$ behavior in their H = 0 T heat capacities, but the high field behaviors are different. Only $CeSn_3$ exhibits type 3 behavior [10-12] and is exemplified by the facts that as the magnetic field is raised the $T^3 \ell nT$ term disappears, γ is lowered and the slope β increases, Fig. 1c. The weakness of the $T^3 \ell nT$ term is probably due to a low T_s (5.7 K). The magnetic susceptibility for $CeSn_3$ is nearly the same as that of Sc (Fig. 2b) except that the curve is displaced to higher temperatures for $CeSn_3$.

3.4 Type 4 Behavior

UAl_2 [3,25] and $U_{0.85}Al_{0.15}Al_2$ [25] are the only known type 4 behavior materials. The $T^3 \ell nT$ term is quite strong and is only partially suppressed even at 17 T, Fig. 1d. γ is lowered, but it has not been established whether or not β increases with field. It is possible that behavior types 3 and 4 differ only in degree. If β remains constant, and that appears to be the case, then UAl_2 (type 4) remains a distinct and different type.

The magnetic susceptibility is shown in Fig. 2c where it is seen to have an inflection point at low temperature. It is obvious that this χ vs. T behavior is intermediate between those shown in Fig. 2b and 2d.

3.5 Type 5 Behavior

In addition to the effects seen in type 4 materials a distinct bump develops in the high field heat capacity of type 5 materials (Fig. 1e). We believe this bump occurs because of an induced moment due to the alignment of the fluctuating electrons by the high magnetic field. The characteristic rapid rise in χ as $T \to 0$ K, see Fig. 2d, suggests that this material is about ready to order. Examples of type 5 materials are $CeSi_x$ (x = 1.9 and 1.85)[9] and $TiBe_2$ [22-24].

3.6 Type 6 Behaviors

The last type of behavior (as observed in Sc_3In) occurs when the interactions of the fluctuating electrons are barely strong enough to give rise to "ferromagnetism". The peak in the heat capacity due to "ferromagnetic" ordering vanishes (see Fig. 1f) as $H \to \sim 12$ T [26-28]. Not evident in Fig. 1f is the development of a bump with increasing field at a temperature slightly higher than the magnetic ordering temperature. The susceptibility and the magnetization curves for Sc_3In are exactly like that expected for a ferromagnet and this is why the material has been called an itinerant "ferromagnet". However, the quenching of "ferromagnetism" by high magnetic fields is unusual and unexpected and that is why we have used quotation marks around the words ferromagnetism and ferromagnet for this material.

4. Summary

The quenching of spin fluctuations in highly exchange enhanced Pauli paramagnets has been observed in approximately 25 different materials. The quenching behaviors, which seem to be correlated with the strength of interactions between the fluctuating spins, have been divided into six types according to the magnetic field dependence of the low temperature heat capacity. The χ vs. T behavior for these six types seems to support the trend of increasing strength of the interactions as the type number increases. The variation of λ_{spin} and S for the six types generally increase on going from type 1 materials to type 6 materials (Table 1). UAl_2 (type 4) is a notable exception in that both are low. Clearly such simple parameters as λ_{spin}, S and T_s in themselves or together cannot explain the observed trends, but they do serve as useful guides.

Although the study of the quenching of spin fluctuations in these highly enhanced systems is slightly more than four years old, some great strides have been made experimentally. The theory, which is 20 years old, is inadequate and needs to be reexamined in light of the recent experimental studies.

References

1. K.Ikeda, K.A.Gschneidner,Jr.: Phys. Rev. Lett. **45**, 1341 (1980)
2. R.J.Trainor, M.B.Brodsky, H.V.Culbert: Phys. Rev. Lett. **34**, 1019 (1975)
3. G.R.Stewart, A.L.Giorgi, B.L.Brandt, S.Foner, A.J.Arko: Phys. Rev. B **28**, 1524 (1983)
4. K.A.Gschneidner,Jr., K.Ikeda: J. Magn. Magn. Mater. **31-34**, 265 (1983)
5. H.Hasegawa: J. Phys. Soc. Japan **38**, 107 (1975)
6. J.J.M.Franse, P.H.Frings, F.R.deBoer, A.Menovsky, C.J.Beers, A.P.J.vanDeursen, H.W.Myron, A.J.Arko: Phys. Rev. Lett. **48**, 1749 (1982)
7. F.Acker, R.Huguenin, J.L.Smith, G.R.Stewart: Phys. Rev. B **27**, 5883 (1983)
8. H.Yashima, N.Sato, H.Mori, T.Satoh: Solid State Comm. **43**, 595 (1982)
9. S.K.Dhar, K.A.Gschneidner,Jr.: unpublished results
10. K.Ikeda, K.A.Gschneidner,Jr.: Phys. Rev. B **25**, 4623 (1982)
11. T.-W.E.Tsang, K.A.Gschneidner,Jr., O.D.McMasters, R.J.Stierman, S.K.Dhar: Phys. Rev. B **29**, 4185 (1984)
12. K.A.Gschneidner,Jr., S.K.Dhar, R.J.Stierman, T.-W.E.Tsang, O.D.McMasters: to be published
13. G.R.Stewart, B.L.Brandt: Phys. Rev. B **28**, 2266 (1983)
14. P.Hertel, J.Appel, D.Fay: Phys. Rev. B **22**, 534 (1980)
15. T.Y.Hsiang, J.W.Reister, H. Weinstock, G.W.Crabtree, J.J.Vuillemin: Phys. Rev. Lett. **47**, 523 (1981)
16. G.W.Crabtree: private communication (1983)
17. K.Ikeda, K.A.Gschneidner,Jr., A.I.Schindler: Phys. Rev. B **28**, 1457 (1983)
18. S.K.Dhar, K.A.Gschneidner,Jr., A.I.Schindler: unpublished results
19. K.Ikeda, K.A.Gschneidner,Jr., R.J.Stierman, T.-W.E.Tsang, O.D.McMasters: Phys. Rev. B **29**, 5039 (1984)
20. K.Ikeda, K.A.Gschneidner,Jr., T.-W.E.Tsang, F.A.Schmidt: Solid State Comm. **41**, 889 (1982)
21. R.J.Stierman, K.A.Gschneidner,Jr., T.-W.E.Tsang, F.A.Schmidt, P.Klavins, R.N.Shelton, J.Queen, S.Legvold: J. Magn. Magn. Mater. **36**, 249 (1983)
22. G.R.Stewart, J.L.Smith, A.L.Giorgi, Z.Fisk: Phys. Rev. B **25**, 5907 (1982)
23. G.R.Stewart, J.L.Smith, B.L.Brandt: Phys. Rev. B **26**, 3783 (1982)
24. F.Acker, Z.Fisk, J.L.Smith, C.Y.Huang: J. Magn. Magn. Mater. **22**, 250 (1981)
25. M.S.Wire, G.R.Stewart, W.R.Johannson, Z.Fisk, J.L.Smith: Phys. Rev. B **27**, 6518 (1983)
26. K.Ikeda, K.A.Gschneidner,Jr.: J. Magn. Magn. Mater. **30**, 273 (1983)
27. W.E.Gardner, T.F.Smith, B.W.Howlett, C.W.Chen, A.Sweedler: Phys. Rev. **166**, 577 (1968)
28. L.R.Testardi, L.M.Holmes, W.E.Reed, F.S.L.Hsu: Phys. Rev. B**6**, 3365 (1972)

Quasiparticle Model for High Frequency Inelastic Neutron Scattering from Intermediate Valence Compounds

D.L. Huber

Department of Physics, University of Wisconsin, Madison, WI 53706, USA

1. Introduction

In recent years considerable progress has been made in understanding the properties of intermediate valence impurities in non-magnetic host metals. Important theoretical advances have come about by utilizing renormalization group, Bethe Ansatz, and other field-theoretic techniques. Unfortunately, there has been much less progress in understanding intermediate valence compounds. Intermediate valence compounds, as the term is used here, are stoichiometric materials containing rare earth or actinide ions where the energy difference between the f^n and f^{n-1} configurations lies close to the fermi energy.

The most fundamental unsolved problem pertaining to intermediate valence compounds concerns the nature of the ground state. Some years ago MOTT [1] proposed that electronic states in intermediate valence systems were characterized by a hybridization of the f-states with the conduction band. Because of the hybridization there was a gap in the density of states in the neighborhood of the fermi energy. The Mott picture was criticized by ANDERSON [2] who argued that when allowance was made for the orbital degeneracy of the f-states there would be states in the gap which did not hybridize with the conduction band (see Fig. 1). Recently MOTT [3] responded to the Anderson argument by asserting that the high correlation of the f-states leads to hybridization with no states in the gap.

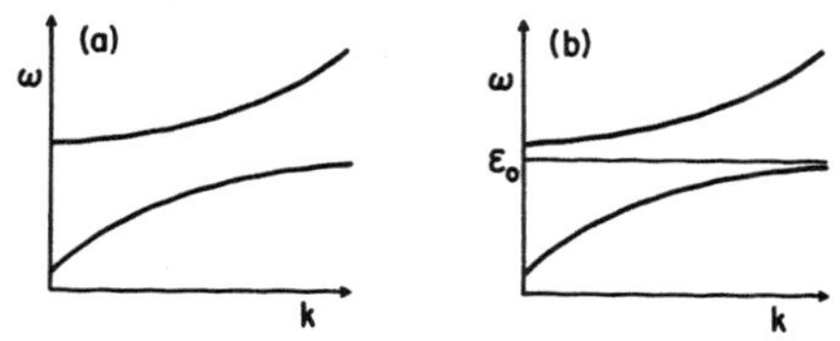

Fig. 1. Schematic effective one-electron states. (a) Mott picture, complete hybridization with conduction band. (b) Anderson picture, horizontal line indicates modes which do not hybridize with conduction band.

The experimental situation is also unclear. The only intermediate valence compound for which there appears to be unambiguous evidence for a hybridization gap is SmB_6 although gaps have been reported for other materials[4]. It is uncertain, however, whether the experimental data can be equally well interpreted in terms of a description closer to the Anderson picture when allowance is made for the difference in the wave functions of the hybridized and non-hybridized modes.

Because of the uncertainties surrounding the Mott and Anderson pictures it is important to undertake studies which emphasize the differences in the two models. In this paper we discuss one such investigation in which we focus on the inelastic neutron scattering. We have carried out an analysis of the high frequency inelastic neutron scattering utilizing an independent quasiparticle model involving hybridized and gap (or non-hybridized) modes [5,6,7] which is predicated on a ground state similar to that proposed by Anderson. Although our approximations appear to be incompatible with the Mott picture we will compare our results with what would have been obtained had we assumed complete hybridization.

2. Dynamic Structure Factor.

In Refs. 5-7 we have outlined a theory for the inelastic neutron scattering from intermediate valence compounds which applies to the high frequency regime, $\omega \gg |\tilde{V}|^2 \rho_o$, where ρ_o is the density of states of the conduction band and $|\tilde{V}|$ is one half of the hybridization gap (see below). It is assumed $|V| \rho_o \ll 1$, and that crystal field splittings are small in comparison with $|\tilde{V}|$. The relevant f states are the Hund's Rule ground states of the f^n and f^{n-1} configurations.

The high frequency tail of the dynamic structure factor, $^{HF}S(q,\omega)$, arises from two-quasiparticle processes: excitation across the hybridization gap from the lower to the upper hybrid mode, intraband scattering involving the hybrid modes, excitation from the gap, or non-hybridized, modes to the upper-hybrid mode, and excitation from the lower-hybrid mode to the gap modes. At the center of the Brillouin zone, q = 0, the intraband scattering disappears, and only vertical transitions, reflecting the conservation of quasimomentum, are allowed (see Fig. 2). We find

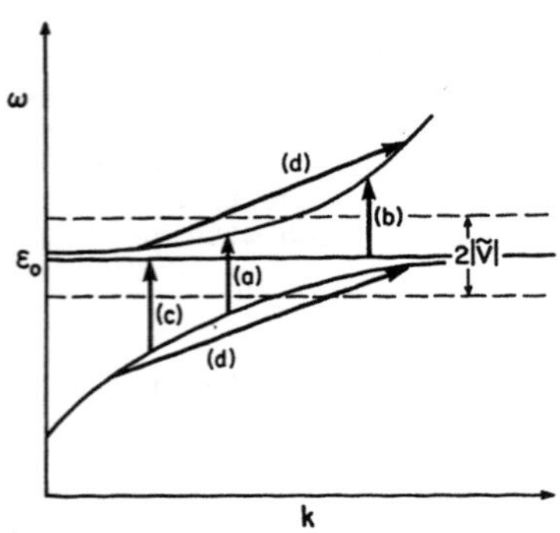

Fig. 2. Schematic diagram of the quasiparticle spectrum. The vertical arrows denote processes contributing to $^{HF}S(0,\omega)$ (a) hybrid-to-hybrid, (b) gap-to-hybrid, (c) hybrid-to-gap. The diagonal arrows (d) represent intraband processes which contribute to the scattering at finite q. In the independent quasiparticle model transitions between the gap modes appear as q-independent elastic scattering when crystal field effects are neglected.

$$
\begin{aligned}
&{}^{HF}S(0,\omega) \\
&= C_o\rho_o R[\omega E(\omega)]^{-1}\Big(f[E(\omega)-\omega]\{1-f[E(\omega)+\omega]\} \\
&\quad + f[-E(\omega)-\omega]\{1-f[-E(\omega)+\omega]\}\Big) \\
&\quad + C_o\rho_o(1-R)\omega^{-2}\{f(0)[1-f(2\omega)] + f(-2\omega)[1-f(0)]\}, \qquad (1)
\end{aligned}
$$

taking the density of states to be constant. In this equation C_o is a constant,

$$E(\omega) = (\omega^2-4|\tilde{V}|^2)^{1/2}, \qquad (2)$$

and

$$f(x) = [e^{\beta(\varepsilon_o + x/2 - \mu)} + 1]^{-1}, \tag{3}$$

with $\beta = 1/kT$. Here ε_o is the (renormalized) energy difference between the ground states of the two configurations and μ is the chemical potential.

The first term in Eq.(1) describes the hybrid-to-hybrid scattering. It is characterized by a threshold singularity at $\omega = 2|\tilde{V}|$, the hybridization gap. The second term is the scattering associated with transitions involving a gap mode and a hybrid mode (see Fig.2). The parameter R is a measure of the relative weight of the hybrid-to-hybrid scattering. Values of R appropriate to various relevant pairs of Hund's rule ground states are given in Table 1.

Table 1. Values of R for various configuration pairs.

$f^n(^{2S+1}L_J)$	$f^{n-1}(^{2S'+1}L'_{J'})$	R
$f^1(^2F_{5/2})$	$f^0(^1S_0)$	5/21
$f^2(^3H_4)$	$f^1(^2F_{5/2})$	0.104
$f^3(^4I_{9/2})$	$f^2(^3H_4)$	0.042
$f^{13}(^2F_{7/2})$	$f^{12}(^3H_6)$	0.159
$f^{14}(^1S_0)$	$f^{13}(^2F_{7/2})$	3/7
$f^1(^2S_{1/2})$	$f^0(^1S_0)^a$	1.0

[a]No orbital degeneracy.

At high temperatures, the coherence between the conduction band states and the f states which is implicit in the hybrid-mode picture will be destroyed. When this happens the decoupled-mode approximation is appropriate, and ${}^{HF}S(0,\omega)$ is given by

$$^{HF}S(0,\omega) = C_o\rho_o\omega^{-2}\{f(0)[1-f(2\omega)] + f(-2\omega)[1-f(0)]\}. \tag{4}$$

3. Finite Wavevector and Hybrid-Mode Damping

The results for the dynamic structure factor given in the preceeding section were obtained at q=0. Provided the q-dependence of C_o is overlooked the scattering in the decoupled mode approximation (Eq.(4)) is q-independent as is the scattering involving the gap modes in Eq.(1). The hybrid to hybrid scattering, however, does depend on wavevector. In particular, the inverse square root singularity in the first term in Eq.(1) disappears at finite q. When the conduction band is treated in the effective mass approximation the factor $(\omega^2-4|\tilde{V}|^2)^{1/2}$ is replaced by

$$[2\omega(\varepsilon_q\varepsilon_o)^{1/2}]^{-1}[(\omega^2-4|\tilde{V}|^2+2\omega(\varepsilon_q\varepsilon_o)^{1/2})^{1/2} - (\omega^2-4|\tilde{V}|^2-2\omega(\varepsilon_q\varepsilon_o)^{1/2})^{1/2}],$$

where $\varepsilon_q = \hbar^2q^2/2m^*$, m^* being the effective mass, and the radicals are evaluated only for positive values of their argument. In deriving this equation it is assumed $(\varepsilon_q\varepsilon_o)^{1/2} << |\tilde{V}|$. The behavior of ${}^{HF}S(q,\omega)$ for

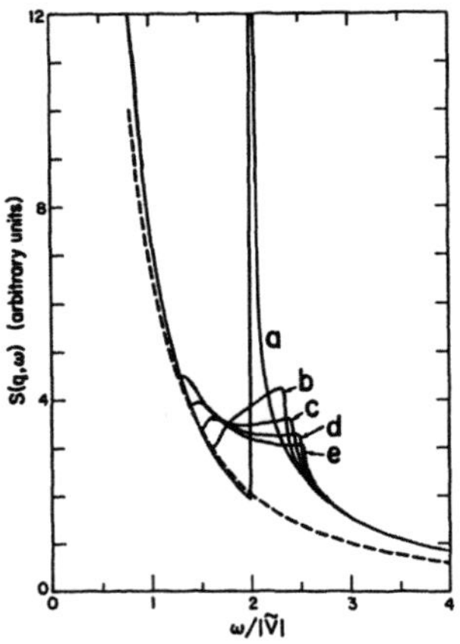

Fig. 3. $S(q,\omega)$ vs $\omega/|\tilde{V}|$. Broken curve: Eq. (4) with $\varepsilon_o = \mu$, T=300K. Solid curves: Eq. (1) (generalized to finite q) with $\varepsilon_o = \mu$, T = 5K, V = 20 meV, R = 5/21; $\varepsilon_q = \hbar^2 q^2/2m^*$.

(a) $(\varepsilon_q \varepsilon_o)^{\frac{1}{2}} = 0$,
(b) $(\varepsilon_q \varepsilon_o)^{\frac{1}{2}} = 0.316|\tilde{V}|$;
(c) $(\varepsilon_q \varepsilon_o)^{\frac{1}{2}} = 0.447|\tilde{V}|$;
(d) $(\varepsilon_q \varepsilon_o)^{\frac{1}{2}} = 0.548|\tilde{V}|$;
(e) $(\varepsilon_q \varepsilon_o)^{\frac{1}{2}} = 0.632|\tilde{V}|$.

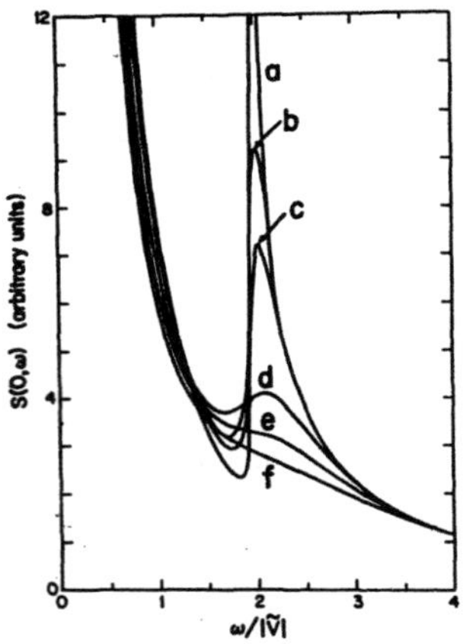

Fig. 4. Effect of hybrid mode damping on $S(0,\omega)$.
(a) $\gamma/|\tilde{V}| = 0.01$; (b) $\gamma/|\tilde{V}| = 0.025$;
(c) $\gamma/|\tilde{V}| = 0.05$; (d) $\gamma/|\tilde{V}| = 0.2$;
(e) $\gamma/|\tilde{V}| = 0.35$; (f) $\gamma/|\tilde{V}| = 0.5$;
other parameters as in Fig. 3.

larger values of q is shown in Fig.3. Note that since $(\varepsilon_q \varepsilon_o)^{1/2} \sim 2q(\text{eV-Å}^{-1})$ for ε_o=1eV and m* equal to the free electron mass, the threshold singularity at $\omega = 2|\tilde{V}|$ broadens unusually rapidly.

The damping of the hybridized modes arises from both extrinsic (e.g. involving phonons and impurities) and intrinsic processes. The extrinsic processes contribute a term to the damping rate which is on the order of the scattering rate for conduction electrons in the absence of hybridization. As will be shown elsewhere [8], the intrinsic damping arises from the quasi-elastic scattering of the quasiparticles. In order to gain some insight into the effects of the damping on the structure factor we have calculated the hybrid-to-hybrid contribution to $^{HF}S(0,\omega)$ for various values of $\gamma/|\tilde{V}|$, γ being the decay rate of the hybridized mode. This was done by replacing $\delta(\omega - \omega^U_{Hy} + \omega^L_{Hy})$, ω^U_{Hy} and ω^L_{Hy} being the frequencies of the upper and lower hybridized modes, respectively, by a Lorentzian of width 2γ at appropriate points in the calculation. The effect of this modification is shown in Fig. 4.

4. Discussion

In the discussion we comment first on the distinction between the scattering calculated in the Mott and Anderson pictures. We then take up the question of the implications of the theory for various experimental studies. As noted in the Introduction, our results for the high frequency tail of the dynamic structure factor in the hybrid-mode regime are compatible with the Anderson picture in the sense that we include scattering involving the non-hybridizing gap modes as well as excitation across the hybridization gap. The latter gives rise to an asymmmetric peak in $S(0,\omega)$ at $\omega = 2|\tilde{V}|$ which rapidly broadens with increasing wavevector. In contrast, the scattering associated with the gap modes varies approximately as ω^{-2}[6] and is virtually q-independent as is the scattering at high temperatures where the decoupled mode approximation is expected to be appropriate. The presence of hybridization-induced structure superimposed on a ω^{-2} (Lorentzian) background is qualitatively similar to the low-temperature results reported by SCHLOTTMAN [9] for intermediate valence impurities with orbital degeneracy except that in the case of impurities the scattering is independent of wave vector. Just as in the case of intermediate valence compounds, the structure disappears at high temperatures.

In the Mott picture there are no gap modes which in our calculation corresponds to taking R = 1 in Eq.(1). Thus at low temperatures the scattering at q = 0 involves only excitation across the hybridization gap, there being no ω^{-2} background coming from the gap modes. It should be noted that Eq.(1) with R = 1 agrees qualitatively with the results at small wavevectors obtained by FEDRO and SINHA [10] and CZYCHOLL and LEDER [11] for a mixed valence compound with no orbital degeneracy. It is also similar to the results obtained recently by KOJIMA et al [12] for a model appropriate to Ce compounds.

In assessing the relevance of our findings to inelastic neutron scattering measurements it must be emphasized that our calculations pertain to the asymptotic regime $\omega >> |\tilde{V}|^2\rho_0$, which is equivalent to taking ω large in comparison with the high temperature limit of the linewidth of the central peak [13]. We have also neglected the crystal field splitting of the f^n and f^{n-1} manifolds, which appears to be a reasonable approximation as long as the widths of the manifolds are much less than $|\tilde{V}|$. It is important to note that the shape of the threshold singularily in $S(q,\omega)$ is affected by quasiparticle damping even at q = 0 (Fig. 4) in addition to the rapid change that occurs with increasing q.

Because of these factors we believe that searches for hybridization effects in neutron scattering studies of intermediate valence compounds should focus on the low-temperature small-q large-ω regime where such effects are expected to be especially prominent. Care should be taken to distinguish between crystal field transitions and excitation across the hybridization gap. This can probably be done either from estimates of the crystal field parameters or from measurements of the q-dependence of the scattering which should be much less pronounced in the case of crystal field transitions. Since SmB_6 is a system for which there is unambiguous evidence for a hybridization gap from a variety of measurements [14], neutron scattering studies of this material should provide a critical test of the theory either in its original form or with the R = 1 modification appropriate to the Mott picture.

References

*Research supported by the National Science Foundation under the grant DMR-8203704.

1. N. F. Mott, Phil. Mag. 30, 403 (1974).
2. P. W. Anderson, in Valence Fluctuations in Solids, edited by L. M. Falicov, W. Hanke, and M. B. Maple (North-Holland, Amsterdam, 1981) p. 451.
3. N. F. Mott, in Valence Instabilities, edited by P. Wachter and H. Boppart (North-Holland, Amsterdam, 1982), p. 403.

4. G. Travaglini and P. Wachter, Phys. Rev. B29, 893 (1984).
5. D. L. Huber, Phys. Rev. B28, 860 (1983).
6. D. L. Huber, Phys. Rev. B29, 456 (1984).
7. D. L. Huber, J. Appl. Phys. 55, 1928 (1984).
8. G.-J. Hu and D. L. Huber (to be published).
9. P. Schlottman, Phys. Rev. B29, 630, 4468 (1984).
10. A. J. Fedro and S. K. Sinha, Valence Fluctuations in Solids, edited by L. M. Falicov, W. Hanke, and M. B. Maple (North-Holland, Amsterdam, 1981) p. 329.
11. G. Czycholl and H. J. Leder, Z. Phys. B48, 67 (1982).
12. H. Kojima, Y. Kuramoto, and M. Tachiki, Z. Phys. B54, 243 (1984).
13. Y. Kuramoto, Z. Phys. B37, 299 (1980).
14. S. von Molnar, T. Theis, A. Benoit, A. Briggs, J. Floquet, J. Ravex, and Z. Fisk, in Valence Instabilities, edited by P. Wachter and H. Boppart (North-Holland, Amsterdam, 1982) p. 389.

Singlet Ground-State Dimer Systems $Cs_3Cr_2Br_9$ and $Cs_3Cr_2Cl_9$

J.K. Kjems[1] and D. Petitgrand
Laboratoire Léon Brillouin, Saclay, F-91191 Gif-Sur-Yvette, France
R. Feile[2]
Risø National Laboratory, DK-4000 Roskilde, Denmark
B. Leuenberger and H.U. Güdel
Department of Inorganic Chemistry, University, CH-3000 Bern, Switzerland

I. Introduction

Singlet ground-state magnetism has attracted attention partly because of the interest in the nature of phasetransitions involving induced ordering and partly because such materials are of potential use for adiabatic cooling [1]. Similarly, there has been recent work on magnetic dimer states in molecular units and complexes. Initially these studies were motivated by the interest in the relations between the chemical environment and the magnetic interactions which becomes especially transparent in such systems [2]. However, as shown by the work presented here, the magnetic many-body aspects are equally exciting and lead to a new class of singlet ground-state compounds.

The energy spectrum of a magnetic dimer system depends on the strength of the magnetic interaction, J, and on the anisotropy due to the crystalline electric fields at the magnetic sites. For spin-only states like those of Cr^{+3} to be discussed here, the dimer-spectrum becomes exceedingly simple, namely given by [3]

$$E = - 1/2 \; J \; K \; (K+1), \quad K = S_1+S_2. \tag{1}$$

In systems with antiferromagnetic interactions, $J < 0$, the singlet state, $K = 0$, is the lowest with the $K = 1$ triplet as the first excited state, and in the compounds where the dimers are not completely isolated the interaction between dimers may give rise to cooperative phenomena such as exciton dispersion and even magnetic ordering.

This contribution summarizes our recent neutron scattering work on the Cr^{+3}-Cr^{+3} dimer compounds $Cs_3Cr_2Br_9$ and $Cs_3Cr_2Cl_9$ [4-7] which have turned out to be very simple singlet ground-state model systems while the ratio between the inter- and intra-dimer couplings varies from very near a

---

[1] Permanent address Risø National Laboratory DK 4000, ROSKILDE, DENMARK
[2] Present address Department of Physics, J. Gutenberg University D6500 MAINZ

critical value for the bromide compound to quite far from it in the chloride compound. Following this trend one would expect the iodide analogue to have a ratio beyond the critical value, i.e. that magnetic order should occur spontaneously at low temperatures. This seems to be confirmed by recent powder diffraction measurements [8].

2. Experiments

The hexagonal structure of the $Cs_3Cr_2Br_9$, $-Cl_9$ compounds is shown in Fig.1. The dimers form two simple hexagonal sublattices and the dimer-dimer interactions within, J_p, and between, J_c, these sublattices are comparable. This in turn gives rise to a tendency towards incommensurate magnetic ordering and in this respect makes the systems analogous to the rare earth metals Pr and Nd [9-10] and to the insulators $RbFeCl_3$ and $CsFeCl_3$ [11-12].

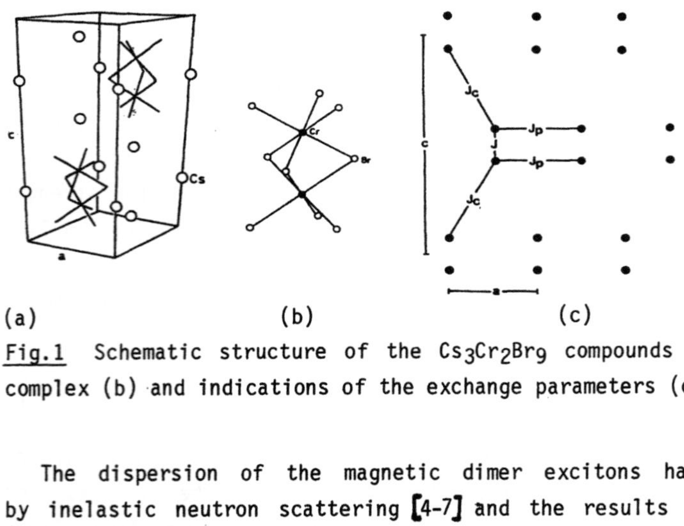

(a) (b) (c)

Fig.1 Schematic structure of the $Cs_3Cr_2Br_9$ compounds (a) with the dimer complex (b) and indications of the exchange parameters (c).

The dispersion of the magnetic dimer excitons have been determined by inelastic neutron scattering [4-7] and the results are shown in Fig.2 together with the fitted dispersion relations resulting from a generalized random phase approximation theory, RPA [5]. In this theory the exciton energies follow from the Hamiltonian

$$H = - J \sum_i \underline{S}_{i1}\,\underline{S}_{i2} - 1/2\, J_p \sum_{i,j} (\underline{S}_{i1}\cdot \underline{S}_{j1} + \underline{S}_{i2}\cdot \underline{S}_{j2}) - 1/2\, J_c \sum_{ij} (\underline{S}_{i1}\cdot \underline{S}_{j'2} + \underline{S}_{i2}\cdot \underline{S}_{j'1}) \qquad (2)$$

and are given by

$$\hbar\omega(q)^{a,o} = (J^2 + JM^2(n_o-n_1)\,(J_p\gamma_p(q)) \pm J_c|\gamma_c(q)|)^{1/2} \qquad (3)$$

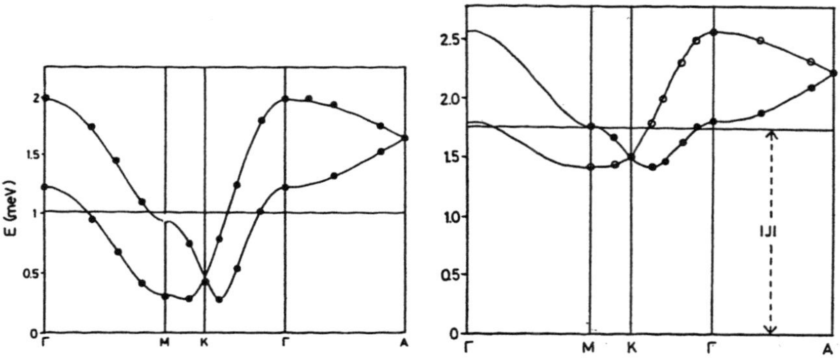

Fig.2 The observed singlet-triplet exciton dispersion in $Cs_3Cr_2Br_9$ at 1.6K (a) and $Cs_3Cr_2Cl_9$ at T = 2K (b). The full lines are model fits described in the text.

where the superscripts refer to the acoustic and the optic branch. $M^2 = 5$ is the square of the singlet-triplet transition matrix element, n_i is the corresponding thermal population factor and γ_i is the Fourier transform of the interactions. As can be seen, the observed dispersion is excellently described with the following parameters, indicated in Fig. 1(c) and given in meV.

	$Cs_3Cr_2Br_9$	$Cs_3Cr_2Cl_9$
J	- 1.03	- 1.75
J_p	- 0.054	- 0.032
J_c	- 0.039	- 0.031
η	0.92	0.36

The critical ratio is defined as $\eta = |M^2 J^*|/J$; $J^* = (J_p\gamma_p(\underline{q}_o) - J_c|\gamma_c(\underline{q}_o)|)$ with $\underline{q}_o$ being the wavevector that corresponds to the minimum energy. With these few parameters it turns out that the same minimum energy is found on a ring of q-points around the K-point whose average radius is $q = J_c/J_p(2\pi\sqrt{3})$ rlu. This is illustrated in the contour plot of the excitation dispersion surface in the basal plane shown for $Cs_3Cr_2Br_9$ in Fig. 3(a). Hence, on the basis of the exciton dispersion alone we cannot predict precisely which q-vector will become the magnetic moment modulation vector in the ordered state which will occur spontaneously in systems for which $\eta > 1$. If $\eta < 1$ magnetic order may be induced by an external perturbation such as an applied magnetic field. The mechanism for the ordering is the soft-mode that results from the Zeeman splitting of the triplet state and the ordering is predicted to occur at the field where the energy of the lowest lying exciton mode vanishes. Using this criterion and the RPA calculation of the exciton energies with the parameters pertinent to $Cs_3Cr_2Br_9$ one arrives at the phasediagram shown in Fig. 3(b).

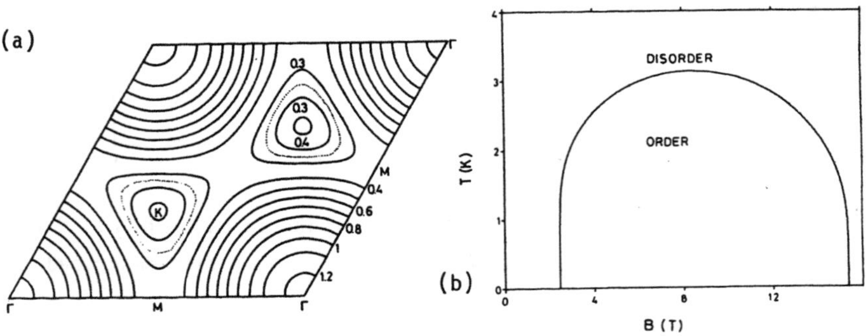

Fig.3 : a) Contour plot of the acoustic exciton energies in the basal plane for $Cs_3Cr_2Br_9$ based on the model parameters given in the text.
b) Calculated phasediagram in an applied field with the same parameters.

Experimental results for the excitation energies observed in $Cs_3Cr_2Br_9$ at q_o = (0.27,0.27,2) at a temperature of 1.7 K and for fields up to 4 Tesla is shown in Fig.4 together with the predictions based on the RPA model. At low fields both the energies and the corresponding intensities agree well with the model calculations, but at fields above 2 Tesla substantial deviations are found. The observed inelastic intensities now

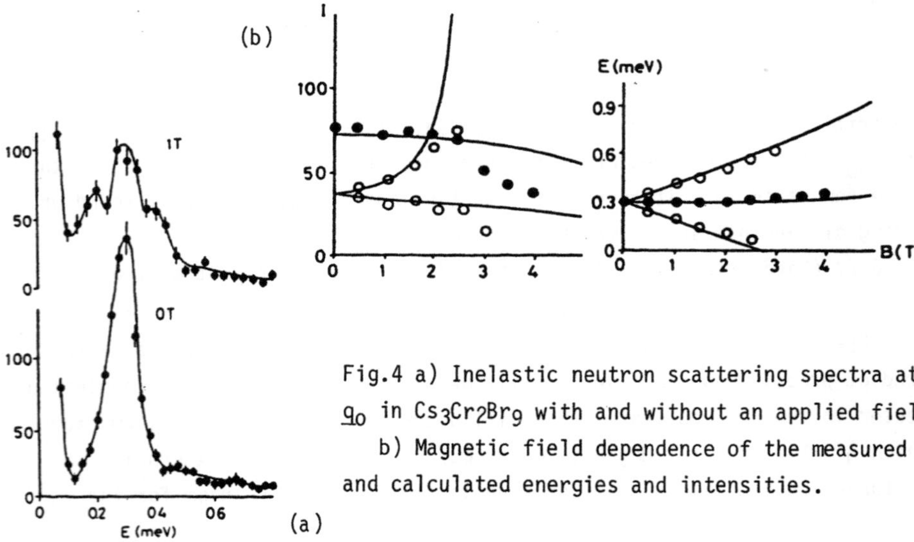

Fig.4 a) Inelastic neutron scattering spectra at q_o in $Cs_3Cr_2Br_9$ with and without an applied field.
b) Magnetic field dependence of the measured and calculated energies and intensities.

fall below the calculated values, but no significant increase in the elastic or quasielastic scattering was observed, neither at $\underline{q}_0$ nor at other values of $\underline{q}$ along the Γ-K-M line. At 4.5 Tesla an abrupt but weak increase in the Bragg scattering at ($\bar{1}\ \bar{1}\ 4$) was observed. The ($\bar{1}\ \bar{1}\ 4$) nuclear Bragg-reflection is extremely weak ($\sim 10^{-4}$ relative to the most intense) and this allowed the detection of the small ferromagnetic component which corresponds to an estimated moment $\langle K^z \rangle^2 \simeq 0.03$.

3. Discussion

The absence of long range order for fields less than 4.5 Tesla indicates that the simple isotropic RPA-model is insufficient for the description of the phasetransitions in these compounds. This shortcoming is either due to the neglect of anisotropy (which is known to be weak for the Cl compound) or due to the more fundamental neglect of fluctuations which may be particularly important because of the competing nature of the interactions. This causes the peculiar degeneracy or near-degeneracy of the excitation energies in the basal plane (Fig. 3(a)) and in this respect these systems become analogous to other incommensurate hexagonal magnets like the light rare earths and $RbFeCl_3$ [13]. It is notable that the soft-mode response in the dimer systems fails to show any significant quasielastic central peak component in the spectra even when the soft mode is driven to energies below 0.1 meV by the application of a magnetic field. This is contrary to the behaviour observed in both the metallic and the non-metallic singlet ground-state systems [13-14] .

To conclude we remark that the neutron scattering studies of the Cr^{+3}-Cr^{+3} dimer compounds have established these as a novel class of singlet ground-state systems in which only few parameters are needed for a microscopic description. This renders the systems well suited for both experimental and theoretical studies. The compounds discussed here have many chemical isomorphs, and an interesting trend with an increasing tendency towards long range magnetic order is found when going through the halogen series. The hexagonal arrangement adds additional richness and/or complexity to the ordering transitions which calls for further study at lower temperatures and higher fields.

References :

1 P. Fulde in Handbook on the Physics and Chemistry of Rare Earths p 296, eds. K.A. Gschneider and L. Eyring, North Holland (1978)

2 See Magneto-Structural Correlations in Exchange-couple Systems, Ed. R.D. Willet, Reidel, Amsterdam (1984)

3 A. Furrer and H.U. Güdel, Phys. Rev. Lett. 39 (1977) 657

4 B. Leuenberger, H.U. Güdel, R. Feile and J.K. Kjems, Phys. Rev. B 28 (1983) 5368

5 B. Leuenberger, A. Stebler, H.U. Güdel, A. Furrer, R. Feile and J.K. Kjems, submitted to Phys. Rev. B

6 B. Leuenberger, H.U. Güdel, J.K. Kjems and D. Petitgrand, to be published

7 B. Leuenberger, H.U. Güdel, R. Feile and J.K. Kjems, submitted

8 P. Fisher, private communication

9 S.K. Burke, W.G. Stirling and K.A. Mc Ewen, J. Phys. C (1981) L 976

10 B. Lebech, K.A. Mc Ewen and P.A. Lingaard, J. Phys. C8 (1975) 1684

11 K. Hirakawa, H. Kadowaki and R. Ubukoshi, Physica 120B (1983) 187

12 D. Petitgrand, B. Hennion, P. Radhakrishna, C. Escribe and S. Legrand in Recent Developments in Condensed Matter Physics 4, Eds. J.T. Devreese et al. (1982) 205

13 P.A. Lindgaard, to be published

14 R.W. Youngblood, G. Aeppli and J.D. Axe, Phys.Rev.Lett. 49 (1982) 1724

Helimagnetism in the Heisenberg Model: New Features and Old Mechanisms

E. Rastelli, L. Reatto, and A. Tassi
Dipartimento di Fisica dell'Università and GNSM, I-43100 Parma, Italy

1. Introduction

Competition between exchange couplings is the well known mechanism [1] giving non collinear spin configuration in an Heisenberg model and here we discuss two different situations. First we consider the case of competition between couplings in the basal plane as is the case, for instance, of compounds of the $NiBr_2$ family. The zero temperature phase diagram can be obtained straightforwardly in the classical approximation [2] and we have shown [3] that on a border line in the parameter space between two different helical configurations interesting new phenomena arise when the transition is not of first order. The ground state has an infinite degeneration corresponding to unequivalent $\vec{Q}$ helix wavevectors that lie on a line of the reciprocal space that we call "degeneration line" and this line is also a "soft line", i.e. the spin wave energy is zero; on the above mentioned border line, due to increased fluctuations due to the soft line, long range order (LRO) is absent at any finite temperature in 3D and even at zero temperature in 2D. The presence of a logarithmic divergence in 3D suggests the existence of an unorthodox low temperature phase, that we call "isotropic helix", without LRO but with algebraic decay of correlations, where the helix wavevector $\vec{Q}$ fluctuates locally through the infinite ground states. Such degeneration lines are present in square, tetragonal, triangular, hexagonal and rhombohedral lattices for suitable values of three in-plane J_1, J_2, J_3 exchange integrals.

Interesting questions arise concerning the stability of the above described behaviour against further interactions such as more distant neighbour interaction and (or) single ion planar anisotropy. We find that further in-plane couplings like J_4 take off the ground state degeneration and the phase transition between unequivalent helices is first order. The energy difference between the different states is extremely small however and one expects that for J_4 small enough the isotropic helix phase could exist at intermediate temperatures following a low temperature ordered phase.

Single ion planar anisotropy D does not lift the infinite degeneration of the ground state on the border line but it destroys the soft line in the magnon dispersion, so that LRO is now stable at low temperature.

Secondly, in view of studying quantum effects and the thermal evolution of the helix wavevector $\vec{Q}(T)$ in helimagnetism, we consider a simpler model in which competition is present only along one crystallographic axis. We call this model ANNNH (anisotropic next nearest neighbour Heisenberg) model since it is the Heisenberg analogous of the ANNNI model studied by FISHER and SELKE [4]. In the classical case the spin wave energy vanishes at $\vec{k}=\pm\vec{Q}$ and this is due to the spin rotation invariance of the hamiltonian. Thus this soft mode is a Goldstone mode that should persist also in the quantum case [5]. We have computed the quantum corrections to the ground state and to the spin wave energy at the lowest order in 1/S and we find that the spin wave energy has indeed a soft mode at $\vec{k}=\pm\vec{Q}$. Finally we consider the ANNNH model in presence of a planar single ion anisotropy. Quantum effects increase the ground state value of Q by only few percents above the classical value. Study of the free energy in the spin wave approximation allows the determination of the temperature variation of Q.

2. Infinite Degeneration of the Ground State, "Soft Lines" and "Isotropic Helix"

The hamiltonian we consider reads

$$\mathcal{H} = -\sum_{\alpha} J_\alpha \sum_{i\delta_\alpha} \vec{S}_i \cdot \vec{S}_{i+\delta_\alpha} \tag{1}$$

where $\vec{S}_i$ is the spin of the site i of a Bravais lattice, $\vec{\delta}_\alpha$ joins the site i with its neighbours of the α-th shell, J_α is the exchange coupling strength between the site i and its neighbours of the α-th shell. We consider the presence of four couplings, the ones with $\alpha=1,2,3$ when $\vec{\delta}_\alpha$ lies in the basal plane and that with $\alpha=0$ when $\vec{\delta}_\alpha$ is out of the basal plane. In this case the helix wavevector $\vec{Q}$ lies in the basal plane.

In order to study statics and dynamics of helical configurations we perform the well known following steps: 1) the spins are represented in a local reference whose ζ-axis of the site i forms an angle $\vec{Q}\cdot\vec{r}_i$ with an in-plane crystallographic axis, 2) the ζ-axis is assumed as a local quantization axis and the bosonic equivalent hamiltonian is obtained by a suitable spin bosontransformation [6], 3) we retain only contribution proportional to S^2 and S and this amounts to a linear approximation of the dynamics. By the steps 1),2),3) we have

$$\mathcal{H} = E_{cl} + \Delta_2 + \sum_k E_k \alpha_k^+ \alpha_k \tag{2}$$

$$E_{cl} = -S^2 N \sum_\alpha J_\alpha \sum_{\delta_\alpha} \cos \vec{Q}\cdot\vec{\delta}_\alpha \ . \tag{3}$$

E_{cl} represents the energy of the model if the spins were classical vectors. Minimization of E_{cl} [2] with respect to $\vec{Q}$ gives the helix wavevector, that we call $\vec{Q}_c$, in the classical approximation. Δ_2 is proportional to S and is the first quantum correction to the ground state energy,

$$\Delta_2 = -\frac{1}{2}\sum_k (A_k - E_k) \tag{4}$$

$$E_k = \sqrt{A_k^2 - B_k^2} \tag{5}$$

with

$$A_k = \sum_\alpha 2J_\alpha S \sum_{\delta_\alpha} \left[\cos\vec{Q}_c\cdot\vec{\delta}_\alpha - \frac{1}{2}\left(1+\cos\vec{Q}_c\cdot\vec{\delta}_\alpha\right)\cos\vec{k}\cdot\vec{\delta}_\alpha\right] \tag{6}$$

$$B_k = -\sum_\alpha 2J_\alpha S \sum_{\delta_\alpha} \frac{1}{2}\cos\vec{k}\cdot\vec{\delta}_\alpha\left(1-\cos\vec{Q}_c\cdot\vec{\delta}_\alpha\right) \tag{7}$$

and α_k^+ is the creation operator of a magnon whose dispersion curve is E_k. The zero temperature phase diagram in classical approximation of our model on a tetragonal lattice is shown in Fig.1 (for details see reference [2]).

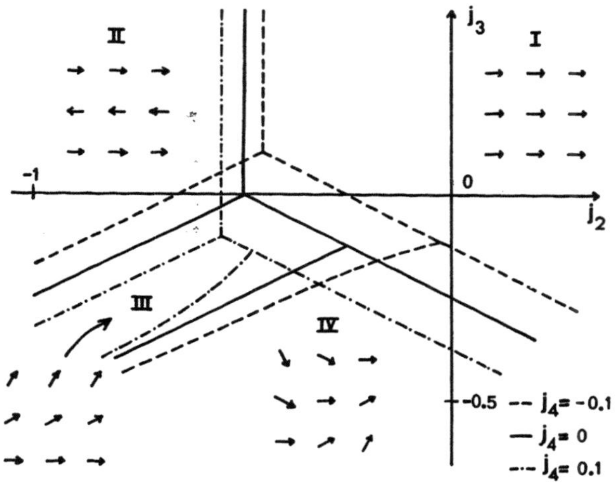

Fig.1. Phase diagram for the square and tetragonal lattice ($j_\alpha = z_\alpha J_\alpha / z_1 J_1$). Continuous curve refers to j_4=0, dashed curve to j_4= -0.1, chain curve to j_4=0.1

The line $J_2=2J_3$ for $J_2 < -J_1/4$ is the phase boundary between two different helices that we call III and IV and a degeneration line is present: any helix whose $\vec{Q}$ satisfies the condition

$$\cos aQ_x + \cos aQ_y = -J_1/2J_2 \,, \tag{8}$$

where a is the lattice constant, is a ground state. Therefore the phase transition at T=0 K between helix III and IV is not of first order. Positive J_0 gives Q_z=0, negative J_0 implies $Q_z = \pi/c$, c being the distance between planes.

From (5) one can see that E_k=0 if k equals any of the infinite ground states wavevectors so that the dispersion curve exhibits a soft line.

The expectation value of the local magnetization reads

$$\langle S_i^{\zeta} \rangle = S - \frac{1}{N}\sum_k \frac{A_k - E_k}{2E_k} - \frac{1}{N}\sum_k \frac{A_k}{E_k}\langle \alpha_k^{+}\alpha_k \rangle \tag{9}$$

which exhibits a logarithmic divergence at T≠0 K arising from fluctuations in the neighborhood of the soft line. In the 2D limit ($J_0=0$) at T≠0 K a power-like divergence is present in addition to the usual logarithmic divergence coming from the long wavelength region. So LRO is destroyed by the fluctuations associated to the soft line whose wavevectors coincide with the ground state $\vec{Q}$ wavevectors. This suggests that at low temperature the system,even if it has not helical order,it has some sort of local helical order changing from place to place and the local $\vec{Q}$ wavevector spans the degeneration line . We call such a state an "isotropic helix". In this case an elastic neutron scattering experiment should find a ring structure in place of the satellites found with a usual helix. In effects ring-like scattering has been found [7] at low temperature in $NiBr_2$ mixed with Zn; unfortunately such experimental data could also be due to inhomogeneity of the sample.

There is some indication that the low temperature phase on the line $J_2=2J_3$ should have algebraic decay of the correlation function. In fact in the spin wave approximation we find

$$\langle \vec{S}_i \cdot \vec{S}_j \rangle \cong \cos \vec{Q}\cdot\vec{r}_{ij} \left(S^2 - cT \ln r_{ij}^{\perp} \right) \tag{10}$$

where $r_{ij}^{\perp}$ is the in-plane projection of $\vec{r}_{ij}$. In other systems with no LRO such logarithmic divergence is the signature of an algebraic decay of correlations and the spin wave approximation gives the low temperature expansion of the exact correlation function [8] . However a full computation of the correlation function in the present case is not yet available.

The previous result related to the presence of a degeneration line is not specific to the tetragonal lattice. In fact such a line is present also for the other lattices that we have studied: triangular,hexagonal and rhombohedral.

We consider now effects which can be of interest in physical compounds,such as more distant neighbour in-plane interactions and planar anisotropy.Figure 1 shows the T=0 K phase diagram of the tetragonal model when a forth neighbour in-plane interaction J_4 ($J_4=\pm 0.05J_1$) is present.In Fig.2 the ground state energy is shown along the line in the reciprocal space where the energy is a minimum in the radial direction for sets of hamiltonian parameters for which the model crosses the border line between helix III and helix IV. As one can see,unlike the case $J_4=0$,new extrema between helix III and IV arise and the transition is now of first order.From quantitative point of view the new extrema are very weak so that the magnon dispersion curve exhibits a structure reminiscent of a soft line.With the choise of hamiltonian parameters performed in Fig.2.(c) and $J_0=J_1=18$ K (which gives a mean field Néel temperature $T_N \sim 50$ K) the ripple in the magnon dispersion curve has an amplitude of nearly 1 K. Therefore for temperature greater than the weak ripple produced by J_4,an isotropic helix phase should take over the low temperature phase with LRO .

Planar anisotropy does not lift the infinite degeneration of the ground state but destroys the soft line of the magnon disper-

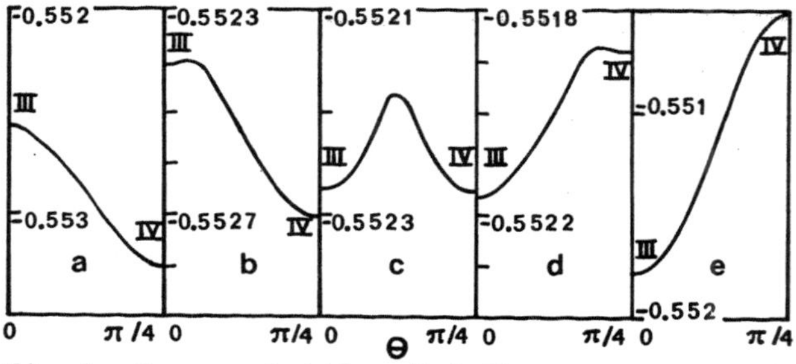

Fig.2. Square lattice T=0 K energy $e_o=E_{cl}/4J_1NS^2$ along the line where it is minimum in the radial direction (θ is the polar angle of this line) for $j_3=-0.25$, $j_4=-0.1$ and $j_2=-0.39$(a), -0.392(b), -0.394(c) (1st order transition), -0.395(d), -0.40(e), respectively

sion curve. In this case LRO is not destroyed at low temperature. If the anisotropy is extremely small the gap is also small and we could again expect the existence of an isotropic helix phase at intermediate temperature.

3. ANNNH Model and Quantum Effects

Quantum effects in helimagnetism have not been studied so far and we have undertaken such a study for the ANNNH model. In this model there is only a n.n. interaction J_o in the basal plane and there is a n.n. interaction J_1 and a n.n.n. interaction J_2 in the z direction. In the classical limit the ground state has helimagnetic order when $J_2/J_1 < -1/4$. We have computed the quantum correction of order 1/S to the spin wave energy E_k. To this order two contributions are present, one due to the four Bose operator terms of the hamiltonian computed in first order perturbation theory and a second contribution comes from the three operators terms (present only in non collinear states) computed to second order. These two contributions do not vanish separately at $\vec{k}=\pm\vec{Q}$ [9] but they exactly compensate each other at $\vec{k}=\pm\vec{Q}$ so that the soft mode behaviour $E_{\pm Q}=0$ persists in the quantum case [5], in agreement with the existence of a Goldstone theorem. We recall that the invariance of the hamiltonian under spin rotation around the axis perpendicular to the plane where the spins are lying is reflected by the soft mode at k=0, while the invariance under rotation of that plane is related to the soft mode at $\vec{k}=\pm\vec{Q}$. This indicates that the 1/S expansion is a valid perturbative approach also for helimagnets. To this order the ground state energy is well defined only for $\vec{Q}=\vec{Q}_c$ so there is no quantum correction of order 1/S to the helix wavevector. This is no more true when an anisotropy is present. However the change of Q is very small, for instance Q increases by 3.3% with respect to Q_c when S=1, $J_o=J_1$, $J_2=-0.5J_1$, $D=0.1J_1$.

Helimagnets show a temperature variation of $\vec{Q}$ and we have investigated this effect for the ANNNH model. In the isotropic case there is no temperature variation within perturbation theory. The situation is different when an anisotropy is present and we have computed Q(T) by minimizing the free energy in the spin wave approximation. This effect is very small; for instance with the same parameters as above we find that Q(T) differs by 4% less than Q(0) for $T=0.9T_N$, where T_N is the mean field Néel temperature.

However, the fact that the temperature renormalization cannot be computed in the isotropic case within the perturbation scheme suggests that other mechanisms should be operating and the previous estimate has to be taken as approximate.

Extension of the computation of quantum effects to the model of section 2 should be interesting but the computation becomes very cumbersome. We expect that quantum effects destroy the presence of a degeneration line throgh a mechanism similar to that due to J_4 but the energy modulation should be very small so that a quasidegeneration should persist.

1 J.Villain: J.Chem.Phys.Solids 11,303 (1957)
2 E.Rastelli,A.Tassi,L.Reatto: Physica 97B,1 (1979)
3 E.Rastelli,L.Reatto,A.Tassi: J.Phys.C 16,L331 (1983)
4 M.E.Fisher and W.Selke: Phys.Rev.Letters 44,1502 (1980)
5 E.Rastelli,L.Reatto,A.Tassi: to be published
6 A.I.Akhiezer,V.G.Bar'yakhtar,S.V.Peletminskii: Spin waves (North Holland,Amsterdam 1968)
7 P.Day,M.W.Moore,C.Wilkinson,K.R.A.Ziebeck: J.Phys.C 14, 3423 (1981)
8 F.Wegner: Z.Phys. 206,465 (1967); V.L.Berezinskii: Sov.Phys. JETP 32,493 (1971)
9 E.Rastelli,L.Reatto,A.Tassi: J.Appl.Phys. 55,1871 (1984)

Magnetic Excitations in the Quasi-2d Planar Magnets $BaM_2(XO_4)_2$ (M = Co,Ni ; X = P,As)

L.P. Regnault, J. Rossat-Mignod, J.Y. Henry[1], R. Pynn[2], and D. Petitgrand[3]

DRF/DN and [1]LETI/CRM, Centre d'Etudes Nucléaires, B.P. 85
F-38041 Grenoble Cedex, France

1. Introduction

In the past few years a great deal of effort has been devoted to studies of the phase transition in two dimensional (2d) systems with a two component order parameter and in particular to the study of 2d magnetic systems with planar anisotropy [1-8]. For this class of systems KOSTERLITZ and THOULESS [1-3] have predicted the existence of a phase transition at a *finite* temperature from a state with free vortices to a state in which the vortices are paired. This situation leads to an exponential divergence of the correlation length according to the relation :

$$\xi(T) = A \exp B(T/T_{KT}-1)^{-1/2} \quad (A \sim 1 \text{ Å} , B \sim \pi/2) \tag{1}$$

but without any longrange order down to T = 0. NELSON and KOSTERLITZ [6] have shown that the unbinding of the vortex-antivortex pairs must induce a universal *jump* in the superfluid density of superfluid systems or in the stiffness constant of hydrodynamic spin wave modes. The existence of such a type of transition has been established for ^4He superfluid and superconducting films [9,10]. Actually 2d magnetic systems are quasi-2d systems because it exists always a small interlayer coupling J' which induces a 3d long range ordering at a finite temperature T_c approximatively given by:

$$k_B T_c \simeq z'|J'|S^2 \xi^2(T_c) \tag{2}$$

where $\xi(T)$ is the correlation length. Nevertheless the behaviour of such a system *must reflect* that of the pure 2d system, if all the undesirable interactions are minimized. In particular the quasi 2d planar system must have an in-plane anisotropy as small as possible, with a symmetry $p > 4$ in the notation of JOSE et al. [8].

The situation described above seems not to be dependent on the sign of J (the in-plane interaction). In fact very recent works have shown fundamental differences between ferromagnetic and antiferromagnetic 2d-XY systems [11-13] essentially when some *frustration* is present. In a same way, the KT picture is certainly modified if exchange integrals of different signs are introduced.

In this paper we describe some properties of the magnetic excitations of the isomorphous compounds $BaM_2(XO_4)_2$ (M=Co,Ni ; X=P,As). These compounds crystallize in a structure which can be described as (001) magnetic layers

[2]ILL, B.P. 156, Centre de Tri, F-38402 Grenoble Cedex, France

[3]LLB, Centre d'Etudes Nucléaires, Saclay,
F-91191 Gif-Sur-Yvette Cedex, France

in which the M^{2+} ions are located in a *honeycomb lattice* with a neighbouring distance $a/\sqrt{3} \sim 2.8Å$, separated by XO_4-Ba-XO_4 layers [14-16]. From the large interlayer distance $c/3 \sim 7.8Å$, a pronounced 2d magnetic character is expected. Note that the symmetry of the lattice corresponds to $p = 6$ [8]. In the following, only the two compounds $BaNi_2(PO_4)_2$ and $BaCo_2(AsO_4)_2$ will be discussed.

2. Magnetic Properties of $BaNi_2(PO_4)_2$

Magnetization and neutron scattering experiments on single crystals [17,18] give evidence for an AF order below $T_N \simeq 23.5 \pm 0.5$ K, characterized by a wavevector $\vec{k} = [000]$. The spins ($S = 1$) are oriented within the layers with a collinear AF arrangement. Precise values of anisotropy and exchange parameters are deduced from the analysis of the dispersion curves of magnetic excitations. Figure 1 shows the experimental spectrum measured by inelastic neutron scattering (INS) (Siloe and Orphee reactors). The presence of two branches in the spectrum originates from the two hexagonal Bravais lattices. For $q \to 0$, the lower branch is linear and displays a small gap $\Delta_{IP} \simeq 0.1$ meV $\simeq 0.025$ THz, reflecting the very weak in-plane anisotropy. The upper branch has a gap $\Delta_{OP} \simeq 2.8$ meV $\simeq 0.7$ THz, reflecting the strong planar (single ion) anisotropy. These data are analysed in terms of the Hamiltonian:

$$H = - \sum_{i,j} J_{ij} \vec{S}_i \vec{S}_j + D \sum_i (S_i^z)^2 \qquad (3)$$

where the J_{ij} are the exchange interactions between first (J_1), second (J_2) and the third (J_3) in-plane nearest neighbours. Solid curves in Fig.1 represent the results calculated from linear spin wave theory [19,20] with the parameters given in Table 1. So the strong planar anisotropy ($H_a/H_e \simeq 0.11$) combined with the weakness of the in-plane anisotropy ($H_a^{IP} \simeq 40$ Oe) and the low value of J' ($\simeq -8.10^{-4}$K), make $BaNi_2(PO_4)_2$ a very good prototype of the 2d planar model. The behaviour of the correlation length predicted for that model has been fitted to the experiment (c.f. Fig.2). In the range $10^{-2} \lesssim T/T_N - 1 \lesssim 0.5$ the data are well reproduced by (1) with $A \simeq 0.6Å$, $B \simeq 1.6$ and taking $T_{KT} \simeq (0.96 - 0.97)T_N$. This upward shift of 3-4% in the transition temperature could be accounted for by the interlayer coupling on basis of the formula:

$$T_N \simeq T_{KT} + B^2|\tilde{J}|\, S^2(\ln|J'/\tilde{J}|)^{-2} \qquad (4)$$

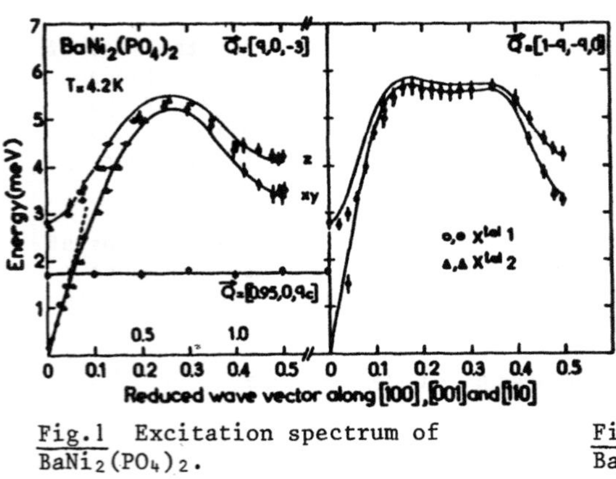

Fig.1 Excitation spectrum of $BaNi_2(PO_4)_2$.

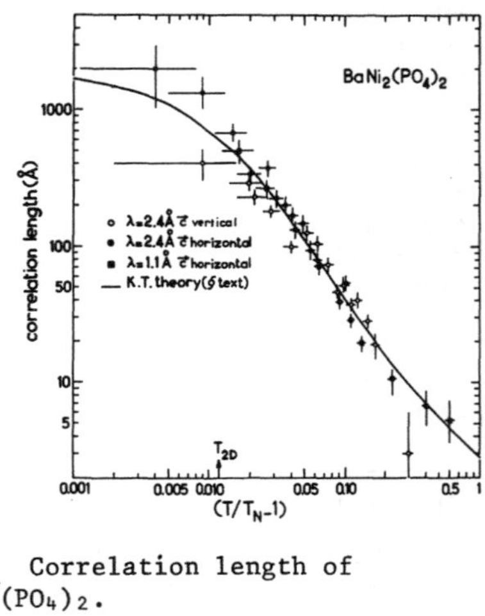

Fig.2 Correlation length of $BaNi_2(PO_4)_2$.

Table 1. Exchange and anisotropy parameters of $BaM_2(XO_4)_2$.

	J_1(K)	J_2(K)	J_3(K)	D	$J^{\parallel}/J^{\perp}$	$z'J'/z\tilde{J}$
$BaNi_2(PO_4)_2$	-2.2	-0.3	-8.8	7.3	-	$4 \cdot 10^{-4}$
$BaCo_2(AsO_4)_2$	38	1.5	-10.3	-	0.39	$\ll 10^{-3}$

obtained from combining (2) and (1). In (4), $\tilde{J}$ is an *effective* interaction between *first* neighbours defined by $\tilde{J} = -E(\vec{k})/3S^2$, where $E(\vec{k})$ is the energy of the classical ground state. With this definition we get $k_B T_{KT}/|\tilde{J}|S^2 \simeq 2.2$, very close to the theoretical value [1,2]. To further test the applicability of the KT model, we have measured the temperature dependence of the excitations at different q-values. For small q, the theory predicts a discontinuous decrease of the spinwave energy [6]. Experiments (c.f. Fig.3) show that for $q \lesssim 0.05\ a^*$ the excitations are strongly renormalized ($\omega(T \simeq T_N)/\omega(T \simeq 0) \simeq 0.5$) but are observed up to T_N, where they disappear suddenly, due to the existence of free vortices. Excitations of larger q-values persist above T_N, but they are also renormalized and damped. These results provide strong indications for the existence of a KT-type transition in $BaNi_2(PO_4)_2$ which appears to be a much better prototype of the 2d planar model than e.g. K_2CuF_4 [21]. To our opinion the good agreement with the KT picture arises from the fact that in $BaNi_2(PO_4)_2$ $\vec{k} = 0$ and the J_i are negative, so that the frustration is irrelevant in this system.

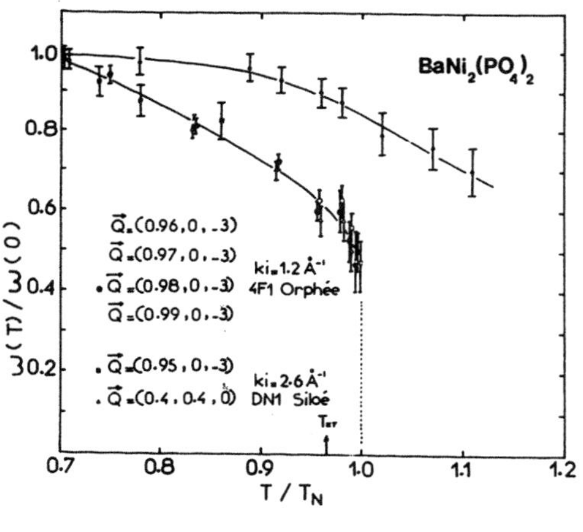

Fig.3 Temperature dependence of excitations in $BaNi_2(PO_4)_2$.

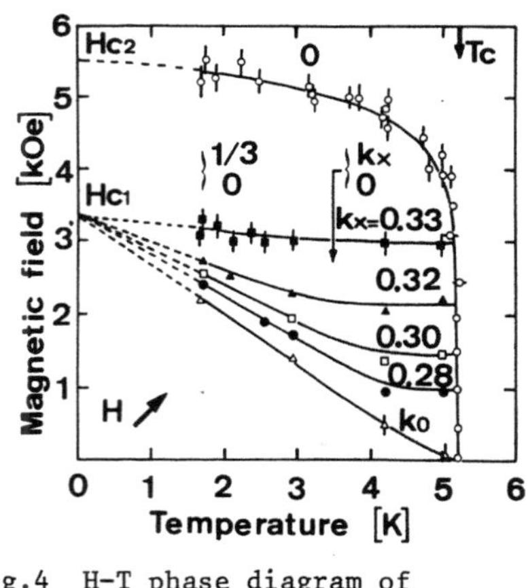

Fig.4 H-T phase diagram of $BaCo_2(AsO_4)_2$ in increasing fields.

3. Magnetic Properties of $BaCo_2(AsO_4)_2$

Magnetization, specific heat and neutron scattering experiments on $BaCo_2(AsO_4)_2$ give evidence of a *helimagnetic* order below $T_N \simeq 5.33 \pm 0.05$ K characterized by an incommensurate wave vector $\vec{k} \simeq [0.27,0,-1.33]$ [22-24]. The magnetic structure can be viewed within a (001) layer as a stacking of quasi ferromagnetic "chains" weakly coupled ($2\pi k_x \simeq 94°$). The peculiarity of this structure is to disappear very suddenly between 5 K and T_N, whereas the specific heat

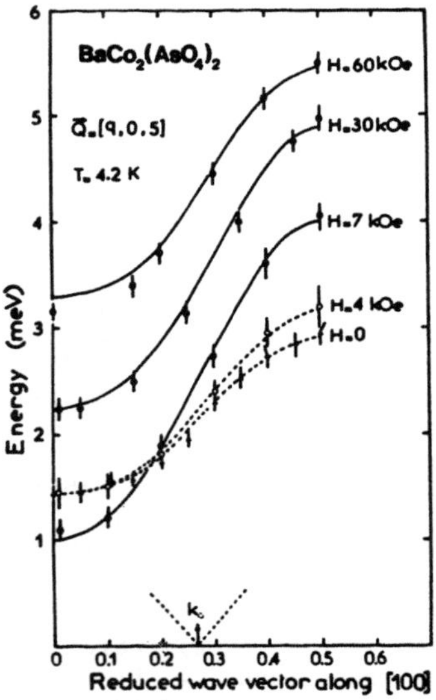

Fig.5 Excitation spectrum of $BaCo_2(AsO_4)_2$ versus H.

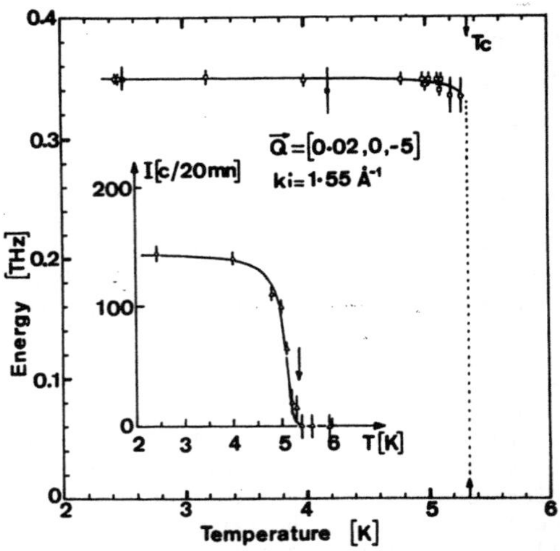

Fig.6 Temperature dependence of the energy and intensity of the excitations in $BaCo_2(AsO_4)_2$.

displays a sharp λ-type anomaly at T_N [16,23]. The analysis of magnetization and susceptibility data indicates a strong planar character, described by the anisotropic Hamiltonian:

$$H = - \sum_{i,j} (J_{ij}^{\perp}(s_i^x s_j^x + s_i^y s_j^y) + J_{ij}^{//} s_i^z s_j^z) + g_{\perp} \mu_B H \sum_i s_i^y \qquad (5)$$

with $J^{//}/J^{\perp} \simeq 0.4$ and where $\vec{s}_i$ is an effective spin 1/2. Figure 4 shows the H-T phase diagram in increasing fields. Two phase transitions at $H_{c1} \simeq$ 3kOe and $H_{c2} \simeq$ 5.5 kOe occur , with an intermediate *quasi-2d* collinear ferrimagnetic ...++-++-... structure and a final saturated paramagnetic structure [24]. In decreasing fields metastability and irreversibility take place as discussed in [24], due to the possibility for the spin within a chain to turn easily. In Figure 5 we give the low energy excitation spectrum measured (at the Siloe and ILL reactors) in the [100] direction for several values of the applied field (H). As in $BaNi_2(PO_4)_2$ two branches are observed in $BaCo_2(AsO_4)_2$. The upper branch (not plotted on Fig.5) is weakly q-dependent and is approximatively given by $\omega(q) \simeq 12.5 + 0.03$ H (ω in meV, H in kOe). The lower branch is more difficult to understand. For $H > H_{c2}$ the linear spin wave theory developed from (5) reproduces quite well the experimental data with the exchange and anisotropy parameters given in Table 1. Note that for this compound J_1 and J_3 are of the same order of magnitude, but with opposite signs, in such a way that important frustration effects are expected. Yet the "simple" spin wave theory cannot explain the low energy excitation spectrum observed when $H < H_{c2}$. This spectrum is characterized by an isotropic gap $\Delta_o \simeq 1.45$ meV $\simeq 0.35$ THz (XY fluctuations). No linear branch merging from $\vec{q} = \vec{k}$ has been detected, although the zero field magnetic specific heat gives evidence for a large T^2 contribution at low T [23]. Moreover in $BaCo_2(AsO_4)_2$ (and $BaCo_2(PO_4)_2$), the low-T specific heat appears to increase anomalously fast, with an exponential law [23] : $C_m = AT^2 + Bf(T)\exp(-\Delta/T)$, with A $\simeq$ 30 mJ/K^3/mole and $\Delta \simeq 30$ K. The important low-T specific heat, combined with the helimagnetic character of the structure and with the large in-plane susceptibility (and the small value of T_N), make the in-plane anisotropy not responsible for the gap Δ_o. These results suggest the existence of *another type* of magnetic excitations with a small creation energy in addition to the spin waves. But it is difficult to be more precise without the help of a quantitative theory. Further information is obtained by measuring the temperature dependence of the low q-excitations. Figure 6 gives, for instance, the temperature dependence of the energy at $\vec{q} = 0.02\ \vec{a}^*$. In contrast to $BaNi_2(PO_4)_2$, a very small renormalization of the energy with T occurs. The main effect of the temperature is seen on the damping and on the intensity of these excitations, which reflect the disappearance of the staggered magnetization. Note that the higher q-excitations persist above T_N (up to 2-3 T_N), with a renormalization and a damping which increase with T [16]. Correlatively a quasi-elastic peak centered around $q = 0$ and $\omega = 0$ is growing up, with an integrated intensity in energy maximum around 6-6.5 K [16]. This contribution persists at low temperatures, where it has a resolution-limited energy width (c f. Fig.7). The effect of the temperature is to produce a fast increase of the energy width Γ_ω above 5 K. Such effects are not observed at $\vec{q} = \vec{k}$, where the scattering intensity remains resolution-limited in energy. These results suggest the existence of fast XY fluctuations around $q = 0$ above T_N. Note that a central peak is effectively predicted around $q = 0$ and $\omega = 0$ for the ferromagnetic 2d planar model, due to the (slow) motion of free vortices [25]. Nevertheless it seems that the simple mechanism involved in the KT transition is modified by the existence of several exchange interactions in $BaCo_2(AsO_4)_2$. It is easy to show that the energies of the structures of the H-T phase diagram (Fig.5) are given by: $E/NS^2 \approx 2(J_1 + J_2) + e_c$, with e_c (the interaction energy between chains) $\ll 2(J_1+J_2)$ (ratio $\sim$ 1/10 to 1/20). As a consequence, the energy of a vortex such as that of Fig.8 is considerably reduced and it can be viewed as an assembly of π-solitons (i.e. a delocalized spin-flip) weakly coupled [26,20]. Such a π-soliton has an energy $E_o \approx 2.2\ J_1S^2 \approx 20K$, which has the good order of magnitude ($\Delta_o \approx$ 16-17K, $\Delta \approx$ 30K). Actually these low energy excitations are responsible for the small value of the ordering temperature in $BaCo_2(AsO_4)_2$ ($k_BT_N/J_1S^2 \approx 0.75$) which is only 1/3 the classical KT-value.

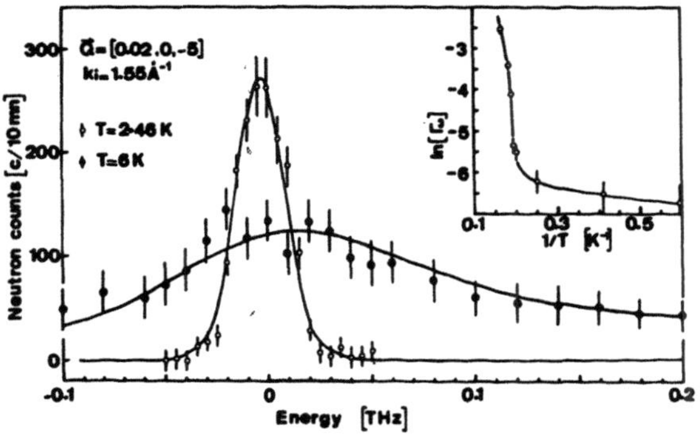

Fig.7 Temperature dependence of the central peak in $BaCo_2(AsO_4)_2$. In insert Γ_ω versus T.

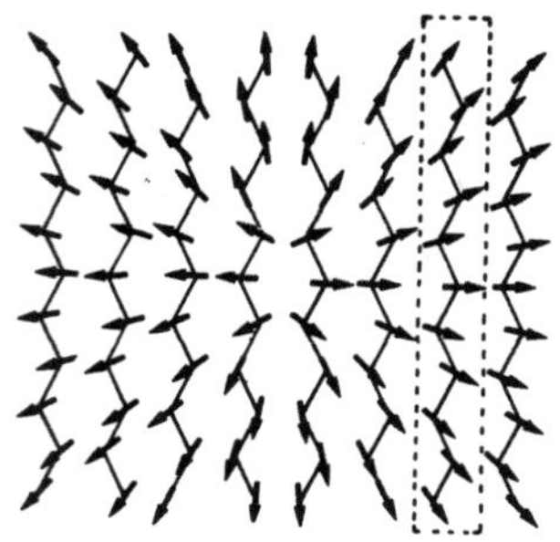

Fig.8 Relation between vortex and π-soliton.

4. Conclusion

The analysis of the experimental data strongly supports that the transition in $BaNi_2(PO_4)_2$ or $BaCo_2(AsO_4)_2$ is closely connected with the KT transition and with the existence of vortex-type excitations. The KT theory appears to apply better for $BaNi_2(PO_4)_2$ than for the quantum system $BaCo_2(AsO_4)_2$ in which quite unusual effects take place. This situation agrees with recent theoretical studies, which show the importance of both the quantum character [27] and the frustration [11-13]. But at this time, more theoretical calculations are needed to understand quantitatively the magnetic excitation properties of this compound (in particular for the honeycomb lattice).

References

1. J.M. Kosterlitz, D.J. Thouless: J. Phys. C6, 1181 (1973).
2. J. Villain: J. de Phys. 35, 27(1974); 36, 581 (1975).
3. J.M. Kosterlitz: J. Phys. C7, 1047 (1974).
4. V.L. Berezinski: Sov. Phys. JETP 32, 493 (1971); 34, 610 (1972).
5. L.J. De Jongh, A.R. Miedema: Adv. Phys. 23, 1 (1974).
6. D.R. Nelson, J.M. Kosterlitz: Phys. Rev. Lett. 39, 1201 (1977).
7. V.L. Pokrovski, G.V. Uimin: Adv. Phys. 28, 595 (1979).
8. J.V. Jose, L.P. Kadanoff, S. Kirpatrick, D.R. Nelson: Phys. Rev. B16, 1217 (1977).
9. D.J. Bishop, J.D. Reppy: Phys. Rev. Lett. 40, 1727 (1978); I. Rudnick: Phys. Rev. Lett. 40, 1454 (1978).
10. A.F. Hebard, A.T. Fiory: Phys. Rev. Lett. 44, 291 (1980).
11. S. Teitel, C. Jayaproskash: Phys. Rev. B27, 598 (1983).
12. D.H. Lee, J.D. Joannapoulos, J.W. Negele, D.P. Landau: Phys. Rev. Lett. 52, 433 (1984).
13. S. Miyoshita, H. Shiba: J. Phys. Soc. Jpn (in press).
14. S. Eymond, A. Durif: Mat. Res. Bull. 4, 595 (1969).
15. J.Y. Henry: Ph.D. thesis (University of Grenoble, France, 1976).
16. L.P. Regnault, J. Rossat-Mignod, J.Y. Henry: J. Phys. Soc. Jpn 52, 1 (1983).
17. L.P. Regnault, J. Rossat-Mignod, J.Y. Henry, L.J. De Jongh: JMMM 31-34, 1205 (1983).
18. L.P. Regnault, J.Y. Henry, J. Rossat-Mignod: JMMM 15-18, 1021 (1980).
19. E. Rastelli, L. Reatto, A. Tassi: Physica B-C 97, 1 (1979).
20. L.P. Regnault: Ph.D. thesis (University of Grenoble, France, 1981).
21. K. Hirakawa, H. Yoshizawa, J.D. Axe, G. Shirane: J. Phys. Soc. Jpn 52, 4220 (1983).
22. L.P. Regnault, P. Burlet, J. Rossat-Mignod: Physica 86-88B, 660 (1977).
23. L.P. Regnault, J. Rossat-Mignod, J. Villain, A. De Combarieu: J. de Phys. 39, 579 (1978).
24. L.P. Regnault, J. Rossat-Mignod: JMMM 14, 194 (1979).
25. D.L. Huber: Phys. Lett. 76A, 406 (1980); J. Appl. Phys. 53, 1899 (1982).
26. J. Villain: (unpublished results 1976).
27. D.D. Betts, S.B. Kelland: J. Phys. Soc. Jpn 52, 11 (1983).

Spin Dynamics of Rare Earth Ions in Crystal Fields Probed by Muon Spin Rotation

P.M. Richards

Sandia National Laboratories, Albuquerque, NM 87185, USA

Muon spin rotation (μSR) has been used to probe the dynamics of rare-earth ions in situations where the crystal field splittings have important effects. This paper outlines the theory of spin dynamics in such cases and discusses its application to μSR studies in two specific systems: erbium in noble metals and holmium in $LuRh_4B_4$. The former shows novel effects of a muon-induced axial crystal field in the otherwise cubic lattice, and the latter exhibits a dramatic change in temperature dependence because of small admixtures from the tetragonal field.

I. INTRODUCTION

Muon spin rotation (μSR) can act as a probe of spin dynamics [1] in a manner which complements the more widely-used techniques of NMR, ESR and neutron scattering. Attention here is focused on the spin dynamics of rare earth ions in situations where the complex crystal field levels produce novel features which are particularly amenable to study with μSR.

The examples selected are from μSR studies of erbium in noble metals [2] and of holmium in $LuRh_4B_4$ [3]. The former displays dynamics quite different than probed by ESR [4], and the latter has some unusual crystal field effects aside from the interest in its superconductivity and ferromagnetism [5].

II. THEORY OF MUON DEPOLARIZATION AND SPIN DYNAMICS

The spectral density $\langle J_\alpha J_{\alpha'} \rangle_\omega$ of a fluctuating rare-earth ion can be probed via the μSR depolarization rate Λ. (Λ^{-1} is equivalent to the NMR T_1 for the present work.) Here J_α is a component of the ion angular momentum and $\langle\ \rangle_\omega$ indicates transform at the angular frequency ω of the thermal equilibrium time correlation function $\langle J_\alpha(t) J_{\alpha'}(0)\rangle$. This is given by the memory function technique [6] which has been applied previously [7] to rare-earth ions in crystal fields, although in somewhat less complex situations than some of the ones we consider. The method is briefly described below and an expression presented in a form suitable for application. Attention is confined to single-ion dynamics in metals where the dominant relaxation mechanism is coupling to conduction electrons.

The Hamiltonian is

$$\mathcal{H} = V_{cf} + g_J \mu_B \vec{J} \cdot \vec{H} - 2I(g_J - 1)\vec{J} \cdot \vec{\sigma} + \mathcal{H}_e \qquad (1)$$

where V_{cf}, the Hamiltonian for the rare-earth ion in a static crystal field, is written in terms of operators O_n^m which are functions of J as described, for example, in ABRAGAM and BLEANEY [8]. The conduction-electron spin σ is coupled to the rare-earth spin via the interaction constant I.

The conduction-electron Hamiltonian $\mathcal{H}_e$ is assumed to be describable by non-interacting Fermions with a density of states N(0) (per spin state per atom) at the zero-temperature Fermi level E_F, and we assume $E_F \gg k_B T$ at the temperature T so that knowledge of N(0) suffices.

The equation of motion for a component of $\vec{J}$ is iterated assuming rapid fluctuation of the conduction electrons. A convolution integral results whose Laplace transform leads to the system of equations

$$\begin{aligned}(s-i\omega_{mn})\alpha_{mn} + \lambda \sum_{pq} [G_{mp,pq} f(\omega-\omega_{pn})\alpha_{qn} + G_{pq,qn} f(-\omega+\omega_{mq})\alpha_{mp} \\ -G_{mp,qn}(f(\omega-\omega_{pn}) + f(-\omega+\omega_{mq}))\alpha_{pq}] = \langle m|J_\alpha|n\rangle - \langle J_\alpha\rangle_{AV}\delta_{mn}\end{aligned} \tag{2}$$

where $\lambda = 2\pi\,[I(g_J-1)N(0)]^2\, k_B T/\hbar$, $f(x) = \beta x/(1-e^{-\beta x})$ ($\beta = 1/k_B T$), $G_{k\ell,mn} = \sum_\gamma \langle k|J_\gamma|\ell\rangle\langle m|J_\gamma|n\rangle$ ($\gamma = x,y,z$) and $\omega_{mn} = (E_m-E_n)/\hbar$ with $|m\rangle$ an eigenstate of the ion Hamiltonian (first two terms in Eq.(1)) with energy E_m. The Laplace transform at $s = \varepsilon + i\omega$ is written as α_{mn} and $\beta\varepsilon \ll 1$ has been assumed in writing the f's. The term $\langle J_\alpha\rangle_{AV}$ is the thermal average. Its inclusion guarantees that $\langle J_\alpha(t)J_{\alpha'}(0)\rangle \rightarrow \langle J_\alpha\rangle_{AV}\langle J_{\alpha'}\rangle_{AV}$ for $t \rightarrow \infty$. Isotropy of conduction-electron correlations, $\langle\sigma_\alpha(t)\sigma_{\alpha'}(0)\rangle = \langle\sigma_\alpha(t)\sigma_\alpha(0)\rangle\delta_{\alpha\alpha'}$ independent of α has been assumed. The Laplace transform of $\langle J_\alpha(t)J_{\alpha'}(0)\rangle$ is obtained from the solution to (2) by using

$$\langle J_\alpha J_{\alpha'}\rangle_s = \sum_{mn} e^{-\beta E_m}\alpha_{mn}\langle n|J_{\alpha'}|m\rangle / \sum_m e^{-\beta E_m}. \tag{3}$$

The above Eqs.(2) and (3) have been checked to give the same results obtained in Ref. [7] when applied to Ce^{3+} in cubic symmetry as studied there. In that work various symmetries were employed to make the equations more compact, whereas Eq.(2) assumes no particular symmetry. They are in a form which makes the values of the coefficients clear; so they can be programmed immediately. The number of equations can sometimes be reduced by noting, at least for the systems considered here, that if $\langle m|J_\alpha|n\rangle = 0$ at $t = 0$, it remains zero, so the number of equations needed for determining $\langle J_\alpha J_{\alpha'}\rangle_s$ is the number of nonzero matrix elements $\langle m|J_\alpha|n\rangle$. With low symmetry, as for an applied field at an arbitrary angle, however, all matrix elements exist and there is no reduction of the full system of $(2J+1)^2$ equations.

III. ERBIUM IN GOLD AND SILVER

By doping the noble metals with paramagnetic impurities it has been possible to observe μSR depolarization and thereby measure muon diffusion rates [2]. Above about 150K and 300K for Au and Ag, respectively, the rapidly-hopping muon makes many encounters with rapidly fluctuating Er^{3+} spins, and Λ becomes a measure of the spectral density. Er^{3+} in the fcc noble metals has a Γ_7 ground state [8] which has been well-studied by ESR [4] below about 20K. The temperatures of interest here are comparable to the overall splittings of about 100K and 200K for Au and Ag, respectively [9], so there are interesting multi-level effects. A more important feature is that the muon which diffuses to the immediate vicinity of the impurity can produce an axial field with interaction $-DJ_z^2$ (z is now the axis of the muon-ion direction) which distorts the levels. This is known [10] to occur in Cu and Al, and if the same electric field gradient exists here as was measured in Cu, it would convert [8] to a large $|D|/k_B$ = 18K for Er^{3+}.

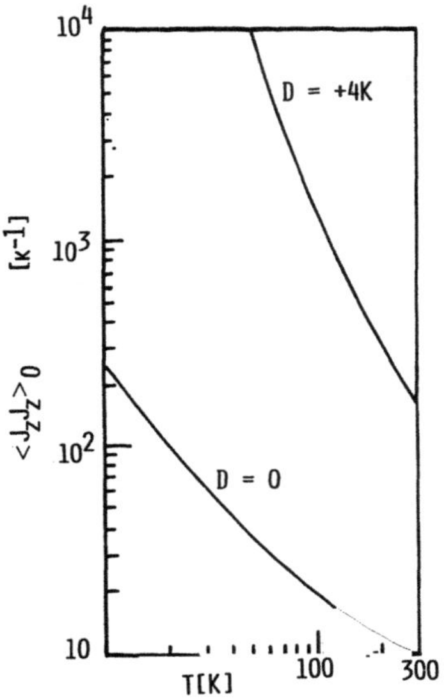

Fig. 1. Zero frequency spectral density for Ag:Er in cubic host (D=0) and with addition of muon-induced axial field for D=+4K.

A positive D>0 has a particularly strong effect since it favors a $|\pm 15/2>$ ground doublet in which no relaxation would occur. The slowed relaxation combined with a larger $<J_z^2>$ should thereby enhance the $\omega = 0$ spectral density. This qualitative argument is borne out by a calculation of $<J_zJ_z>_0$ shown in Fig. 1 both for D=0 and D/k_B = 4K. The curves are for the cubic field parameters appropriate [9] to Ag:Er and IN(0) = 0.014 consistent with the low temperature ESR [4]. Calculations were also done for Au:Er which show an even greater enhancement for the same D because of the smaller cubic field in Au [9]. The order of magnitude increase is in agreement with the data which had seemed to require either an enhanced spectral density or nondipolar muon-ion coupling. Note that 4K is a very conservative estimate for D in view of the above-mentioned results [10]. Unfortunately the sign of D could not be determined in ref.[10], and the sign is crucial. A negative D would produce the opposite effect of diminishing $<J_\alpha J_\alpha>_0$.

IV. HOLMIUM in $LuRh_4B_4$

The rare-earth rhodium borides have received much attention because they exhibit both superconductivity and ferromagnetism [5]. The interest here, however, is more on some novel features of the single-ion dynamics in the crystal field of tetragonal symmetry. One cannot hope to understand how the ferromagnetism and superconductivity influence the spin fluctuations unless he understands the simpler single-ion problem.

To a good first approximation the tetragonal component can be ignored and the crystal field [11] regarded as purely axial, in which case the eigenstates are simply $|\pm m>$ quantized along the z symmetry axis. For J = 8 Ho^{3+} z is an easy axis so that $|\pm 8>$ lies lowest. The first excited state $|\pm 7>$ is at 54K and the overall $|\pm 8>$ to $|0>$ splitting is about 120K [12]. Thus Ho^{3+} forms an excellent Ising system and relaxation within the ground doublet should be highly forbidden. The ion relaxation time would then be Arrhenius with an activation energy of at least 55K.

Small tetragonal crystal field terms are present [11,12] which produce $\Delta m = \pm 4$ admixtures. These, however, do not change the Ising character, so there would be no effect if it weren't for the fact that a zero-field splitting is produced in the non-Kramers ion. Although quite small, about 0.006 K, it leads to relaxation since J_z has matrix elements between the split eigenstates. The situation is the same as described several years ago for observing ESR of non-Kramers ions with Ising ground states in the ethyl sulphates [13]. The doublet relaxation rate which results from Eq. (2) for this Ising system is

$$\Gamma = \Gamma_e[x^2\sin^2\theta(1+x^2\cos^2\theta)/(1+x^2(3\cos^2\theta-1)+x^4\cos^2\theta)] \tag{4}$$

where $\Gamma_e = (g_{11}/g_J)^2\lambda$, $x = E/\Gamma_e\hbar$, $\theta=\tan^{-1}(\Delta/E)$ in which Δ is the zero-field splitting and $E = (g_{11}^2\mu_B^2H^2+\Delta^2)^{1/2}$ is the splitting in a field whose z-component is H with g_{11} the z-axis g-factor for an effective spin 1/2 [8,13] in the absence of the non-axial terms. The limiting forms are $\Gamma = \Gamma_e\sin^2\Theta$ and $\Gamma = E^2\sin^2\Theta/\Gamma_e\hbar^2$ for $\Gamma_e \ll E/\hbar$ and $\Gamma_e \gg E/\hbar$, respectively. Note that $\Gamma=0$ in the absence of zero-field splitting ($\Theta=0$) as anticipated and that for large damping compared with the splitting there is an "inverted" temperature dependence.

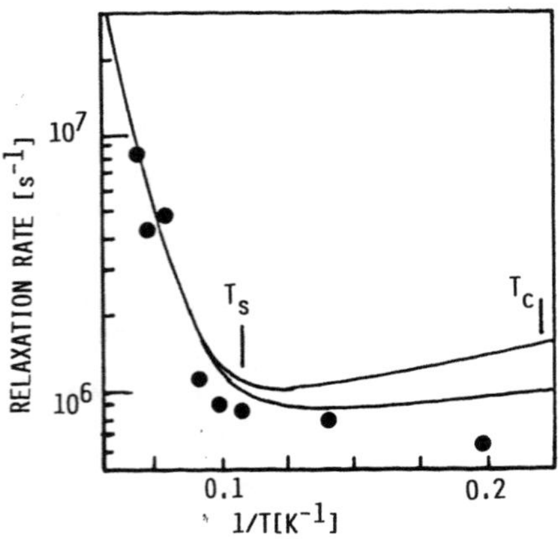

Fig. 2. Muon depolarization rate Λ(●) in $Ho_{0.7}Lu_{0.3}Rh_4B_4$ from ref. 3 and theoretical curves for slow decay rate of Ho spin fluctuations Γ_0. Two branches represent lowest (±1/2) and highest (±7/2) Ho^{165} nuclear spin states contribution to hyperfine field (upper branch for ±1/2).

The above shows that there is a finite relaxation within the ground doublet which can take over from the activated process at sufficiently low temperature. The cross-over temperature depends on details of the crystal field parameters and the full Eq.(2), which in this instance contains 73 simultaneous equations for determination of $\langle J_zJ_z\rangle_s$, must be used. The calculated values of the slow decay rate Γ_0 (obtained by fitting $\langle J_zJ_z\rangle_s = A/(\Gamma_0+s)$ for $s \to 0$) are shown in Fig. 2 along with data from Ref.[3] in the region where the quasi-static condition [14] is thought to apply. I used crystal field parameters supplied by DUNLAP [12] and IN(0)=0.08 from NMR data [15]. The applied field was zero but a hyperfine field was included which can vary between about 230 and 1610 Oe depending on the quantum state of the I=7/2 Ho^{165} nuclear spin which is 100% abundant [16]. Curves are shown for each of these values (the upper corresponding to 230 Oe) where there is an appreciable difference. There are thus no adjusted parameters in the theoretical curves, in view of which the agreement seen in Fig. 2 is quite remarkable. No account was taken of the onset of superconductivity at 7.7K, i.e. the same N(0) was used throughout without introduction of a gap. Data [3] show little concentration dependence in the quasi-static region; so the single-ion Korringa process dominates over spin-spin relaxation, and the present method applies even at the high concentration shown.

Although the agreement may be fortuitous, it seems safe to conclude that the small zero-field splitting produces a significant modification which can explain the departure from strongly activated behaviour. Possible effects due to superconductivity and ferromagnetism should be analysed keeping this in mind.

ACKNOWLEDGEMENTS

This work was done in collaboration with S. Dodds (Rice) and R. Heffner and M. Schillaci (Los Alamos). Work at Sandia National Laboratories was supported by the U. S. Department of Energy under Contract No. DE-AC04-76-DP00789.

REFERENCES

1. For a review of the technique,see A. Schenck,in Nuclear and Particle Physics at Intermediate Energies, edited by J. B. Warren (Plenum, N.Y., 1967).
2. J. A. Brown et al., Phys. Rev. Lett. 47, 261 (1981); R. H. Heffner, Hyperfine Int. 8, 665 (1981); and unpublished analysis of Ag:Er data.
3. R. H. Heffner et al., J. Appl. Phys. 55, 2007 (1984).
4. C. Rettori, D. Davidov and H. M. Kim, Phys. Rev. B 11, 5335 (1973).
5. See Proc. Internatl. Conf. on Ternary Superconductors, ed. by G. K. Shenoy, B. D. Dunlap and F. Y. Fradin (North Holland, N.Y., 1981) for several references to the $RERh_4B_4$ system (RE = rare-earth).
6. H. Mori, Progr. Theor. Phys. 33, 423 (1965).
7. K. W. Becker, P. Fulde and J. Keller, Z. Phys. B 28, 9 (1977).
8. A. Abragam and B. Bleaney: Electron Paramagnetic Resonance of Transition Ions (Oxford Univ. Press., N.Y., 1970).
9. G. Williams and L. I. Hirst, Phys.a Rev. 185, 407 (1969).
10. A. Schenck, Helv. Phys. Act. 54, 471 (1981).
11. B. D. Dunlap and D. Niarchos, Solid State Commun. 44, 1577 (1982).
12. Unpublished values for Ho^{3+} supplied by B. D. Dunlap.
13. J. M. Baker and B. Bleaney, Proc. Roy. Soc. A245, 156 (1958).
14. R . Kubo and T. Toyabe, in Magnetic Resonance and Relaxation, ed. by R. Blinc (North Holland, N.Y., 1967), p. 810.
15. K. Kumagai and F. Y. Fradin, Phys. Rev. B 27, 2770 (1983).
16. The hyperfine constant was taken from Table 5.5 of Ref. 8.

Ground-State Properties of $s=1$ Easy-Plane Ferromagnetic Heisenberg Chains

T. Schneider

IBM Zürich Research Laboratory, CH-8803 Rüschlikon, Switzerland

1. Introduction

One-dimensional (1-d) magnetic systems have been the subject of intensive experimental [1] and theoretical studies [2-5]. Since exact results are scarce for all but $s = 1/2$ models [2-5] and the Ising-Heisenberg ferromagnet for arbitrary s [6], theoretical predictions and the interpretation of experimental data have been based on the classical approximation ($\hbar \to 0$, $s \to \infty$, $\hbar s$ = const). Another striking feature is the large number of mappings between quantum spin chains and a wide assortment of 1 - d and 2 - d models [2-4,7]. Up to now, such isomorphisms have been established almost exclusively for $s = 1/2$ and $s >> 1$ only.

Here, we review the ground-state properties of $s = 1$ ferromagnetic Heisenberg chains with planar xy behavior. The results include the mapping onto the classical $d = 2$ xy model in zero field and onto an anisotropic ferromagnet in a sufficiently strong field, applied perpendicular to the easy plane. We consider $s = 1$ Heisenberg chains defined by

$$\mathcal{H} = \sum_{\ell} \left[- \frac{J}{2} (S_{\ell}^{+} S_{\ell+1}^{-} + S_{\ell+1}^{-} S_{\ell}^{+}) - \Delta J\, S_{\ell}^{z} S_{\ell+1}^{z} + D(S_{\ell}^{z})^2 \right] - h \sum_{\ell} S_{\ell}^{z} , \qquad (1)$$

where the parameters J, D and h are positive. In this case, the D-term represents an (xy) easy-plane single-ion anisotropy. This model Hamiltonian is of particular interest since it is thought to describe $CsNiF_3$ [1] with the parameters

$$J = 23.6\text{ K},\ \Delta = 1,\ D = 9\text{K},\ s = 1 . \qquad (2)$$

This compound has been studied rather extensively using inelastic neutron scattering.

The paper is organized as follows. In Section 2, we review finite-ring and analytic results for the ground-state properties of Hamiltonian (1) in zero field, revealing interesting critical properties in the Δ, D/J plane. Section 3 is devoted to the case with an applied field parallel to the single-ion anisotropy. For a sufficiently strong field, the ground state becomes ferromagnetic, so that the mapping of $CsNiF_3$ onto Hamiltonian (1) might be tested more precisely.

2. Ground-State Properties in Zero Field

The zero field ground-state properties of Hamiltonian (1) have recently been studied rather extensively by finite-ring calculations [8-10] and the path-integral approach [11]. The phase diagram as obtained by GLAUS and SCHNEIDER [9] is given in Fig. 1. There are three phases: A ferromagnetic phase, separated from the planar phases by a first-order line. An essential singularity line separates the planar xy phases, where the low D/J xy phase is gapless. The finite-ring results suggested [8-10] that this line exhibits KOSTERLITZ-THOULESS [12] behavior. This conjecture was recently confirmed by the path-integral approach [11]. Accordingly, close to the onset of the gapless phase, the gap between the ground state and the $\sum_\ell S_\ell^z = \pm 1$ manifold vanishes exponentially as

$$G(\Delta,D) \sim \exp\left[-a\left(\frac{J}{D^*(\Delta)} - \frac{J}{D(\Delta)}\right)^{-\sigma}\right] . \tag{3}$$

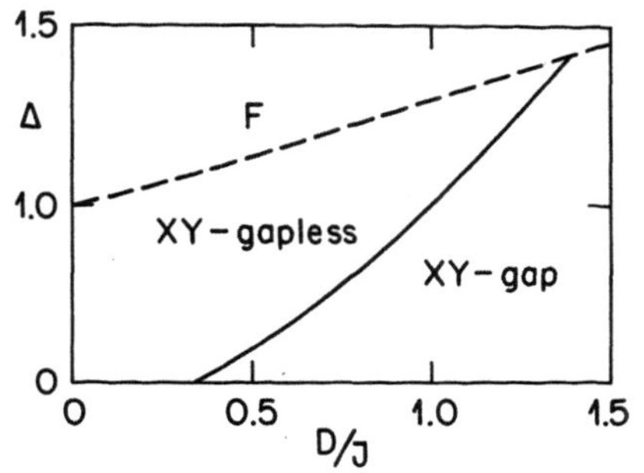

Fig. 1. Finite-ring estimate for the phase diagram of Hamiltonian (1) in zero field [9]. The planar xy phases are separated from the ferromagnetic phase by the first-order transition line (---). The essential singularity line (——), with critical properties equivalent to those of the Kosterlitz-Thouless transition in the classical d = 2 xy model, separates the xy phases with and without gap

Moreover, at $D^*(\Delta)$, the spin-spin correlation function falls off as a power of the distance between the spins,

$$\langle S_\ell^x \, S_0^x \rangle \sim |\ell|^{-\eta} . \tag{4}$$

From KOSTERLITZ's [13] analysis of the d + 1 = 2 - dimensional analog, we infer

$$\sigma = 1/2, \ \eta = 1/4 \ \gamma = 2 - \eta = 7/4 , \tag{5}$$

γ being the critical exponent of the susceptibility. In the gapless xy regime, there is similar behavior, with the important difference that

$$\eta = \eta(D,\Delta) . \tag{6}$$

These results have important implications for the dynamics, namely, also for the dynamic form factor (DFF), $S_{xx}(q,\omega)$. To derive this result, we note that the exponent z, describing scaling between wave number and frequency in $S_{xx}(q,\omega)$ is given by [8,9,11]

$$z = 1 . \tag{7}$$

Moreover, (4) and (6) imply at criticality and in the gapless xy phase,

$$S_{xx}(q) \sim q^{-1+\eta(\Delta,D)} . \tag{8}$$

Invoking then the dynamic scaling hypothesis, we obtain for small frequencies and wave numbers,

$$S_{xx}(q,\omega) \sim (\omega^2-\omega^2(q))^{-1+1/2\eta(\Delta,D)} , \tag{9}$$

where $\omega(q) \sim q^2$. Thus, $S_{xx}(q,\omega)$ probes transitions from the ground state to the $\Sigma_\ell S_\ell^z = \pm 1$ continuum, and it exhibits a singularity along the bottom of this continuum. For the parameters (2) corresponding to $CsNiF_3$, finite-ring calculations yield the estimate

$$\eta(\Delta = 1,\ D/J = 0.38) = 0.08\ . \tag{10}$$

A qualitative understanding of the excitation spectrum might be obtained by invoking the $O(1/s)$ approximation of the Villain representation of spin operators, and by neglecting the boson interaction. On this basis, one expects $S_{xx}(q,\omega)$ to be dominated by the boson pair continuum, and $S_{zz}(q,\omega)$ by the boson δ resonance. The boundaries of the continuum are determined by

$$\omega_c(q) = \omega_1(\tfrac{q}{2} + k) - \omega_1(\tfrac{q}{2} - k),\ 0 \le k \le \pi\ , \tag{11}$$

where ω_1 is the boson frequency,

$$\omega_1(q) = JS\Big(2(1-\cos q)\ (2(1-\Delta\cos q) + 2\,\tfrac{D}{J})\Big)^{1/2}\ . \tag{12}$$

Finite-ring results for parameters corresponding to $CsNiF_3$ (2) are shown in Fig. 2. The results are consistent with the singular behavior of $S_{xx}(q,\omega)$ for small q-values at the bottom of the continuum, and also reveals that the dominant features of $S_{zz}(q,\omega)$ are equivalent to a resonance with dispersion $\omega_1(q)$, becoming broader by approaching the zone boundary.

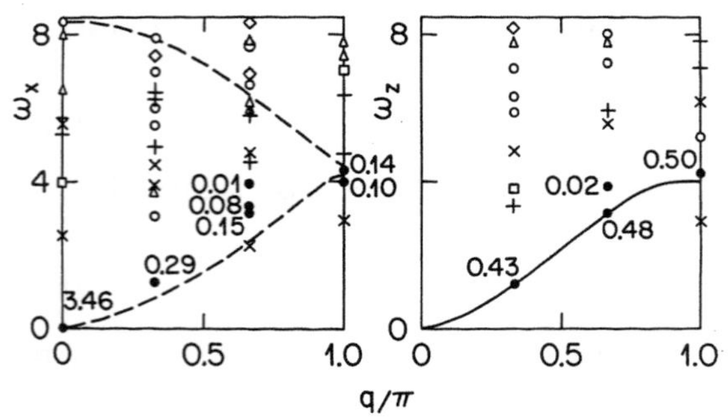

Fig. 2. Finite-ring results for $S_{xx}(q,\omega)$ and $S_{zz}(q,\omega)$ for $s = 1$, $D/J = 0.38$, and $N = 6$. Numbers and symbols give the value of $S_{AA}(q,\omega)$, where x : 10^{-3}, + : 10^{-4}, Δ : 10^{-5}, and O : 10^{-6}. Dashed lines in $S_{xx}(q,\omega)$ denote the estimates for the upper and lower boundaries of the boson pair continuum (11). The full line in $S_{zz}(q,\omega)$ is the boson frequency (12)

In the ferromagnetic phase the DFF's are exhausted by the magnon resonance,

$$S_{xx}(q,\omega) = S_{yy}(q,\omega) = \frac{1}{2}\,\delta(\omega-\omega_1(q))\ , \tag{13}$$

where

$$\omega_1(q) = -\,D + 2J(\Delta-\cos q)\ , \tag{14}$$

and $S_{zz}(q,\omega)$ vanishes. Moreover, there is also an exchange and single-ion two-magnon bound state, which can be calculated explicitly, but does not enter the DFF's, accessible in neutron scattering.

Finally, we turn to the planar xy phase, where $S_{xx}(q,\omega)$ is dominated for small q and ω by the excitation corresponding to the gap between the ground state and the $\sum_\ell S_\ell^z = \pm 1$ manifold (3).

Clearly, our understanding of the zero-field ground-state properties of Hamiltonian (1), apart from the critical properties, is still rather qualitative and incomplete. The main difficulties stem from the crucial importance of quantum fluctuations, invalidating most approximate treatments. Even the 1/s-expansion results are in doubt, because they do not signal the absence of the essential singularity line separating the planar xy phases for s = 3/2. In this view, it would be desirable to test the mapping of $CsNiF_3$ onto Hamiltonian (1) in a simpler limiting case. For this purpose, we next apply a magnetic field parallel to the anisotropy.

3. Ground-State Properties in an Applied Field Parallel to the Single-Ion Anisotropy

By applying a magnetic field parallel to the anisotropy (1), the ground state becomes ferromagnetic for $h > h_c$. The critical field h_c might be determined from

$$\frac{1}{NJ} \langle \uparrow\uparrow\uparrow \ldots | \mathcal{H} | \uparrow\uparrow\uparrow \ldots \rangle = - \Delta + \frac{D}{J} - \frac{h}{J} , \tag{15}$$

so that

$$h_c = D . \tag{16}$$

For

$$h \geq h_c , \tag{17}$$

the problem with one flipped spin leads to a cyclic matrix with eigenvalues

$$\omega_1(q) = 2 J(\Delta - \cos q) + h - h_c \tag{18}$$

relative to the fully aligned ground state. For $h = h_c$, the dynamic form factors are exhausted by this magnon resonance. In fact,

$$S_{xx}(q,\omega) = S_{yy}(q,\omega) = \frac{1}{2} \delta(\omega - \omega_1(q)) \tag{19}$$

and

$$S_{zz}(q,\omega) = 0 : \Delta > 1 . \tag{20}$$

For $\Delta = 1$, however,

$$S_{xx}(q,\omega) = S_{yy}(q,\omega) = S_{zz}(q,\omega) .$$

Thus, for $h \geq h_c$, where for $CsNiF_3$ $h_c/J = 0.38$, the excitation spectrum probed by the DFF's is very simple, and even at finite but low temperatures exhausted by the magnon resonance. Accordingly, a sufficiently strong magnetic field parallel to the single-ion anisotropy should provide accurate tests of the mapping of $CsNiF_3$ on model (1).

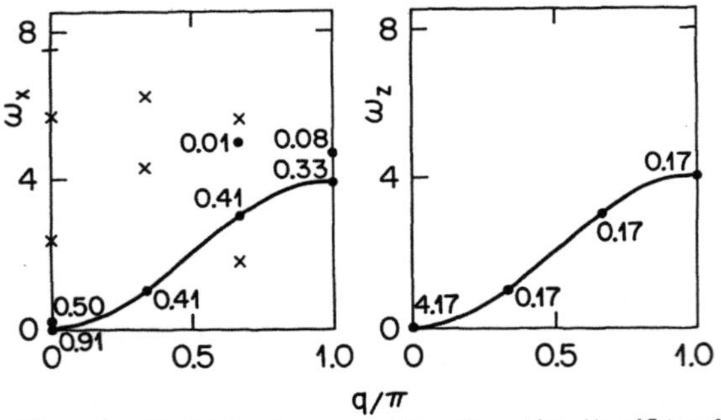

Fig. 3. Finite-ring results for the Hamiltonian (1) for $s = 1$, $D/J = 0.38$, $\Delta = 1$, $h/J = 0.342$, and $N = 6$. Numbers and symbols give the value of $S_{xx}(q,\omega)$ and $S_{zz}(q,\omega)$, respectively, where $x : 10^{-3}$. The full line is the magnon frequency (18)

To illustrate the crossover to the ferromagnetic ground state, in Fig. 3 we show finite-chain results for $h/J = 0.342$, being close to $h_c/J = 0.38$. The main feature is obviously well described by the magnon resonance (19), but there are still marked deviations in $S_{xx}(q,\omega)$ from the resonance strength 1/2, valid for $h \geq h_c$. Moreover, $S_{zz}(q,\omega)$ does not yet vanish.

1. M. Steiner: Physics in One Dimension, ed. by J. Bernasconi and T. Schneider (Springer, Heidelberg 1981), p. 140
2. J.C. Bonner, H.W. Blöte, H. Beck and G. Müller: Physics in One Dimension, ed. by J. Bernasconi and T. Schneider (Springer, Heidelberg 1981), p. 115
3. L.A. Takhtadzhan and L.D. Fadeev: Russ. Math. Surveys 34, 5 (1979)
4. M.P.M. Den Nijs: Phys. Rev. B 23, 6111 (1981)
5. T. Schneider and E. Stoll: J. Appl. Phys. 53, 1850 (1982)
6. T. Schneider and E. Stoll: Phys. Rev. B 25, 4721 (1982)
7. J.B. Kogut: Rev. Mod. Phys. 51,659 (1979)
8. T. Schneider, U. Glaus, and E.P. Stoll: J. Appl. Phys.
9. U. Glaus and T. Schneider: (unpublished)
10. R. Botet, R. Julien, and M. Kolb: (unpublished)
11. O.F. de Alcantara Bonfim and T. Schneider: (unpublished)
12. J.M. Kosterlitz and D.J. Thouless: J. Phys. C 6, 118 (1973)
13. J.M. Kosterlitz: J. Phys. C 7, 1046 (1974)

Bound State of Phonons and CEF-Excitations in $CeAl_2$

P. Thalmeier

Hamburger Synchrotronstrahlungslabor HASYLAB at DESY, Notkestraße 85
D-2000 Hamburg 52, Fed. Rep. of Germany

1. Introduction

The coupling of phonons to the crystalline electric field (CEF)-states of Rare Earth (RE) compounds with integer RE-valence leads to various phonon anomalies for temperatures smaller than the CEF-splitting ($T \lesssim 100$ K). In the long wavelength limit, the T-dependece of sound velocities, i.e. elastic constants is due to this "magnetoelastic coupling" which has been shown by numerous ultrasonic experiments [1]. A similar mechanism has been found to explain temperature, magnetic field and RE-concentration dependence of optical phonons in RE-halides and their corresponding La-diluted alloys as observed in Raman scattering experiments [2]. More recently inelastic neutron scattering experiments [3,4] revealed that intermetallic $CeAl_2$ is an exceptionally striking example for the effects produced by a strong coupling of phonons to CEF-excitations.

2. Neutron Scattering Results for $CeAl_2$

In the Laves phase $CeAl_2$ the Ce^{3+}-ions are sitting on places with tetrahedral site symmetry, forming a diamond-type sublattice. Accordingly the $4f^1$ ($J = 5/2$) ground state of Ce^{3+} should split into $\Gamma_7(0)$ doublet and $\Gamma_8(\Delta)$ quartet CEF-states where Δ is the splitting energy. From measuring the elastic constant $c_{44}(T)$ $\Delta = 100$ K was concluded [5]. However, inelastic neutron scattering [3] clearly resolved *two* inelastic peaks in the magnetic scattering cross section at $\Delta_1 = 180$ K, $\Delta_2 = 100$ K instead of a single $\Gamma_7 - \Gamma_8$ transition at Δ. Furthermore, these peaks showed almost no dispersion, their total intensity corresponds to the expected value for a single $\Gamma_7 - \Gamma_8$ transition and they are unchanged in the diluted $La_{1-x}Ce_xAl_2$. The origin of this splitting which seemed to be in conflict with tetrahedral Ce-site symmetry was unknown until additional experiments concentrating on lattice vibrations of $CeAl_2$ showed a sharp phonon peak at about $h\omega_0 = 140$ K which corresponds to the average $1/2(\Delta_1 + \Delta_2)$ of the excitation energies in the magnetic cross section. This suggests that interaction with phonons is responsible for the existence of two peaks even more so since the phonons themselves show a strong T-dependence [4] which is absent in $LaAl_2$ and hence must be a consequence of coupling between phonons and 4f-states. A theoretical model based on this coupling mechanism and explaining the neutron scattering results in $CeAl_2$ was given in [6,7,8] and its essence is described in the following.

3. Model Hamiltonian for the Interaction of Phonons and CEF-Excitations

The relevant phonons correspond to vibrations of the diamond type sublattice. The peak in the phonon density of states (PDOS) at $\hbar\omega_0 = 140$ K ($T = 296$ K) is due to TA, TO modes which are almost degenerate at the zone boundary [4]. Distortion produced by these phonon modes change the CEF-potential at Ce-sites leading to a coupling of phonons to CEF-excitations. Assuming that only distortion of Γ_5-type are important this model is described by $H = H_0 + H_1$ with

$$H_0 = \sum_{\alpha n} \epsilon_\alpha |\Gamma^n_\alpha\rangle\langle\Gamma^n_\alpha| + \sum_{\underline{k}s} \hbar\omega_{\underline{k}s}(a^+_{\underline{k}s}a_{\underline{k}s}+1/2)$$

$$H_1 = -\sum_{\substack{\underline{k}s,\mu \\ 1\tau}} g^\mu_0(\underline{k}s,\tau)(a_{\underline{k}s}+a^+_{-\underline{k}s})O_\mu(\underline{R}_{1\tau})e^{-i\underline{k}\underline{R}1\tau} \quad . \qquad (1)$$

Here $|\Gamma^n_\alpha\rangle$, ϵ_α are CEF-states and energies (α = multiplet, n = degeneracy index), i.e. $\epsilon_7 = 0$, $\epsilon_8 = \Delta$. Phonon branches are denoted by $(\underline{k}s)$ and Ce-sites by (1τ) with $\tau = 1,2$. g^μ_0 = coupling constant. The quadrupolar operators O_μ ($\mu = 1,2,3$) with $O_1 = J_yJ_z+J_zJ_y$ etc. transform like a Γ_5 triplet. Only nondiagonal matrix elements in the combination $\gamma^2_0 = \sum_{\mu,m} |\langle\Gamma^n_7|O_\mu|\Gamma^m_8\rangle|^2$ are important. The $(\underline{k}s)$-dependence of the coupling constant $g^\mu_0(\underline{k}s,\tau)$ is not known and it will be replaced by an average g_0 later on.

4. Bound State and the Magnetic Excitation Spectrum

The magnetic scattering cross section of a CEF-system is proportional to

$$S(\omega) = (1-e^{-\beta\omega})\,\mathrm{Jmu}(\omega) \qquad (2)$$

where $\beta = 1/kT$ and $u(\omega)$ is the dynamical dipolar single ion suceptibility. For a noninteracting system, $\mathrm{Jmu}(\omega)$ would have a single δ-peak at $\omega = \Delta$ corresponding to the $\Gamma_7-\Gamma_8$ excitation. However, the interaction with phonons and conduction electrons will change this simple spectrum. To study the effect of coupling to phonons, ABRIKOSOV'S pseudofermion technique [9] is most convenient. For the noninteracting system H_0 in (1) $\Delta = 1/2(\Delta_1+\Delta_2)$ is assumed and the unrenormalized phonons (which nearly correspond to those measured at room temperature) are described by a triangular PDOS:

$$D(\omega) = \begin{cases} (\frac{2}{\omega_b})^2\ (\omega-\omega_1) & \omega_1<\omega<\omega_0 \\ (\frac{2}{\omega_b})^2\ (\omega_2-\omega\ ) & \omega_0<\omega<\omega_2 \end{cases} \qquad (3)$$

with $h\omega_1$ = 120 K, $\hbar\omega_2$ = 160 K, $\hbar\omega_0$ = 140 K and a bandwidth $\hbar\omega_b = \hbar(\omega_2-\omega_1)$ = 40 K. The pseudofermion Green's function of CEF-state $|i\rangle$ ($i = \alpha n$) renormalized by the interaction H_1 is given by

$$\bar{G}(i,\omega) = [\omega-\epsilon_i-\bar{\Sigma}(i,\omega)+i\eta]^{-1}$$

$$\bar{\Sigma}(i,\omega) = \sum_{\substack{\underline{k}s \\ j\mu}} \frac{g^2_0\,|\langle i|O_\mu|j\rangle|^2}{\omega-\epsilon_j-\omega_{\underline{k}s}+i\eta} \quad . \qquad (4)$$

The bare CEF-energy ϵ_i is therefore renormalized by a self energy $\bar{\Sigma}$ which is due to virtual excitations of phonons, shown in Fig. 1.

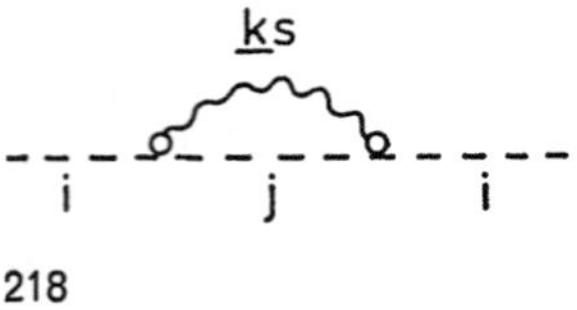

Fig. 1 Self-energy diagram for pseudofermion Green's function of CEF-states (broken lines). Wavy lines represent phonons and circles the coupling constant g_0.

For small coupling g_0 the self energy $\bar{\Sigma}(i,\omega)$ can be approximated by a constant $\bar{\Sigma}(i,\varepsilon_i)$ which simply leads to a shift in the CEF-energy $\varepsilon_i'=\varepsilon_i+\bar{\Sigma}(i,\varepsilon_i)$ leaving the pole structure of $\bar{G}$ unchanged. Therefore $S(\omega)$ would be similar to the noninteracting case: a simple peak at $\omega = \Delta'$. For larger coupling g_0 however, the dynamical nature of $\bar{\Sigma}(\omega)$ cannot be neglected and a qualitatively different $S(\omega)$ may result. For the Γ_8 self energy one obtains:

$$\bar{\Sigma}(\Gamma_8,\omega) = g_0^2\gamma_0^2 H_1(\omega) - \frac{i}{\pi} g_0^2 \gamma_0^2 D(\omega)$$

$$H_1(\omega) = \int \frac{D(\omega')}{\omega-\omega'} d\omega' \quad . \tag{5}$$

This function is shown in Fig. 2.

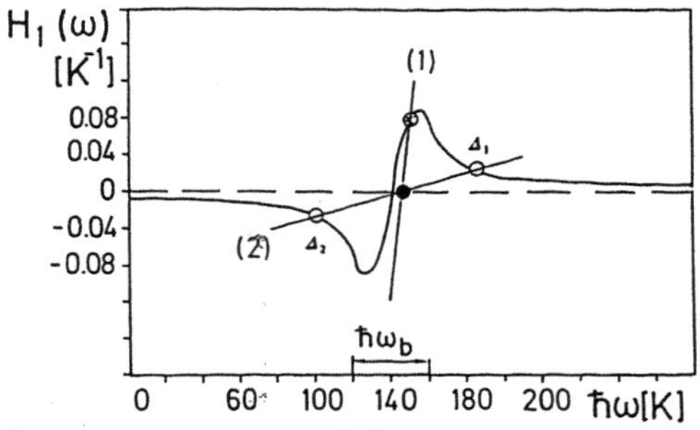

Fig. 2 Plot of $H_1(\omega)$ and graphic solution of (6) for different g_0. Full circle: position of bare Δ. (1) small g_0: no bound states. (2) large g_0: bound and antibound states (open circles) at Δ_2,Δ_1.

The possible poles of $\bar{G}(\Gamma_8,\omega)$ which determine structure of $S(\omega)$ are solutions of

$$\frac{1}{g_0^2\gamma_0^2}(\omega-\Delta) = H_1(\omega) \quad . \tag{6}$$

They can be found by graphical means (Fig. 2). For $CeAl_2$ the original Δ is inside the phonon continuum of width $\hbar\omega_b$. For small g_0^2 (case 1) only a shift of Δ and a broadening occurs, for large g_0 (case 2) however, bound (Δ_2) and antibound (Δ_1) states emerge outside the phonon band and are denoted by $\tilde{\Gamma}_{82}$ and $\tilde{\Gamma}_{81}$. These new poles in $\bar{G}(\Gamma_8,\omega)$ also appear in $u(\omega)$ [8] and therefore lead to the observed double peak structure in $S(\omega)$. The essential result is that for finite $\hbar\omega_b$ a threshold coupling strength g_0^c is given by $g_0^c = (\lambda/\gamma_0)\omega_b$; with $\lambda = 1/2(2\ln 2)^{-1/2}$ and $\gamma_0 = (40)^{1/2}$ $\hbar\omega_b = 40$ K this means $g_0^c = 2.7$ K. If one describes $CeAl_2$ approximately by a dispersionless model, i.e. $\hbar\omega_b \to 0$, one obtains analytical solutions of [7]

$$\Delta_{1,2} = 1/2(\Delta+\hbar\omega_0)\pm[1/4(\Delta-\hbar\omega_0)^2+g_0^2\gamma_0^2]^{1/2} \quad . \tag{7}$$

Additional inclusion of exchange interaction $-I_{ex}(g_J-1)\underline{\sigma}\cdot\underline{J}$ with conduction electrons then leads to the broadened double peak spectrum of Fig. 3 which agrees qualitatively with the spectrum of $CeAl_2$. The only adjustable parameter is g_0; chosing $g_0 = 6.3$ K leads to peaks in $S(\omega)$ at the proper positions. This value for g_0 agrees reasonably with the magnetoelastic coupling strength determined by elastic constant measurements [5].

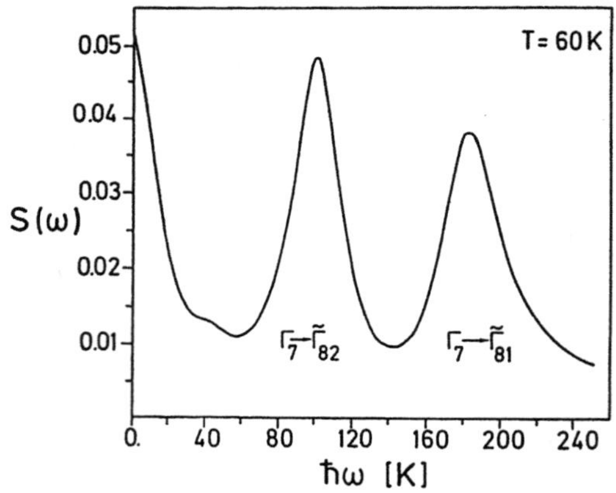

Fig. 3 Magnetic structure function for $CeAl_2$ calculated for the dispersionless model with g_0 = 6.3 K and an exchange parameter $I_{ex}N(0)$ = 0.06. (N(0) = cond. el. density of states).

5. Temperature Dependence of Phonons in $CeAl_2$

At room temperature phonon energies are hardly changed by magnetoelastic interactions. Therefore the unrenormalized $\hbar\omega_0$ corresponds to the peak in the PDOS measured at 295 K. At lower temperatures (77 K, 4 K) some phonon branches have also been measured [4] and a strong softening was found which was absent in $LaAl_2$. The softening of $CeAl_2$ phonons is due to a process shown in Fig. 4.

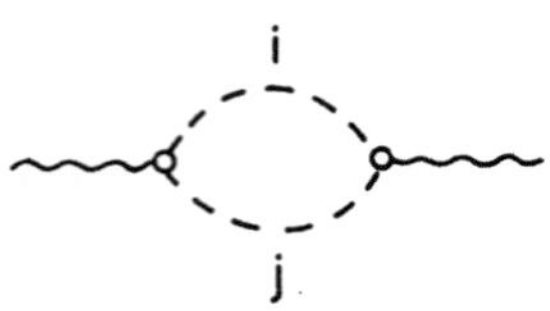

Fig. 4 Phonon self energy diagram. The bubble corresponds to $u_Q(\omega,T)$; i,j denote CEF-states.

This leads to a renormalized phonon Green's function

$$D(\omega)^{-1} = (2\omega_0^2)^{-1}(\omega^2-\omega_0^2)-g_0^2u_Q(\omega,T)\,. \tag{8}$$

Here $u_Q(\omega,T)$ is the quadrupolar susceptibitly

$$u_Q(\omega,T) = \sum_{i,j} O_{ij}O_{ji}\frac{p_i-p_j}{\omega+\Delta_{ji}+i\eta}$$

$$O^{\mu}_{ij} = \langle i|O_\mu|j\rangle,\quad \Delta_{ji} = \varepsilon_j-\varepsilon_j \tag{9}$$

where p_i are the thermal occupations of CEF-states. This RPA-expression leads to a softening which is too strong and must be modified by including a phenomenological linewidth, i.e. $\Delta \to \Delta + i\Gamma$ in (9). Using g_0 = 6.3 K as before the renormalized average phonon frequency $\hbar\omega_{av}$ after (8) is shown in the following figure.

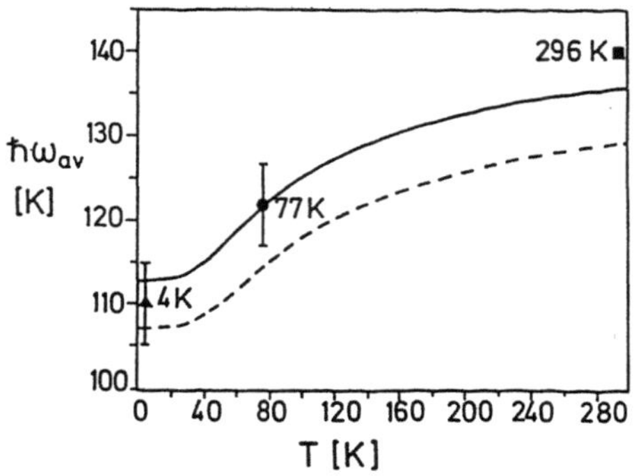

Fig. 5 Average phonon frequency $\hbar\omega_{av}$. Broken line: RPA result (Γ =0). Solid line: linewidth =25 K. Experimental points from [4].

6. Conclusions

$CeAl_2$ is the first RE-compound where a bound state of phonons and a CEF-excitation has been found. Its origin lies in an extraordinary strong magnetoelastic coupling in this compound. This is also responsible for a pronounced phonon softening at low temperatures. Other $REAl_2$ do not show such anomalies due to a much weaker coupling of phonons and CEF excitations. Recent Raman scattering experiments [10] on $CeAl_2$ confirm the existence of a bound state; however discrepancies with respect to the phonon softening were found to exist.

References

1. B. Lüthi, Dynamical Properties of Solids, Vol. 3, (North-Holland, Amsterdam, 1978)
2. K. Ahrens and G. Schaack, Phys. Rev. Lett. 42, 1488 (1979)
3. M. Loewenhaupt, B.D. Rainford and F. Steglich, Phys. Rev. Lett. 25, 1709 (1979)
4. W. Reichardt, to be published
5. B. Lüthi and C. Lingner, Z. Phys. B34, 157 (1979)
6. P. Thalmeier and P. Fulde, Phys. Rev. Lett. 49, 1588 (1982)
7. P. Thalmeier, J. Appl. Phys. 55, 1916 (1984)
8. P. Thalmeier, J. Phys. C, in press
9. P. Fulde and I. Peschel, Adv. Phys. 21, 1 (1971)
10. G. Güntherod, A. Jayaraman, G. Batlogg, M. Croft and E. Melczer, Phys. Rev. Lett. 51, 2330 (1983)

Local Magnetic Field Distributions in Ising Models

M.F. Thorpe and M. Thomsen[1]

Physics and Astronomy Department, Michigan State University
East Lansing, MI 48824, USA

D. Sherrington and T.C. Choy

Physics Department, Imperial College, London SW7 2BZ, United Kingdom

We show that the statistical mechanics of Ising models can be conveniently reformulated in terms of the local magnetic field probability distribution function P(h). Analytic results are presented for the 2D square net Ising ferromagnet and Monte Carlo results for the 3D simple cubic Ising ferromagnet. We also give results for the SK Ising spin glass which are analytic above T_{sg} and from Monte Carlo simulations below T_{sg}.

A spin ½ Ising system is defined with the Hamiltonian

$$H = -\tfrac{1}{2} \sum_{ij} J_{ij} \, \sigma_i \sigma_j \tag{1}$$

where the J_{ij} are arbitary and the factor ½ is to prevent double counting. This Hamiltonian describes a very general spin ½ Ising system with $\sigma_i = \pm 1$ and an arbitary range for the parameters J_{ij}. We introduce the probability distribution $P_i(h)$ for the local magnetic field at site i by

$$P_i(h) = \langle \delta(h - h_i) \rangle \tag{2}$$

where $h_i = \sum_j J_{ij} \sigma_j$ and $\langle \cdot \cdot \cdot \cdot \rangle$ denotes the usual thermal average. The probability distribution for the whole system P(h) is then

$$P(h) = \frac{1}{N} \sum_i P_i(h) \tag{3}$$

where there are N sites. We only have room in this short note to summarize our main results. Full results can be found in references [1] and [2].

The function P(h) is an extremely useful function because both the thermodynamic properties and the neutron scattering law $S(\underline{k},\omega)$ can be obtained from it. It also has an easy physical interpretation. For discussion of its previous use and misuse see [1] and [2]. The energy E of the system is given by

$$E = -\tfrac{1}{2} N \int h \tanh \beta h \, P(h) dh \tag{4}$$

and the magnetization by

Work supported by the N.S.F.

[1]Exxon Fellow

$$M = N \int \tanh \beta h \, P(h) dh \tag{5}$$

and we see that E couples to the symmetric part of P(h) whereas M couples to the antisymmetric part of P(h). The inelastic neutron scattering cross section [1] is given by

$$S(\underline{k},\omega) = \frac{1}{2\pi} \int_{-\infty}^{+\infty} e^{-i\omega t} dt \sum_i < \sigma_i^x \, \sigma_i^x(t) > \tag{6}$$

which can be written as [1]

$$S(\underline{k},\omega) = \tfrac{1}{2}N \, [P(\omega/2) + P(-\omega/2)]/[1 + \exp(-\beta\omega)] \, . \tag{7}$$

Note that $S(\underline{k},\omega)$ is independent of $\underline{k}$ because there is no spin dynamics, and that $S(\underline{k},\omega)$ couples to the symmetric part of P(h).

An analytic calculation of $P_i(h)$ involves all possible correlations among the spins interacting with site i. This can be seen most directly by noting that the distribution can be reconstructed from its moments, $<h_i^n>$. Many of the required correlation functions are available in the literature while others can be expressed in terms of them using various transformations [1]. Thus, for many two dimensional Ising ferromagnets P(h) may be calculated analytically while in three dimensions we can employ Monte Carlo techniques [2]. Note that for these systems, $P(h) = P_i(h)$.

We now present results for the nearest neighbor ferromagnetic Ising square net and simple cubic. These are shown in Fig. 1 where the field h is given in units of J.

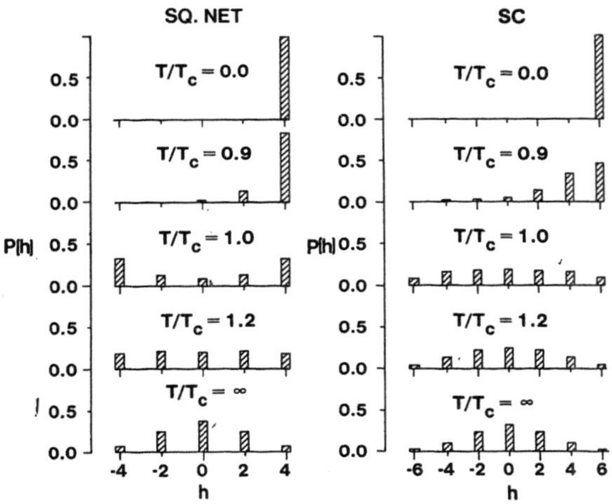

Fig. 1. The distribution P(h) for the square net and simple cubic lattices at various temperatures

It can be seen that P(h) consists of a series of delta functions, whose integrated weight sums to unity. At high temperatures the "envelope" is roughly Gaussian while at low temperatures there is a single peak. This behaviour is common to both lattices in contrast to the behaviour around T_c which is quite different in the two cases. In two dimensions a distinct dip is seen at T_c whereas in three dimensions there is some flattening but no dip. We have found all two dimensional lattices to have a similar dip [1] and all three dimensional lattices to have a similar flattening [2]. These effects are independent of the number of nearest neighbors and depend only on dimensionality. We believe that the reason for the difference is that the spins are correlated over a larger region in two dimensions at T_c and a domain picture is appropriate. In three dimensions, because of the greater connectivity, less correlation is present at T_c and the domain picture is less appropriate.

In principle the symmetric part of P(h) could be measured as a function of temperature using inelastic neutron scattering. This has not been done to our knowledge. In practice most Ising-like systems such as K_2CoF_4 and $KCoF_3$ also have small Heisenberg terms in the Hamiltonian. However these are small enough that the peaks in $S(k,\omega)$ should not overlap and our calculation of the weights as a function of temperatures should be good.

We have also used the P(h) formalism to study the SK Ising spin glass [3]. The Hamiltonian is given by Eq. (1) but every spin interacts with every other spin so that the model has no dimensionality. The J_{ij} are distributed randomly (each bond is independent) with mean zero and variance $\tilde{J}/\sqrt{N}$ where N is the number of spins in the system. This scaling of the exchange is adopted so that the model behaves reasonably in the thermodynamic limit. The model can be solved (at least for $T > T_{sg}$) and exhibits a spin glass transition at $k_B T_{sg} = \tilde{J}$.

We have managed to calculate P(h) exactly in the paramagnetic phase using the method of TAP[4] in which N is replaced by z, the coordination of a Bethe lattice, and the interactions are nearest neighbor only. We find that P(h) for $T > T_{sg}$ is given by [5]

$$P(h) = \exp(-\beta^2\tilde{J}^2/2 - h^2/2\tilde{J}^2)\cosh(\beta h)/(\sqrt{2\pi}\,\tilde{J}) \tag{8}$$

which is shown in Fig. 2.

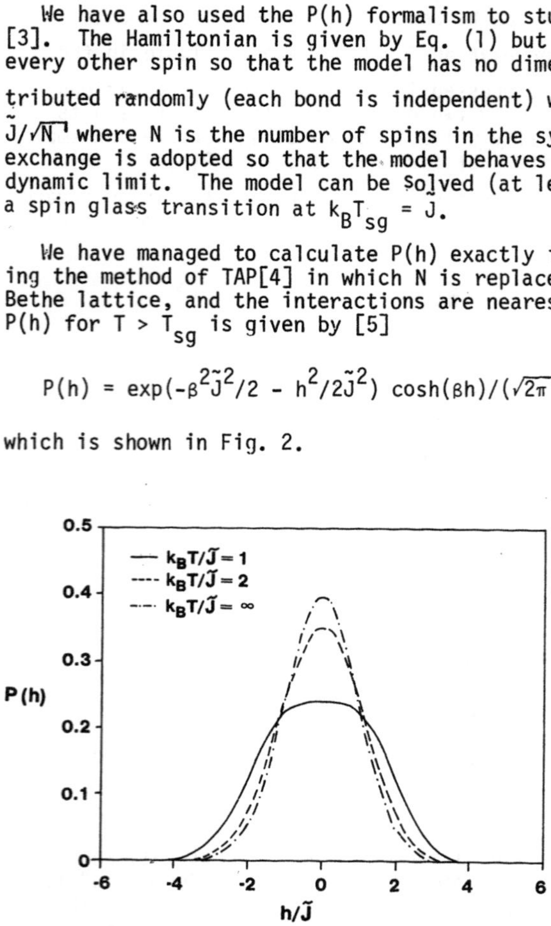

Fig. 2. The distribution P(h) for the SK Ising model in the paramagnetic phase ($k_B T_{sg} = \tilde{J}$)

At infinite temperatures, $P(h)$ is Gaussian. As the temperature is lowered, the distribution flattens and near T_{sg} behaves like

$$P(h) = P(0) + a(T_{sg} - T)h^2 + O(h^4) \tag{9}$$

for small h. We have not been able to obtain an analytic solution for $T < T_{sg}$ but in Fig. 3 we show the result of a Monte Carlo simulation.

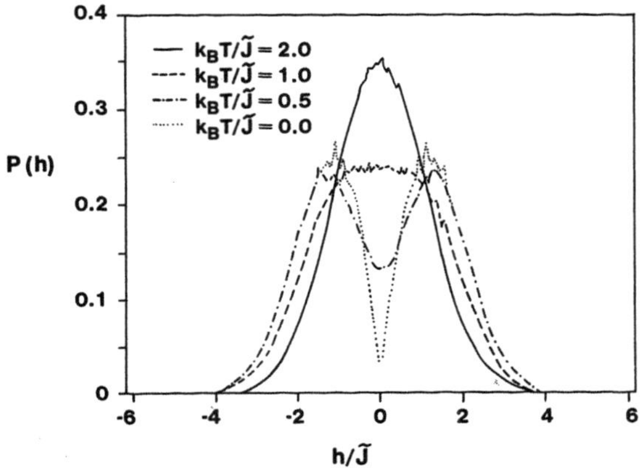

Fig. 3. The distribution $P(h)$ for the SK Ising model obtained by Monte Carlo simulation averaged over four different 1020 spin configurations. The zero temperature data has been symmetrized to lower the noise level.

It can be seen that a dip develops just below T_{sg} and as the temperature approaches zero our results are consistent with the heuristic TAP predictions

$$P(h) \sim 0.307h/\tilde{J}^2 \ . \tag{10}$$

In this short paper we have tried to show that $P(h)$ is a useful quantity in that the energy, magnetization and $S(\underline{k},\omega)$ can be obtained directly from it. Caution must be taken in reading the literature as other similar quantities like

$$P_i'(h) = \delta(h_i - \langle h_i \rangle) \tag{11}$$

are not as useful as Eq. (2) in our opinion as they do not lead to observables [4] - [6]. They may exhibit similar behaviour at very low temperatures. The quantity $P(h)$ is easier to visualise physically, as it involves a local field h, than say a global quantity like the free energy.

These concepts can clearly be extended to vector spins and classical Heisenberg models. It is much harder to see how to extend them to quantum systems as the whole concept of a local field is a rather "classical" notion.

References

1. M. Thomsen, M.F. Thorpe, T.C. Choy and D. Sherrington, Phys. Rev. B 1 July 1984.
2. T.C. Choy, D. Sherrington, M. Thomsen and M.F. Thorpe, to be submitted to Phys. Rev.
3. D. Sherrington and S. Kirkpatrick, Phys. Rev. Letters 35, 1792 (1975).
4. D.J. Thouless, P.W. Anderson and R.G. Palmer, Phil. Mag. 35, 593 (1977).
5. M. Thomsen, M.F. Thorpe, T.C. Choy and D. Sherrington, to be submitted to Phys. Rev.
6. C.M. Soukoulis and G.S. Grest, Phys. Rev. B 28, 1495.

List of Contributors

Allia, P. 110

Balucani, U. 46,51
Banavar, J.R. 115
Birgeneau, R.J. 121
Bonner, J.C. 55
Borsa, F. 2
Boucher, J.P. 6
Bouillot, J. 6
Bullough, R.K. 80

Capel, H.W. 86
Choy, T.C. 222
Cieplak, M. 115
Cieslewicz, W. 150
Collins, M.F. 12
Cottam, M.G. 153
Cowley, R.A. 121

de Groot, H.J.M. 21
de Jonge, W.J.M. 16,27
de Jongh, L.J. 21
De Raedt, B. 172
De Raedt, H. 172
Dhar, S.K. 177

Feile, R. 189
Fivez, J. 172
Fogedby, H. 92
Fowler, M. 99

Gaulin, B.D. 12
Gerling, R.W. 61
Gschneidner, Jr., K.A. 177

Güdel, H.U. 189

Hagen, M. 121
Henry, J.Y. 6,201
Hone, D.W. 126
Hoogerbeets, R. 27
Huber, D.L. 183

Johnson, M.D. 99

Kamieniarz, G. 137
Kjems, J.K. 189
Kontos, D.E. 153
Kopinga, K. 16,27

Lagendijk, A. 172
Landau, D.P. 61
Leuenberger, B. 189
Lindgård, P.-A. 159
Lockwood, D.J. 33
Loly, P.D. 165
Lovesey, S.W. 67

Magyari, E. 102
Meech, G.P. 67

Parkinson, J.B. 55
Petitgrand, D. 189,201
Phaff, A.C. 27
Pilling, D.J. 80
Pini, M.G. 2,46,51,72
Pires, A. 6
Pynn, R. 201

Rastelli, E. 195
Reatto, L. 195
Regnault, L.P. 6,201
Reiter, G. 77
Renard, J.P. 6
Rettori, A. 2,46,72
Rezende, S.M. 126
Richards, P.M. 207
Rossat-Mignod, J. 6,201

Schneider, T. 212
Sherrington, D. 222
Shirane, G. 121
Southern, B.W. 165
Stasch, A. 150
Steiner, M. 39
Stirling, W.G. 6
Swüste, C.H.W. 27

Tassi, A. 195
Thalmeier, P. 217
Thiel, R.C. 21
Thomas, H. 102
Thomsen, M. 222
Thorpe, M.F. 222
Timonen, J. 80
Tinus, A.M.C. 16
Tognetti, V. 2,46,51

Vaia, R. 46
van Duyneveldt, A.J. 27
Villain, J. 142
Vinai, F. 110

Wright, N.F. 99

Yoshizawa, H. 121

MIX
Papier aus verantwortungsvollen Quellen
Paper from responsible sources
FSC® C105338

If you have any concerns about our products,
you can contact us on
ProductSafety@springernature.com

In case Publisher is established outside the EU,
the EU authorized representative is:
Springer Nature Customer Service Center GmbH
Europaplatz 3, 69115 Heidelberg, Germany

Printed by Libri Plureos GmbH
in Hamburg, Germany